T0344983

Artificial Intelligence in Digital Holographic Imaging

Artificial Intelligence in Digital Holographic Imaging

Technical Basis and Biomedical Applications

Inkyu Moon
Department of Robotics and Mechatronics Engineering
Daegu Gyeongbuk Institute of Science & Technology (DGIST)
Daegu, South Korea

This edition first published 2023

Registered Office
John Wiley & Sons, Inc., 111 River Street, Hoboken, NJ 07030, USA

Editorial Office
111 River Street, Hoboken, NJ 07030, USA

For details of our global editorial offices, customer services, and more information about Wiley products visit us at www.wiley.com.

Wiley also publishes its books in a variety of electronic formats and by print-on-demand. Some content that appears in standard print versions of this book may not be available in other formats.

Library of Congress Cataloging-in-Publication Data applied for
Hardback ISBN: 9780470647509

Cover Design: Wiley
Cover Images: © KATERYNA KON/SCIENCE PHOTO LIBRARY/Getty Images

Set in 9.5/12.5pt STIXTwoText by Straive, Pondicherry, India

SKY10037282_102522

Contents

Preface

Quantitative label-free optical imaging technique represents a new, highly promising approach to identify cellular biomarkers, particularly when it is combined with artificial intelligence (AI) technologies for scientific, industrial, and biomedical applications. Among several new optical quantitative imaging techniques, digital holographic microscopy (DHM) has recently emerged as a powerful new technique well suited to non-invasively explore cell structure and dynamics with a nanometric axial sensitivity and the ability to identify new cellular biomarkers. This book provides detailed explanations for using DHM to perform label-free phenotypic cellular assays, thus allowing the non-invasive isolation of different specific cellular phenotypes. Practically, phenotypes related to the monitoring of cell responses and cytotoxicity profiling upon interaction with drugs are presented. Thus, promising theragnostic cellular biomarkers can be successfully explored. This book further provides explanations of AI and deep learning pipelines for the development of an intelligent DHM that can perform optical phase measurement, phase image processing, feature extraction, and classification. Multiple biophysical single-cell features such as morphological parameters, optical loss characteristics, and protein concentration are automatically measured in individual biological cells. These biophysical measurements form a hyper-dimensional feature space in which supervised learning can be performed for cell analysis. This technology is undergoing clinical testing for blood screening and live cardiomyocytes analysis as well as for studying neuronal activities in mental diseases including psychiatry disorders. Furthermore, combining DHM with stem-cell technology including induced pluripotent stem cell (iPSC) approaches paves the way to develop personal medicine, considering that iPSCs derived from a patient can be differentiated a priori into any types of cells, including cardiac cells, neural cells, and so on. However, the development of systems and methods for performing high-throughput analysis of holographic images in a large volume are indispensable to have a DHM-based automated high-content screening approach aiming at identifying theragnostic cellular biomarkers. By combining DHM with AI technology, including recent deep learning approaches, this system can achieve a record-high accuracy in non-invasive, label-free cellular phenotypic screening. It opens up a new path to data-driven diagnosis. Specifically, AI is one of the most rapidly evolving subjects in computing and engineering fields, with a special emphasis on creating intelligent automated systems or applications. These AI algorithms are becoming essential for developing intelligent systems. The main goal of this book is to explain key concepts and algorithms of AI to show how to build intelligent DHM systems drawing on techniques from artificial neural networks (ANN), convolutional neural networks (CNN), and generative adversarial network (GAN). Principles behind these techniques are explained by showing how various techniques can be implemented for intelligent DHM systems design. Depending on problems to be solved, AI algorithms can be applied to recognition, classification, regression, and prediction problems. This book describes representative algorithms for each problem with some good examples and how to implement intelligent DHM systems with ANNs, CNNs, and GANs on the computer. Furthermore, this book gives details of the deep learning CNN to automatically reconstruct the best focused images in DHM. It describes GAN models to eliminate superimposed twin-image noise in the phase image of Gabor holography. It also introduces a deep learning model to compute an unwrapped phase solution in DHM. This book brings together the literature addressing biomedical applications of DHM combined with AI algorithms (e.g. drug safety testing and compounds selection as a new paradigm for drug toxicity screening) to present recent achievements in this interesting field. For readers with various backgrounds, this book provides a detailed discussion of the use of intelligent DHM in biomedical fields with great potential for biomedical application. This book provides two representative examples of applying intelligent DHM in biomedical fields. The first example describes how instant phenotypic assessment of red blood cells (RBCs) storage lesion can be automatically performed by AI based-DHM, which has the potential to lead to new efficient tools for safe transfusions as well as measurement of stored RBC quality. The second example demonstrates that relevant dynamic parameters of cardiomyocytes can be obtained by DHM phase signal analysis based on AI algorithms to characterize the physiological state of live cardiomyocytes. This finding opens the possibility of automated quantitative analysis of cardiomyocytes suitable for further monitoring some specific drug mediated effects on the dynamics of cardiomyocytes, which represents a promising label-free approach for drug discovery. Therefore, my intention in writing this book, is to introduce AI-based DHM background with detailed description of these two examples in biomedical fields to make it easier for readers with diverse backgrounds to read this book. I hope that readers will be able to gain an understanding of the basics for implementing AI in DHM designs and connecting practical biomedical questions that arise from the use of DHM with various AI algorithms in intelligence models.

Part I

Digital Holographic Imaging

1

Introduction

Biomedical imaging technologies promise opportunities for effective diagnostics and treatments as well as new economic perspectives for the medical technology industry. Increasing demand for accurate early diagnosis and the growth of an aging population are major forces driving the biomedical imaging market. An increase in the number of patients suffering from chronic diseases, such as cancer, and a demand for early diagnosis and treatment are also major factors fueling the growth of medical diagnostics and medical therapeutics.

Cellular imaging remains one of the most important techniques to solve major challenges in life sciences and medicine. Optical microscopy is one of the most productive scientific tools in cell imaging. The identification of microorganisms and cells was explored for the first time in the nineteenth century, which could be considered the beginning of the emergence of modern biology and medicine. However, optical microscopy still faces two major problems. One is resolution limitation due to Abbe's law. Another is the lack of quantitative information due to the inherent limits of conventional optical microscopes. Conventional intensity-based imaging techniques are not robust enough to provide detailed quantitative information for cell morphology. They provide low-contrast images, especially when investigating cells with transparent or semi-transparent features, which makes it difficult to analyze cells.

Consequently, several optical imaging modalities based on contrast mechanisms were developed to overcome these limitations. Among many contrast-generating modes, the Zernike phase contrast (PhC) mode and Nomarski differential interference contrast (DIC) are widely used for live-cell imaging. Unlike fluorescence-based imaging techniques, PhC and DIC can visualize transparent specimens, particularly subcellular structures of living cells, without using a specific staining contrast agent. However, these two non-invasive modes cannot provide a direct or quantitative measure of phase shift or the optical path length over a cell area. Since

Artificial Intelligence in Digital Holographic Imaging: Technical Basis and Biomedical Applications, First Edition. Inkyu Moon.

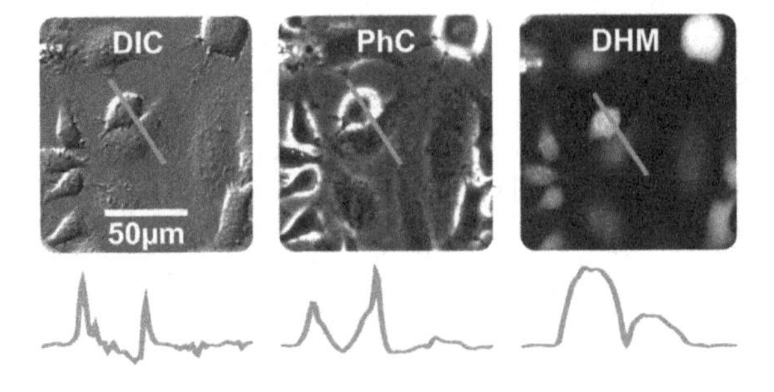

Figure 1.1 Quantification comparison of digital holographic microscopy and other optical imaging modalities.

PhC or DIC signals contain only qualitative information, it is difficult to further perform quantitative analysis of the biophysical properties of live cells with reconstructed cell images (see Figure 1.1).

On the other hand, interference microscopy can provide a direct quantitative measurement of the optical path length based on interference between the reference wave and the object wave that passes through the specimen. Interference microscopy was introduced in the 1950s. Gabor proposed the concept of holography in 1948, which enabled lens-less imaging by reproducing the exact wavefront emerging from the observed specimen. However, due to the non-availability of coherence light sources such as lasers and the high cost of opto-mechanical designs, only a few studies of imaging of live cells are reported in the literature.

In recent years, fluorescence microscopy in a confocal configuration, and its extension into multi-photon fluorescent excitation, have been widely used for cell imaging in biology among various contrast-generating modes. However, they cannot provide any information about dielectric properties in terms of the underlying biological functions of live cells. Fortunately, these dielectric properties can be measured using new digital holographic imaging approaches that have recently emerged as promising techniques for the accurate, quantitative visualization of cell structure and dynamics in a non-invasive manner. To accurately observe live cells without disturbing them is tremendously important. For example, when the aim is to assess drug-mediated cellular effects.

The rapid development of computing technologies and scientific advances in light sources has opened up a new opportunity in the field of holography and interferometry. Integrating techniques in holography with numerical processing has led to the development of digital holographic microscopy (DHM) with a nanometric axial sensitivity that provides a reliable and quantitative phase signal observed in live cells [1–5]. Therefore, the fusion of DHM and information technologies offers an automatic, low-cost, and reliable tool to identify various cell types, including protozoa, bacteria, plant cells, blood cells, nerve cells, stem cells, and cardiomyocytes. DHM allows scientists [2] to observe the growing process of the cell as close to natural conditions as possible, whether in a cell-culture flask or the tissue environment. Moreover, by numerical reconstruction, DHM offers unique possibilities including an extended depth of focus and a posteriori numerical refocusing, which allow quantitative and non-invasive analysis of cell structure, contents, and dynamics with different time scales that vary from a few milliseconds to several days. Another advantage of DHM is that images of both single cells and populations can be obtained. However, application of DHM to the field of cell biology is still in its beginning stages. There are few studies on the automated quantitative analysis of large-scale holographic cell datasets. Therefore, systems and methods for performing intelligent analyses of large-scale holographic images are becoming more important as digital holographic information rapidly increases with the development of DHM technology. The development of automated procedures to study various cell types using DHM can significantly benefit cell biology studies. Until now, most experiments using DHM in its single mode were performed to prove that the technique is useful. Practically, DHM can measure relevant biophysical cell parameters including absolute volume, dry mass density, nanoscale membrane fluctuations, and biomechanical properties. Furthermore, the development of DHM in a multimodal platform with the fusion of DHM and other cell imaging methods such as fluorescence confocal imaging opens up the possibility to simultaneously measure a large number of relevant and specific cell parameters, which can help scientists understand cellular processes including cell differentiation, cell cycle, apoptosis, and cell migration. For example, DHM systems integrated with fluorescence microscopy can add valuable information about cell morphology and motility with a broad variety of fluorescence labeling tools to study cell function. Measurements of multiple cell parameters resulting from multimodal imaging could even enable high throughput cellular screening assays to identify cell biomarkers for various diseases. The time has come to develop intelligent DHM in a multimodal platform and apply multimodal holographic cell imaging informatics to medical and biological research.

To summarize, the DHM enables label-free, quantitative assessment of biological specimens. There is recent growth in the study of techniques and applications of DHM to address important biomedical questions that cannot be solved with conventional optical imaging techniques. This rapidly emerging field enables scientists to investigate cells and tissues in terms of their morphology and dynamics at a nanoscale resolution over temporal scales ranging from milliseconds to days. Quantitative measurements of intrinsic optical, chemical, and mechanical properties are likely to yield a new understanding of cell and tissue pathophysiology.

Before explaining key concepts of DHM and its success in biomedical imaging and applications, Chapter 2 provides an introductory background on DHM with a quick review and summary of scalar diffraction theory, coherent imaging, and diffraction-limited imaging. In Chapter 3, fundamental definitions and concepts in lateral and depth resolutions in optical imaging are described. Phase information obtained by DHM provides principal values wrapped in a range of $-\pi$ to π, which can cause 2π phase jumps due to phase periodicity of trigonometric functions. A phase unwrapping process must be conducted to remove 2π phase discontinuities in the image and obtain an estimate of the true continuous phase image. Chapter 4 provides a review of phase unwrapping algorithms to solve challenging problems such as phase discontinuities. Advanced unwrapping algorithms can be categorized into three types: global algorithms, region algorithms, and path-following algorithms. This chapter further gives a detail explanation of the key concepts of quality-guided path-following algorithms. Chapter 5 introduces off-axis DHM in an example transmission of the quantitative visualization of phase objects such as living cells. This chapter also presents a detailed description of the numerical reconstruction procedure in DHM. In addition, this chapter shows that the transverse resolution is equal to the diffraction limit of the imaging system. Chapter 6 introduces Gabor DHM with a simple optical setup as a promising tool for measuring the distribution(s) of particles in a liquid solution with large depths. It demonstrates that the Gabor DHM can resolve the locations of several thousand particles and measure their motions and trajectories with time-lapse imaging.

References

1. Marquet, P., Rappaz, B., Magistretti, P. et al. (2005). Digital holographic microscopy: a noninvasive contrast imaging technique allowing quantitative visualization of living cells with subwavelength axial accuracy. *Opt. Lett.* 30: 468–470.
2. Moon, I., Daneshpanah, M., Javidi, B., and Stern, A. (2009). Automated three-dimensional identification and tracking of micro/nanobiological organisms by computational holographic microscopy. *Proc. IEEE* 97: 990–1010.
3. Cuche, E., Marquet, P., and Depeursinge, C. (1999). Simultaneous amplitude and quantitative phase contrast microscopy by numerical reconstruction of Fresnel off-axis holograms. *Appl. Opt.* 38 (34): 6994–7001.
4. Anand, A., Moon, I., and Javidi, B. (2017). Automated disease identification with 3-D optical imaging: a medical diagnostic tool. *Proc. IEEE* 105: 924–946.
5. Moon, I., Daneshpanah, M., Anand, A., and Javidi, B. (2011). Cell identification computational 3-D holographic microscopy. *Opt. & Photon. News* 22: 18–23.

2

Coherent Optical Imaging

This chapter provides a basic understanding of the scientific principles associated with optical image formation based on Fourier optics [1–3] for numerous biomedical applications. Digital holographic microscopy (DHM) is a powerful method for three-dimensional (3D) and quantitative sensing, imaging, and measuring of biological and microscopic samples. Methods explored in this chapter form the basis of DHM systems used in biomedical imaging.

2.1 Monochromatic Fields and Irradiance

A monochromatic scalar field with a single-frequency propagating in free space can be described by (2.1):

$$u(x,y,z,t) = A(x,y,z)\cos\left[2\pi ft - \varphi(x,y,z)\right], \tag{2.1}$$

where A is the amplitude, φ is the phase at a position (x, y, z), and f is the temporal frequency. A highly monochromatic light source, like lasers, can provide a specific form of Eq. (2.1), which is a plane wave that propagates in the z direction:

$$u(z,t) = A\cos[2\pi ft - kz],$$

The wavenumber k is defined as $k = \frac{2\pi}{\lambda}$, where λ is the light wavelength. This plane wave can be interpreted as extending infinitely in x and y directions. If the monochromatic light is propagating in a linear medium such as air, the temporal frequency of the resulting light field will be unchanged. Therefore, the temporal term can be ignored. Note that replacing the cosine function with a complex exponential form leads to a function that expresses the spatial distribution of the light field:

$$U(x,y) = A(x,y)\exp\left[j\phi(x,y)\right].$$

Artificial Intelligence in Digital Holographic Imaging: Technical Basis and Biomedical Applications, First Edition. Inkyu Moon.

It is assumed that the field in the (x, y) plane is located at some position on the z-axis and that z is the fundamental propagation direction. Optical sensors respond to the time-averaged, squared magnitude of the light field, which is expressed by

$$I(x,y) = U(x,y)U(x,y)^* = |U(x,y)|^2.$$

2.2 Analytic Expression for Fresnel Diffraction

Light propagation from a 2D source plane can be indicated by the coordinate variables ξ and η (see Figure 2.1). At the source plane, an area Σ defines the illuminated aperture and $U_1(\xi, \eta)$ denotes the field distribution. The Huygens–Fresnel principle predicts the field $U_2(x, y)$ in the observation plane:

$$U_2(x,y) = \frac{z}{j\lambda}\int\int_{\Sigma} U_1(\xi,\eta)\, exp\, \frac{(jkr_{12})}{r_{12}^2} d\xi d\eta. \tag{2.2}$$

where λ is the light wavelength, k is the wavenumber—which is equal to $\frac{2\pi}{\lambda}$ for free space—z is the distance between the centers of the source and observation planes, and r_{12} is the distance between a position on the source plane and a position on the observation plane; r_{12} is expressed as

$$r_{12} = \sqrt{z^2 + (x-\xi)^2 + (y-\eta)^2}. \tag{2.3}$$

Note that integral limits in Eq. (2.2) correspond to the region of the source Σ. This principle supposes that the source acts as an infinite set of virtual point sources, each generating a spherical wave at any position (ξ, η). These spherical waves are summed up at the observation position (x, y), giving rise to interference.

In general, Eq. (2.2) can be expressed as a convolution integral

$$U_2(x,y) = \int\int U_1(\xi,\eta) h(x-\xi, y-\eta) d\xi d\eta \tag{2.4}$$

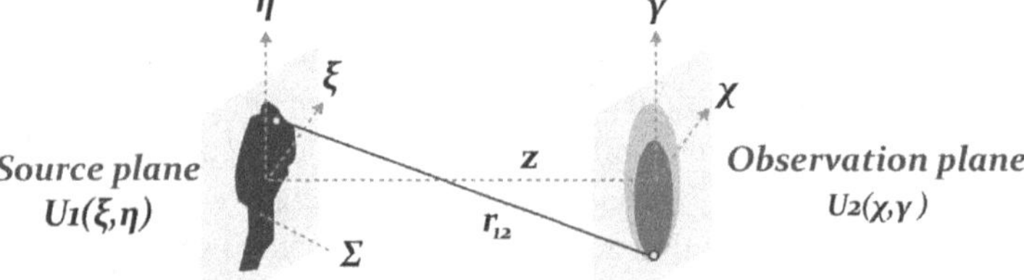

Figure 2.1 Propagation geometry between source and observation planes Adapted from [1, 2].

where the impulse response function in the Rayleigh–Sommerfeld diffraction formula is $h(x,y) = \frac{z}{j\lambda}\frac{exp\,(jkr)}{r^2}$, and where $r = \sqrt{z^2 + x^2 + y^2}$. If the Fourier convolution theorem is applied to Eq. (2.4), resulting equation is

$$U_2(x,y) = IFT\{FT[U_1(x,y)]FT[h(x,y)]\}$$

where *FT* and *IFT* are Fourier and inverse Fourier transforms, respectively.

The Rayleigh–Sommerfeld diffraction expression in Eq. (2.4) can give an accurate diffraction solution since the distance between the source and the observation planes in most optical systems is much greater than a wavelength. To obtain a more manageable diffraction form, the distance r_{12} in Eq. (2.3) can be rewritten using binomial expansion:

$$r_{12} \approx z\left[1 + \frac{1}{2}\left(\frac{x-\xi}{z}\right)^2 + \frac{1}{2}\left(\frac{y-\eta}{z}\right)^2\right].$$

This approximation can be applied to the distance term in the phase of the exponential in Eq. (2.2), which amounts to assuming a parabolic radiation wave rather than a spherical wave. Furthermore, using the approximation $r_{12} \approx z$ in the denominator of Eq. (2.2), the Fresnel diffraction expression is

$$U_2(x,y) = \frac{e^{jkz}}{j\lambda z}\iint U_1(\xi,\eta)\,exp\left\{j\frac{k}{2z}\left[(x-\xi)^2 + (y-\eta)^2\right]\right\}d\xi d\eta.$$

This expression is also a convolution integral, where the impulse response function is

$$h(x,y) = \frac{e^{jkz}}{j\lambda z}\,exp\left[\frac{jk}{2z}(x^2 + y^2)\right].$$

The transfer function is

$$H(f_X,f_Y) = \Im[h(x,y)] = e^{jkz}\,exp\left[j\pi\lambda z(f_X^2 + f_Y^2)\right].$$

Another useful form of the Fresnel diffraction expression can be obtained by moving the quadratic phase term that is a function of x and y outside integrals:

$$U_2(x,y) = \frac{exp\,(jkz)}{j\lambda z}\,exp\left[j\frac{k}{2z}(x^2 + y^2)\right]$$
$$\times \iint\left\{U_1(\xi,\eta)\,exp\left[j\frac{k}{2z}(\xi^2 + \eta^2)\right]\right\}exp\left[-j\frac{2\pi}{\lambda z}(x\xi + y\eta)\right]d\xi d\eta.$$

Along with the amplitude and chirp multiplicative factors out front, this expression is a Fourier transform of the source field times a chirp function. The frequency variable substitutions used for the transform are

$$f_\xi \rightarrow \frac{x}{\lambda z}, \qquad \text{and} \qquad f_\eta \rightarrow \frac{y}{\lambda z} \tag{2.5}$$

Fraunhofer diffraction in far-field can be obtained by assuming that the diffraction pattern is the result of very long propagation from the diffracting object. In other words, $z \gg \left(\frac{k(\xi^2+\eta^2)}{2}\right)_{max}$ as follows

$$U_2(x,y) = \frac{exp\,(jkz)}{j\lambda z}\, exp\left[j\frac{k}{2z}(x^2+y^2)\right] \times \iint U_1(\xi,\eta)\, exp\left[-j\frac{2\pi}{\lambda z}(x\xi + y\eta)\right] d\xi d\eta.$$

The Fraunhofer diffraction pattern is also viewed at the focal plane of an imaging lens. The Fraunhofer expression can be considered a scaled version of the Fourier transform of the source field with the variable substitutions in Eq. (2.5).

2.3 Lens Transmittance Function

Light sources in the previous section are assumed to be apertures illuminated by a plane wave. They are modeled with a zero-phase component. In this section, the spherical lens transmittance function that alters the magnitude and phase of an initial field is presented. An optical imaging system generally uses a converging or a diverging optical beam. As shown in Figure 2.2, a beam with a spherical wavefront can converge toward the point z_f on the z-axis.

We can derive an equation to find the converging phase function in the x–y plane at $z = 0$. This is written as

$$\phi_s(x,y) = -k\sqrt{z_f^2 + x^2 + y^2}. \tag{2.6}$$

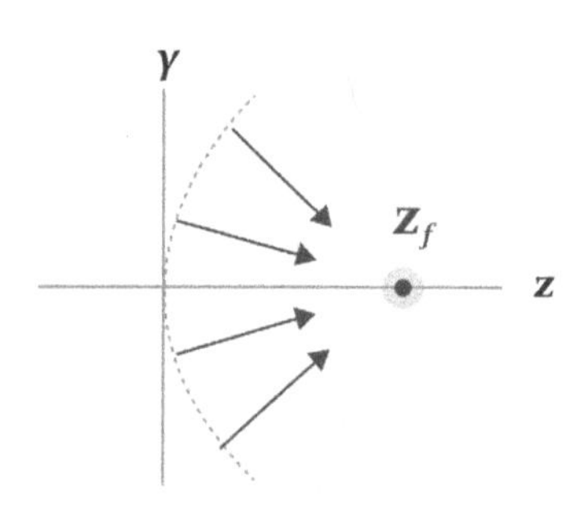

Figure 2.2 A beam with a spherical wavefront [2] / with permission of SPIE.

A converging wavefront has a negative sign in Eq. (2.6) as shown in Figure 2.2, whereas a diverging wavefront has a positive sign. The application of the binomial approximation gives a parabolic phase front that approximates the spherical phase front:

$$\phi(x,y) = -\frac{k}{2z_f}(x^2+y^2).$$

Therefore, the transmittance function for a converging optical beam can be written as

$$t_A(x,y) = exp\left[-j\frac{k}{2z_f}(x^2+y^2)\right].$$

This is a phase chirp function of the same form for the Fresnel impulse response function h, although the exponent sign is negative. A lens is an optical element that uses refraction to focus or diverge light. Therefore, the transmittance function for an ideal, simple lens can be written as

$$t_A(x,y) = P(x,y)\, exp\left[-j\frac{k}{2f}(x^2+y^2)\right], \tag{2.7}$$

where f is known as the focal length and $P(x, y)$ is the pupil function. A positive focal length produces a converging wavefront from a plane-wave input, whereas a negative focal length produces a diverging wavefront. The pupil function accounts for the physical size of the lens. For example, the most common lens pupil function is a circle:

$$P(x,y) = \text{circle}\left[\frac{\sqrt{x^2+y^2}}{w_L}\right],$$

where w_L is the radius of the lens aperture. It is not always practical to implement the transmittance function of Eq. (2.7) in Fresnel propagation. However, if the field incident on the lens is $U_1(x_1, y_1)$, then the field exiting the lens is $U_1(x_1, y_1)t_A(x_1, y_1)$. Insert this into the Fresnel diffraction expression and set $z=f$, the chirp function in the integral cancels and

$$\begin{aligned} U_2(x_2,y_2) = {} & \frac{exp\,(jkf)}{j\lambda f}\, exp\left[j\frac{k}{2f}(x_2^2+y_2^2)\right] \\ & \times \iint U_1(x_1,y_1)P(x_1,y_1)\, exp\left[-j\frac{2\pi}{\lambda f}(x_2x_1+y_2y_1)\right] dx_1 dy_1. \end{aligned} \tag{2.8}$$

The expression in Eq. (2.8) shows that the field at the focal plane of an ideal positive lens is simply the Fraunhofer pattern of the incident field with $z=f$.

2.4 Geometrical Imaging Concepts

In general, the purpose of optical imaging is to reproduce the irradiance distribution of an object at an image plane. To obtain an image of an object, we must collect and focus the optical beam from an arbitrary object point at the image plane. Geometrical optics and diffraction theory are used extensively in lens and optical imaging system design. For an optical imaging situation, the lens law describes the relationship needed under the paraxial condition (small ray angles relative to the optical axis) for the best focus imaging:

$$\frac{1}{z_1} + \frac{1}{z_2} = \frac{1}{f},$$

where f is the lens focal length, z_1 is the distance from the object to the front principal plane of the lens, and z_2 is the distance from the back principal plane to the image location.

To form a real image, z_1 and z_2 are positive and the lens focal length f is also positive. A positive lens converges light rays, whereas a negative lens diverges rays. Practical imaging systems use combinations of lenses to control aberrations. However, imaging still requires a positive focal length for the combined lens group. The ratio of the image height y_2 to the object height y_1 is known as the transverse magnification M_t. For a single lens system, it is given by

$$M_t = \frac{y_2}{y_1} = -\frac{z_2}{z_1}.$$

The minus sign indicates an inverted image. An imaging system is characterized by its pupils. Pupils are virtual apertures that indicate the opening available to collect light from the object (entrance pupil, EP) and the opening from which the collected light exits on its way to form an image (exit pupil, XP). These pupils are known as the aperture stop, which limits the collection of light. The lens is the stop for the system in Figure 2.3. The stop generates fundamental diffractive effects in the image. These diffractive effects are generated due to the stop that represents the fundamental performance limit of an imaging system.

Figure 2.4 illustrates that the physical elements of a system (lenses, mirrors, iris, etc.) can be reduced to EP and XP models. The distance from an object point on the

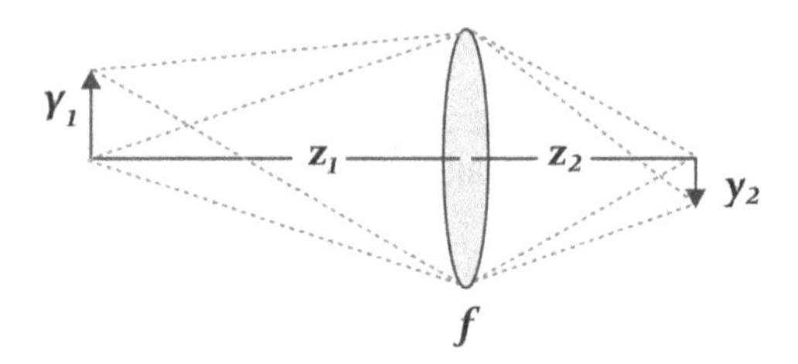

Figure 2.3 Geometrical imaging with a thin, positive lens of focal length [2] / with permission of SPIE.

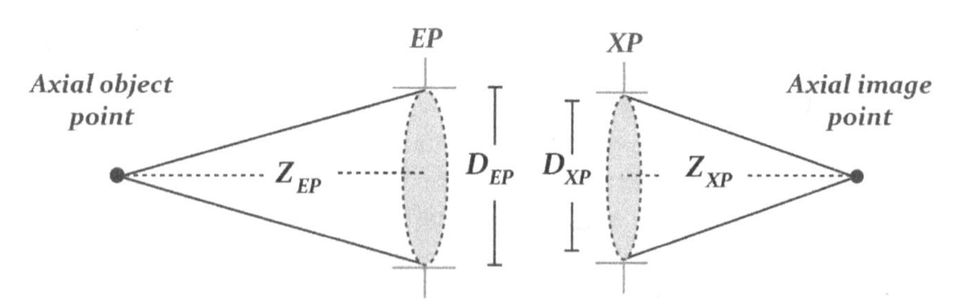

Figure 2.4 Entrance pupil (EP) and exit pupil (XP) model of an imaging system [2] / with permission of SPIE.

optical axis to the EP is z_{EP}, and the distance from the XP to the axial image point is z_{XP}. The EP diameter is D_{EP}, and the XP diameter is D_{XP}. The aperture stop of a system leads to fundamental diffractive effects in the image. For a thin-lens imaging system, $z_1 = z_{EP}$, $z_2 = z_{XP}$, and $D_{EP} = D_{XP}$ = lens diameter.

2.5 Coherent Imaging Theory

Figure 2.5 shows a general imaging arrangement. Imaging with coherent illumination, such as with a coherent laser, can be described with a convolution operation involving the optical field. The process is expressed as

$$U_i(u, v) = h(u, v) \otimes U_g(u, v), \tag{2.9}$$

where u, v are image plane spatial coordinates, U_i is the field at the image plane, h is the impulse response for the imaging system, and U_g is the ideal geometrical-optics predicted image field, which is a scaled copy of the object field $U_o(x, y)$:

$$U_g(u, v) = \frac{1}{|M_t|} U_o\left(\frac{u}{M_t}, \frac{v}{M_t}\right),$$

Note that if M_t is negative, then the resulting image will appear inverted relative to the object. In Eq. (2.9), the ideal geometrical field is blurred through the convolution with the impulse function. In the frequency domain, the corresponding spectra for Eq. (2.9) are related by the equation

$$G_i(f_U, f_V) = H(f_U, f_V) \otimes G_g(f_U, f_V),$$

where H is the coherent image transfer function (or amplitude transfer function). It is defined by $H(f_U, f_V) = P(-\lambda z_{XP} f_U, -\lambda z_{XP} f_V)$, where P is the pupil function of the system. The negative sign in the pupil argument gives a scaled, inverted pupil

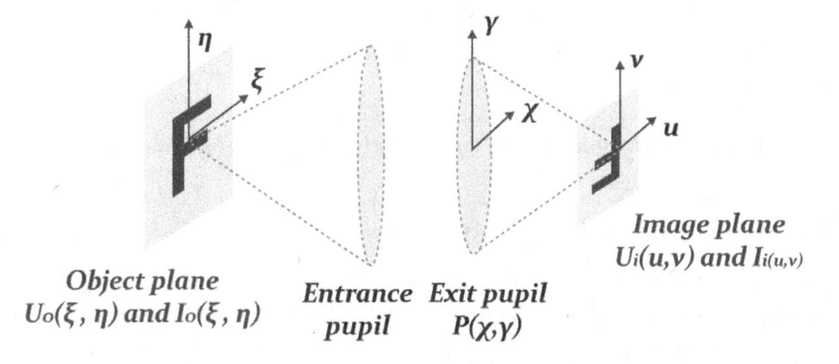

Figure 2.5 Generalized model of an imaging system Adapted from [1, 2].

function. In a system with a perfect pupil function, only boundaries of the pupil are involved in diffractive effects (diffraction limited).

For the first example, a square pupil function is

$$P(x,y) = \text{rect}\left(\frac{x}{2w_{XP}}\right)\text{rect}\left(\frac{y}{2w_{XP}}\right). \tag{2.10}$$

From Eq. (2.10), the coherent transfer function is

$$H(f_U, f_V) = \text{rect}\left(\frac{-\lambda z_{XP} f_U}{2w_{XP}}\right)\text{rect}\left(\frac{-\lambda z_{XP} f_V}{2w_{XP}}\right).$$

Since the rectangle function is symmetric, the negative sign can be ignored. The coherent cutoff frequency along the u or v direction is defined as

$$f_0 = \frac{w_{XP}}{\lambda z_{XP}}, \tag{2.11}$$

Spatial frequencies with values greater than f_0 will not be preserved in the image plane field. A second example is the circular pupil function

$$P(x,y) = \text{circ}\left(\frac{\sqrt{x^2+y^2}}{w_{XP}}\right).$$

The coherent transfer function is

$$H(f_U, f_V) = \text{circ}\left(\frac{\sqrt{(-\lambda z_{XP} f_U)^2 + (-\lambda z_{XP} f_V)^2}}{w_{XP}}\right) = \text{circ}\left(\frac{\sqrt{f_U^2 + f_V^2}}{f_0}\right),$$

where f_0 is again the coherent cutoff frequency as defined in Eq. (2.11). Unlike the square aperture, the cutoff frequency in this case is the same radially (in all directions) in the frequency plane. To observe or record a coherent image, the irradiance given by $I_i = |U_i|^2$ is measured. As a result of the squaring operation, the irradiance image can theoretically gain up to twice the frequency content of the field. Think about the fact that $cos^2(2\pi bx) = \frac{1}{2}[1 - cos(2\pi 2bx)]$. Thus, when an irradiance image is formed, the following cutoff should be considered $2f_0 = \frac{2w_{XP}}{\lambda z_{XP}}$.

Simulating coherent imaging on the computer based on Eq. (2.9) can be implemented as

$$U_i(u,v) = IFT\{H(f_U, f_V) FT[U_g(u,v)]\}.$$

For more information on optical image formation based on Fourier optics, perhaps the best source is Joseph Goodman's excellent book [1] or David Voelz's book [2].

References

1 Goodman, J. (1996). *Introduction to Fourier Optics*. New York: McGraw-Hill.
2 Voelz, D. (2011). *Computational Fourier Optics: A MATLAB Tutorial*. SPIE Press.
3 Hecht, E. (2002). *Optics*, 4e. Boston, MA: Addison-Wesley.

3

Lateral and Depth Resolutions

In the field of optics, the numerical aperture (NA) of an optical system is a dimensionless parameter that characterizes the range of angles over which the optical system can accept or emit light. By incorporating the index of refraction in the definition, NA has the following property: when there is no refractive power at the interface, it is constant for a beam as it travels from one material to another. In most areas of optics, especially in microscopy, the NA of an optical system such as an objective lens is defined as

$$NA = n\sin\theta,$$

where n is the refractive index of the medium surrounding the lens and θ is the maximal half-angle of the cone of light that can enter or exit the lens.

Optical systems, including digital holographic microscopy, form a 3D image of the object by reconstructing the wave plane scattered by the object. Resolution characterizes the quality of the image formed by this optical system. Resolution has two types: lateral resolution and axial resolution.

3.1 Lateral Resolution

In 1835, Airy reported that the diffracted light at a circular aperture exhibited a concentric ring in which the maximum and minimum intensities alternated [1]. Light passing through a circular aperture such as a lens interferes with itself, which creates this ring-shaped diffraction pattern known as the Airy pattern, when the wave front of the transmitted light over the aperture is spherical or flat. The Airy pattern and Airy disk describe the optimally focused light spot that a perfect lens with a circular aperture can create, which is limited by the diffraction

Artificial Intelligence in Digital Holographic Imaging: Technical Basis and Biomedical Applications, First Edition. Inkyu Moon.

of the light. The Airy pattern is observable in the far field. Thus, the intensity of the Airy pattern follows the Fraunhofer diffraction pattern of a circular aperture:

$$I(x,y) = I_0 \left| \frac{2J_1\left(\frac{kaq}{L}\right)}{\frac{kaq}{L}} \right|^2, \tag{3.1}$$

where (x, y) are the coordinates of the observation point at the image plane, I_0 is the maximum of the intensity at the Airy disk center, J_1 is the Bessel function of the first kind of order one, k is the wavenumber, a is the radius of the aperture, $q = (x^2 + y^2)^{1/2}$ is the radial distance from the observation point to the optical axis, and L is the distance from the detector to the aperture. In 1882, Abbe provided a formula of lateral resolution limits:

$$\boldsymbol{R}^{Abbe}_{Lateral} = \frac{\lambda}{2n\sin\theta_{max}} = \frac{\lambda}{2NA},$$

where $\theta_{\max}$ = the maximum of the scattering angle detected by the optical system, n = the refractive index ($n = 1$ for vacuum or air), and NA = the numerical aperture of the system [2].

The two point sources are regarded as barely resolved when the principal diffraction maximum of the Airy disk of one light source coincides with the first minimum of the diffraction pattern of the Airy disk of the other [3, 4] as illustrated in Figure 3.1a. In Eq. (3.1), the first minimum of the diffraction pattern occurs when the argument of the Bessel function is 3.83 [5]. It is measured from the direction of incoming light. Thus, the Rayleigh resolution criterion for incoherent light is given by

$$\boldsymbol{R}^{Rayleigh}_{Lateral} \simeq 0.61\frac{\lambda L}{a} = 0.61\frac{\lambda}{NA},$$

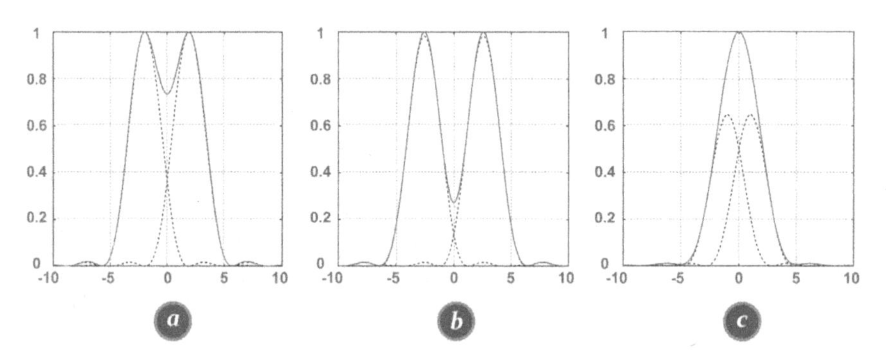

Figure 3.1 Images of two point-sources and resolution criteria. Images of two-sources are (a) barely resolved, (b) resolved, and (c) unresolved.

where λ is the wavelength of the light. The effect of light coherence on lateral resolution can be found in optical textbooks [4, 6]. The two point sources are regarded as resolved when the zero-order diffraction maximum of one source's diffraction pattern coincides with the first-order diffraction maximum of the other source's diffraction pattern, which occurs when the argument of the Bessel function in Eq. (3.1) is 5.14 [5], as shown in Figure 3.1b:

$$R_{Lateral}^{Rayleigh,coherent} \simeq 0.82\frac{\lambda L}{a} = 0.82\frac{\lambda}{NA}.$$

The object consists of two point sources positioned symmetrically from the optical axis and on a line parallel to the screen, as shown in Figure 3.2. The scattered spherical waves emerge from two objects, O_1 and O_2, at positions $\mathbf{r}_1$ and $\mathbf{r}_2$, respectively, which are vertically spaced d apart, causing interference. An interference pattern of the intensity is displayed on the screen D away from the two points. The total wave field is

$$O_T(\mathbf{r}) = A_1\frac{exp\,[ik|\mathbf{r}-\mathbf{r}_1|]}{|\mathbf{r}-\mathbf{r}_1|} + A_2\frac{exp\,[ik|\mathbf{r}-\mathbf{r}_2|]}{|\mathbf{r}-\mathbf{r}_2|},$$

where A_1 and A_2 are complex amplitudes of two scattered spherical waves, k is the wavenumber, and $\mathbf{r}$ is the position vector to the observation on the source screen. The intensity of the contrast image is

$$I_T(\mathbf{r}) = \frac{|A_1|^2}{|\mathbf{r}-\mathbf{r}_1|^2} + \frac{|A_2|^2}{|\mathbf{r}-\mathbf{r}_2|^2} + A_1A_2^*\frac{exp\,[ik(|\mathbf{r}-\mathbf{r}_1|-|\mathbf{r}-\mathbf{r}_2|)]}{|\mathbf{r}-\mathbf{r}_1||\mathbf{r}-\mathbf{r}_2|}$$
$$+ A_1^*A_2\frac{exp\,[-ik(|\mathbf{r}-\mathbf{r}_1|-|\mathbf{r}-\mathbf{r}_2|)]}{|\mathbf{r}-\mathbf{r}_1||\mathbf{r}-\mathbf{r}_2|},$$

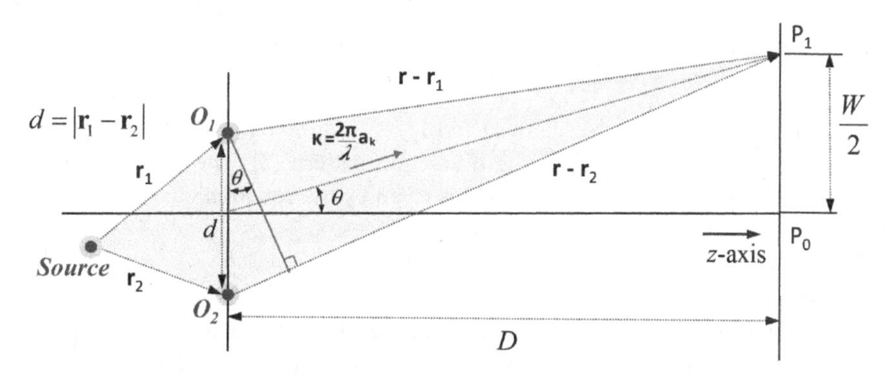

Figure 3.2 Schematics of lateral resolution.

$$I_T(\mathbf{r}) \simeq \frac{|A_1|^2}{|\mathbf{r}-\mathbf{r}_1|^2} + \frac{|A_2|^2}{|\mathbf{r}-\mathbf{r}_2|^2} + \frac{2|A_1||A_2|}{|\mathbf{r}-\mathbf{r}_1||\mathbf{r}-\mathbf{r}_2|} \cos\left[k(|\mathbf{r}-\mathbf{r}_1| - |\mathbf{r}-\mathbf{r}_2|) + \varphi\right], \tag{3.2}$$

where φ is the relative phase between A_1 and A_2. For simplicity, we assume that the relative phase is zero.

The cosine term in Eq. (3.2) produces the modulation of the holographic interference pattern, which has the maxima when its argument is $2\pi n$. To resolve two objects, that is, to distinguish between them, it must be ensured that at least the zero-order and first-order maxima of the interference pattern are reliably recorded. The phase at point P_0 where the zero-order maximum of the interference pattern is located is shown below because the point P_0 is the same distance away from the two objects:

$$\varphi_{P_0} = k(|\mathbf{r}-\mathbf{r}_1| - |\mathbf{r}-\mathbf{r}_2|)_{P_0} = 0.$$

On the other hand, since D is sufficiently larger than d, the phase at point P_1 where the first-order maximum of the interference pattern is located can be approximated by

$$\varphi_{P_1} = k(|\mathbf{r}-\mathbf{r}_1| - |\mathbf{r}-\mathbf{r}_2|)_{P_1} \simeq -\frac{2\pi}{\lambda} d \sin\theta.$$

To distinguish between the two objects, the difference between the phases in P_1 and P_0 must be at least 2π: $\varphi_{P_0} - \varphi_{P_1} = \frac{2\pi}{\lambda} d \sin\theta \geq 2\pi$. Therefore, the two points can be resolved laterally if their lateral separation satisfies

$$\boldsymbol{R}_{Lateral} = d = |\mathbf{r}_1 - \mathbf{r}_2| \geq \frac{\lambda}{\sin\theta} = \frac{\lambda}{NA},$$

where $NA = \sin\theta = \frac{\frac{W}{2}}{\sqrt{D^2 + \left(\frac{W}{2}\right)^2}}$ and W is the screen width.

Meanwhile, the diffraction pattern resolution criterion is determined by the highest detectable frequency in the diffraction pattern [5]. The distribution of the scattered wave front in the far field is given by the equation

$$U(\mathbf{r}) = \iiint \exp(ikz_0) U(\mathbf{r}_0) \frac{\exp(ik|\mathbf{r}-\mathbf{r}_0|)}{|\mathbf{r}-\mathbf{r}_0|} d\mathbf{r}_0,$$

where $U(\mathbf{r}_0)$ is the object distribution, $\mathbf{r}_0$ is the position vector of the object, and $\mathbf{r}$ is the position vector of the observation in the far-field domain. The argument of the second exponent in the integral can be expressed as

$$k|\mathbf{r}-\mathbf{r}_0| = k\sqrt{r^2 - 2\mathbf{r}\cdot\mathbf{r}_0 + r_0^2} \approx k\left(r - \frac{\mathbf{r}}{r}\cdot\mathbf{r}_0\right) = kr - \mathbf{K}\cdot\mathbf{r}_0,$$

where $\mathbf{K} = k\frac{\mathbf{r}}{r} = \frac{k}{r}(x\,\hat{\mathbf{x}} + y\,\hat{\mathbf{y}} + z\,\hat{\mathbf{z}}) = K_x\hat{\mathbf{x}} + K_y\hat{\mathbf{y}} + K_z\hat{\mathbf{z}}$ is the scattering vector. Thus, we can obtain

$$U(\mathbf{K}) \propto \iiint exp\,(ikz_0)U(\mathbf{r}_0)\,exp\,(-i\mathbf{K}\cdot\mathbf{r}_0)d\mathbf{r}_0$$

$$= \iiint exp\,(ikz_0)U(x_0, y_0, x_0)\,exp\left[-i(K_x x_0 + K_y y_0)\right]$$

$$\times\ exp\left[-iz_0\left(k^2 - K_x^2 - K_y^2\right)\right]dx_0dy_0dz_0,$$

where the resulting far-field distribution is in (K_x, K_y) coordinates.

A wave front diffracted on two objects with period d will create a peak in its far-field diffraction pattern at $K_{x,y}^d = \frac{2\pi}{d}$. The largest detected $K_{x,y}$ in k-space is acquired at the largest diffraction angle:

$$K_{x,y} = ksin\theta_{max} = \frac{2\pi}{\lambda}\sin\theta_{max}. \tag{3.3}$$

Thus, the lateral resolution is $\boldsymbol{R}_{Lateral}^{diffr.pattern} = d_{x,y} = \frac{2\pi}{K_{x,y}} = \frac{\lambda}{\sin\theta_{max}} = \frac{\lambda}{NA}$.

3.2 Depth (or Axial) Resolution

When a point source is imaged by an optical system, the axial distribution of the intensity of the wave front is given as

$$I(0,0,\Delta z) = I_0\left|\frac{\sin\left(\frac{\beta a^2}{2}\right)}{\frac{\beta a^2}{2}}\right|^2,$$

where $I_0 = \left(\frac{\pi a^2}{\lambda L}\right)^2, \beta = \frac{\pi\Delta z}{(\lambda L^2)}$, Δz is the defocus distance from the in-focus position at L, and a = the radius of the aperture [5].

According to the Rayleigh criterion, two point sources are regarded as barely resolved when the first minimum of diffraction pattern occurs at $\beta\frac{a^2}{2} = \pi$. Therefore, the axial resolution is given by

$$\mathbf{R}_{Axial}^{Rayleigh} = \Delta z = \frac{2\lambda L^2}{a^2} = \frac{2\lambda}{(NA)^2},$$

where $NA \approx \frac{a}{L}$.

Axial resolution for the diffraction pattern can be determined as

$$\mathbf{R}_{Axial}^{Diff.Pattern} = \frac{2\pi}{\Delta K_z},$$

where ΔK_z is the available spread of K_z, which is given by the maximum variation of K_z that can be estimated on the detector [5]. On the optical axis, $K_z = k$. At the largest diffraction angle, $K_z = \sqrt{k^2 - K_{x,y}^2} \approx k - \left(\frac{K_{x,y}^2}{2k}\right)$. The spread of K_z is given as $\Delta K_z = k - \sqrt{k^2 - K_{x,y}^2} \approx \left(\frac{K_{x,y}^2}{2k}\right)$. Using Eq. (3.3), the equation obtained is

$$\mathbf{R}_{Axial}^{Diff.Pattern} \approx \frac{2\pi}{\Delta K_z} = 2k\frac{2\pi}{K_{x,y}^2} = \frac{2\lambda}{sin^2\theta_{max}} = \frac{2\lambda}{(NA)^2}.$$

As shown in Figure 3.3, the laser beam from the source is diffracted by objects at two points O_1 and O_2, which are horizontally spaced d apart, causing interference. An interference pattern of the intensity is displayed on the screen D away from these two points. In this figure, d is very small compared to D. Thus, the angle between the optical axis (the z-axis) and the line $\overline{O_1P_1}$ can be approximated as θ. The phase at point P_0 where the zero-order maximum of the interference pattern is located can be expressed as

$$\varphi_{P_0} = k(|\mathbf{r} - \mathbf{r}_1| - |\mathbf{r} - \mathbf{r}_2|)_{P_0} = \frac{2\pi}{\lambda}d.$$

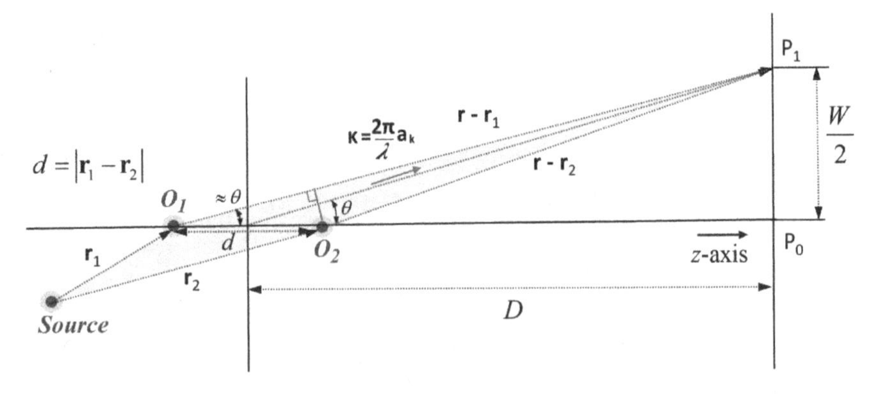

Figure 3.3 Schematics of depth (or axial) resolution.

On the other hand, the phase at point P_1, where the first-order maximum of the interference pattern is located, can be expressed using the binomial expansion of the square root where $1 \gg sin^2\theta$:

$$\varphi_{P_1} = k(|\mathbf{r}-\mathbf{r}_1| - |\mathbf{r}-\mathbf{r}_2|)_{P_1} \simeq \frac{2\pi}{\lambda} dcos\theta$$
$$= \frac{2\pi}{\lambda} d\sqrt{(1 - sin^2\theta)} \approx \frac{2\pi}{\lambda} d\left(1 - \frac{1}{2} sin^2\theta\right).$$

To be able to distinguish two objects, the difference between phases in P_0 and P_1 must be at least 2π, in other words, $\varphi_{P_0} - \varphi_{P_1} = \left(\frac{\pi}{\lambda}\right) d\, sin^2\theta \geq 2\pi$. The lateral resolution can be expressed as $d = |\mathbf{r}_1 - \mathbf{r}_2| \geq \frac{2\lambda}{sin^2\theta} = \frac{2\lambda}{(NA)^2}$, where

$$NA = sin\theta = \frac{\frac{w}{2}}{\sqrt{D^2 + \left(\frac{w}{2}\right)^2}}.$$

References

1 Airy, G. (1835). On the diffraction of an object-glass with circular aperture. *Trans. Cambridge Philos. Soc.* 5: 283–291.

2 Abbe, E. (1882). XI. The relation of aperture and power in the microscope. *J. R. Microsc. Soc.* 2: 460–473.

3 Rayleigh, J. (1879). Investigations in optics, with special reference to the spectroscope. *Philos. Mag. J. Sci.* 8: 261–274.

4 Born, M. and Wolf, E. (1999). *Principles of Optics*, 7e. Cambridge University Press.

5 Latychevskaia, T. (2019). Lateral and axial resolution criteria in incoherent and coherent optics and holography, near- and far-field regimes. *Appl. Opt.* 58: 3597–3603.

6 Goodman, J. (2004). *Introduction to Fourier Optics*, 3e. Roberts and Company Publishers.

4

Phase Unwrapping

Phase information associated with a fringe pattern in an interferogram from digital holographic microscopy (DHM) is calculated by shifting the fringe through different known phase increments or by Fourier transforming the fringe pattern, which is obtained by adding a considerable tilt to the wave front causing carrier fringes [1, 2]. In either case, the phase distribution of a phase image means that principal values are wrapped in a range of $-\pi$ to π, which can cause 2π phase jumps due to phase periodicity (with a phase modulus of 2π) of trigonometric functions. A phase unwrapping process must be conducted to remove 2π phase discontinuities in the image and obtain an estimate of the true continuous phase image. Phase unwrapping consists of detecting the location of the phase jump then connecting adjacent pixels by adding or subtracting multiples of 2π to remove phase discontinuities.

Many phase unwrapping algorithms were proposed to solve challenging problems such as phase discontinuities. Phase unwrapping algorithms can be generally grouped into three major categories: global algorithms, region algorithms, and path-following algorithms [3, 4]. Global algorithms minimize differences between discrete gradients of wrapped and unwrapped phase images [5–11]. The L_P-norm and least-squares algorithms are typical examples of this category. Although these algorithms are generally robust, their computational requirements are huge, making them unsuitable for real-time, live-cell imaging applications.

Region algorithms split an image into smaller ones, unwrap regions with respect to each other, and merge them into larger regions until the whole image is processed. These algorithms are regarded as a compromise between robustness and computational intensiveness. Region algorithms can be subclassified into two groups: region-based algorithms [7, 12–15] and tile-based

Artificial Intelligence in Digital Holographic Imaging: Technical Basis and Biomedical Applications, First Edition. Inkyu Moon.

algorithms [8, 9, 16–18]. Region-based algorithms work by dividing the phase map into sub-regions. Phase unwrapping is first performed on sub-regions. Unwrapped regions are then grown or merged gradually. Tile-based algorithms are a special case of region-based algorithms. Tile-based algorithms divide an image into small local grids (tiles), which are unwrapped by simpler algorithms. Tiles are unwrapped individually and merged to a continuous phase map. This approach has some very attractive properties. Most notably, the tile unwrapping step can be implemented efficiently by parallelization. However, tile-based algorithms are prone to error propagation in cases of failed single tile unwraps as well as phase residues.

Path-following algorithms can unwrap the phase map by detecting 2π phase jumps between adjacent pixels along a path. They operate by using simple linear paths [19], sophisticated branch-cut algorithms [20–23] or by selecting an unwrapping path based on a quality criterion [4, 24]. Path-following algorithms can be classified into path-dependent algorithms, residue-compensation algorithms, and quality-guided path-following algorithms. Among path-dependent algorithms, the simplest algorithm is the Schafer and Oppenheim's unwrapper [25]. This algorithm includes spiral and multiple-scan direction methods. These path-dependent algorithms can detect the position of edges or abrupt phase jumps in the image and use this information to calculate running phase offsets. These algorithms can perform fast unwrapping along a predetermined search path. However, they do not remove noise well [25]. Residue-compensation algorithms search for residues in a wrapped image and generate branch cuts to connect residues of opposite orientation [7, 18, 20, 21, 23, 26–30]. The role of cut lines is to generate an unwrapping barrier and prevent the unwrapping path from going through them. The placement of a particular set of cut lines for any given wrapped-phase map is not unique. They can be placed in many different arrangements and orientations. These algorithms can determine the quality of an unwrapped image according to a cut selection strategy. They are generally computationally efficient (or fast), but not robust [7, 21, 26].

Quality-guided, path-following algorithms are some of the most promising methods. They depend on the assumption that a good quality map, or phase map, will lead to a reasonable unwrapping path while grouping pixels [4, 18, 24, 31–38]. The unwrapping path is determined using pixels' reliability. According to the phase map, the highest-quality pixels with the highest-reliability are unwrapped first while the lowest-quality pixels with the lowest-reliability are unwrapped last to prevent error propagation. Although some unwrapping errors can remain undetected and propagate in a way dependent on the unwrapping path, these algorithms are surprisingly robust in practice [3]. They are generally computationally efficient [7] and robust in real-time applications.

4.1 Branch Cuts

The unwrapping along a loop can be achieved by computing the number of 2π discontinuities, $d(i)$ ($i = 1, 2, ..., N$), between adjacent pixels [23]:

$$d(i) = \left[\frac{\Phi(i) - \Phi(i-1)}{2\pi}\right],$$
$$i = 1, 2, ..., N,$$

where $\Phi(i)$ is the phase at the ith pixel and $[\cdot]$ denotes rounding to the nearest integer (i.e., 1, 0, 1). It can be systematically unwrapped by considering a closed loop around each of the smallest possible units of the phase map (i.e., a square of 4 pixels) as shown in Figure 4.1a. The distribution of the discontinuity source map is computed from $\Phi(m, n)$ given by [23]:

$$s(m,n) = \left[\frac{\Phi(m,n+1) - \Phi(m,n)}{2\pi}\right] + \left[\frac{\Phi(m+1,n+1) - \Phi(m,n+1)}{2\pi}\right] + \left[\frac{\Phi(m+1,n) - \Phi(m+1,n+1)}{2\pi}\right] + \left[\frac{\Phi(m,n) - \Phi(m+1,n)}{2\pi}\right].$$

For such a path (traversed in the clockwise sense), the unwrapping error will always be either −1, 0, or 1, which is called the residue. Due to noise, sampling problems, and characteristics of the object, real phase maps can become logically inconsistent, that is, phase unwrapping can become path dependent. These

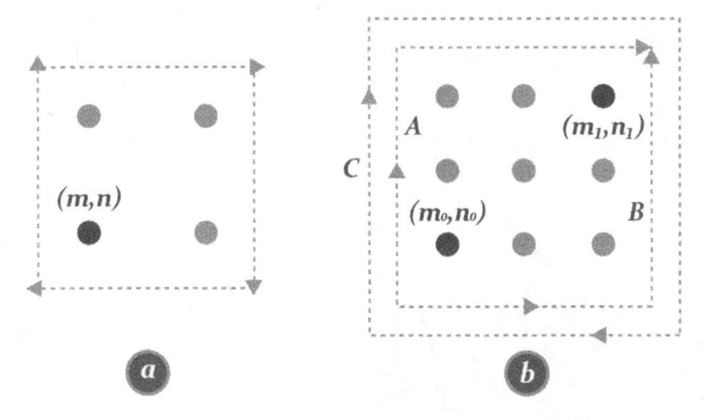

Figure 4.1 (a) Calculation of the distribution of the discontinuity source map. (b) Path A and B are two alternative paths for unwrapping the phase at the point (m_1, n_1), given the phase at the point (m_0, n_0) Adapted from [23].

inconsistencies arise from the existence of discontinuity sources in the phase map. A closed path containing or enclosing one or more discontinuity sources generally has a nonzero value.

Generally, S can be defined for any close loop as $S = \sum_{m,n} s(m, n) = 0$, where the sum is over all pixels enclosed by the loop. In Figure 4.2b, the uniqueness of the unwrapped phase at (m_1, n_1) requires that the total number of discontinuities along the two paths is equal: $S = \sum_{i=1}^{N_A} d_A(i) - \sum_{j=1}^{N_B} d_B(j) = 0$. If path B is reversed, all signs of d_B are changed so that S is the total number of 2π discontinuities around the clockwise, closed loop C: $S = \sum_{k=1}^{N_A+N_B} d_C(k) = 0$.

Each source must be at one end of a cut, with the other end attached to a source of the opposite sign or to the boundary of the phase map [22]. Discontinuity sources (or residues) tend to occur naturally in pairs of opposite signs, although isolated sources can occur near the boundary. Minimizing the length of cut is one criterion used when deciding how to pair sources. A cut is constructed between the two sources (or source and boundary) separated by the shortest distance. Any network of branch cuts that satisfies the criterion—the sum of the residues joined by the branch cuts is zero for all branch cuts—will result in consistent phase unwrapping, in the sense that it is path independent. If branch cuts are constructed by joining groups of residues such that the sum of all joined residues is zero, the unwrapping of the resulting phase will have no inconsistencies so long as the path along which the unwrapping progresses never crosses a branch cut (or discontinuity). For example, path A or B from P_1 to P_2 in Figure 4.2b. On the other hand, since path C crosses a branch cut (see Figure 4.2a), the phase unwrapping along the path will not be free of inconsistencies. When a path crosses a branch cut, 2π multiples must be added to or subtracted from the phase of one of the two pixels that are adjacent to the branch cut on that path.

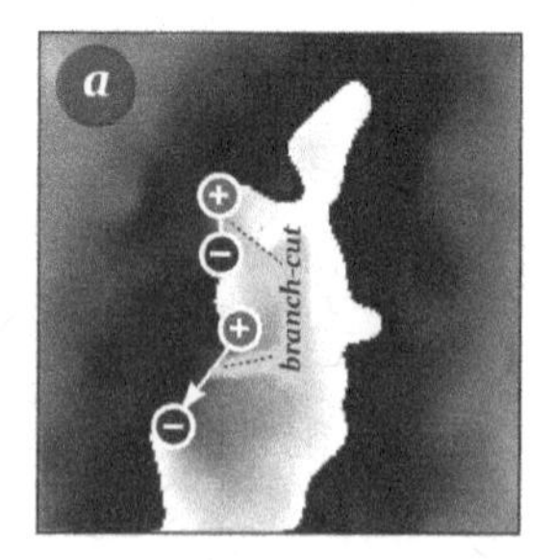

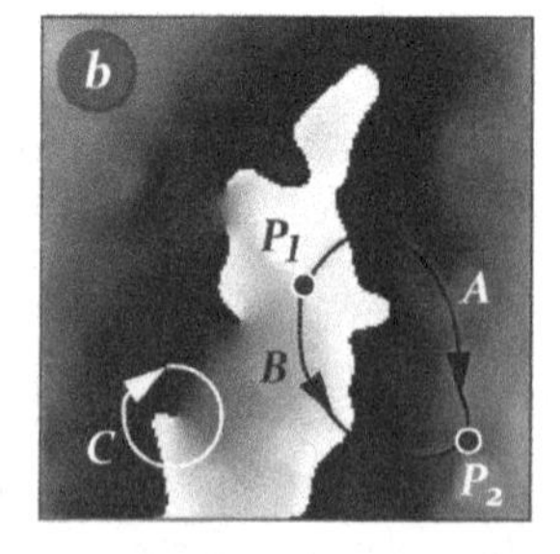

Figure 4.2 Phase map with (a) discontinuity sources and branch cuts, which (b) continue the 2π discontinuity lines [22] / with permission of Optical Society of America.

4.2 Quality-guided, Path-following Algorithms

Quality-guided, path-following algorithms consist of two main concepts: calculation of reliability values and design of the unwrapping path [4, 35, 36]. The algorithm uses criteria to determine the reliability of a point in an image based on gradients or differences between a pixel and its adjacent pixels. Reliability is a criterion that determines the degree of difference between each pixel and its surroundings. The reliability of a pixel is calculated using the second difference between orthogonal and diagonal neighboring pixels. First, the second difference D of the (i, j)th pixel in the 3×3 window is separately calculated (see Figure 4.3a) with

$$D(i,j) = \sqrt{H^2(i,j) + V^2(i,j) + D_1^2(i,j) + D_2^2(i,j)},$$

where

$$H(i,j) = \gamma[\varphi(i,j-1) - \varphi(i,j)] - \gamma[\varphi(i,j) - \varphi(i,j+1)],$$

$$V(i,j) = \gamma[\varphi(i-1,j) - \varphi(i,j)] - \gamma[\varphi(i,j) - \varphi(i+1,j)],$$

$$D_1(i,j) = \gamma[\varphi(i-1,j-1) - \varphi(i,j)] - \gamma[\varphi(i,j) - \varphi(i+1,j+1)],$$

$$D_2(i,j) = \gamma[\varphi(i+1,j-1) - \varphi(i,j)] - \gamma[\varphi(i,j) - \varphi(i-1,j+1)],$$

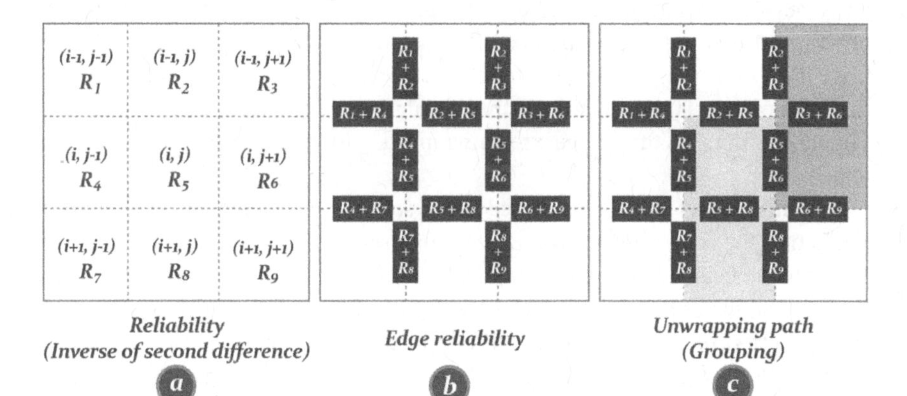

Figure 4.3 Schematics of the quality-guided, path-following unwrapping algorithm. (a) Calculation of reliability, (b) edge reliability, and (c) unwrapping path. Light grey pixels are unwrapped and grouped by the edge with the highest edge reliability; for example, $R_5 + R_8$. Dark grey pixels are unwrapped and grouped by the edge with the second-highest edge reliability; for example, $R_3 + R_6$. Note that the edge $R_5 + R_8$ has the highest edge reliability and the edge $R_3 + R_6$ has the second-highest edge reliability.

where i and j are coordinates of a given pixel in the phase image, H is the horizontal difference, V is a vertical difference, and D_1 and D_2 are diagonal differences [4]. $\gamma(\cdot)$ is a simple unwrapping operation to add or subtract 2π in a phase jump, and φ is the phase value at the corresponding pixel. The second difference can be computed for all pixels except at the borders of an image, where the second difference is set to infinity to be resolved last. Next, the reliability of each pixel in a 3×3 window is separately defined as $R = \frac{1}{D}$.

For simplicity, the reliability of each pixel in the window is represented by $R_1, R_2, \ldots, R_9$, as shown in Figure 4.3a. Initially, no pixel in the phase image is considered to belong to any group. The reliability of edges is estimated by adding the reliability of two facing pixels as shown in Figure 4.3b. The reliability of all edges is sorted and stored in one array. Phase unwrapping is then performed by starting with two facing pixels with the highest reliability. For example, yellow pixels with the highest edge reliability, the (i, j)th and $(i + 1, j)$th pixels in Figure 4.3c, are first unwrapped then joined into a single group.

Phase unwrapping is established by adding or subtracting multiples of 2π to each group. There are three situations in the phase unwrapping process: (i) two selected pixels belong to different groups, (ii) both pixels do not belong to any group, and (iii) one pixel belongs to a group but the other pixel does not belong to any group. In the first case, the pixel in the smallest group is unwrapped with respect to any pixel in the largest group. These two groups are then joined together. In the second case, both pixels are unwrapped with respect to each other, then joined into a single group. In the third case, the pixel that does not belong to any group is unwrapped with respect to the pixel that belongs to the group. The unwrapped pixel then joins the group. The phase unwrapping is performed sequentially in the order of highest edge reliability until all edges in the sorted array are processed. Finally, borders of the image are unwrapped with respect to the rest of the image [4].

The reliability of an edge is defined as the summation of reliabilities of the two pixels that the edge connects [4]. Based on a previous study [4], the unwrapping process is explained using a numerical example as shown in Figure 4.4. In Figure 4.4a, pixels a and b are connected by the edge with the highest edge reliability. Both pixels are unwrapped with respect to each other to construct the first group, I. Pixels c and d connected by the edge with the second-highest edge reliability are unwrapped with respect to each other to construct the second group, II. Pixel e, which does not belong to any group, is connected to pixel c by the edge with the third-highest edge reliability. Thus, they should be unwrapped with respect to each other. However, pixel c is already unwrapped with respect to pixel d. Thus, both pixels belong to the same group II. The 2π multiples required to be added or subtracted to unwrap pixel e can be calculated and added to or subtracted from pixel e, which is joined into group II. Pixels f and g are also connected by the edge

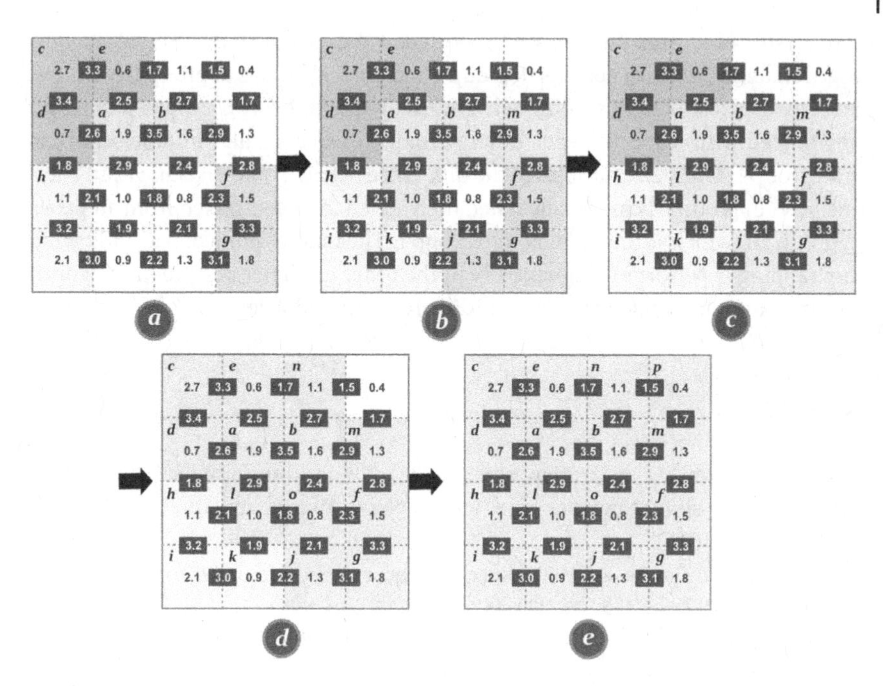

Figure 4.4 Numerical example of the unwrapping path [4] / with permission of Optical Society of America. (a) In order of edge reliability, pixels *a* and *b* (group I), pixels *c* and *d* (group II), pixels *f* and *g* (group III), and pixels *h* and *i* (group IV) are unwrapped with respect to each other and grouped. Pixel *e* is joined into the group II. (b) Pixel *j* is joined into the group III and pixel *k* is joined into the group IV by the third case of the phase unwrapping process. In a similar way, pixels *l* and *m* are joined into the group I. (c) By the first case of the phase unwrapping process, pixels in the group III is unwrapped with respect to pixel *f* in the group I and the group III is joined into the group I. (d) Pixels *n* and *o* are joined into the group I. In a similar way to (c), the group II and the group IV are joined into the group I. (e) Finally, pixel *p* is joined into the group I. The phase unwrapping is complete.

with the third-highest edge reliability. They are unwrapped with respect to each other to construct the third group, III. Pixels *h* and *i*, which are connected by the edge with the fourth-highest edge reliability, are unwrapped with respect to each other to construct the fourth group, IV.

The unwrapping and grouping of pixels *g* and *j* connected by the edge with the fifth highest edge reliability are performed similarly to those of pixels *c* and *e* as shown in Figure 4.4b. The same procedure is performed for pixels *i* and *k*. As a result, pixel *k* is joined into group IV and pixel *j* is joined into group III. Pixels *a* and *l* and pixels *b* and *m* are connected by edges with the same priority edge reliability. However, pixels *a* and *b* are already unwrapped with each other. They belong to the same group I. Consequently, pixels *l* and *m* are joined into the group I.

In Figure 4.4c, pixels *f* and *m* are connected by the edge with the seventh-highest edge reliability. Pixel *m* belongs to group I and pixel *f* belongs to group III. That is, these two pixels belong to different groups. Thus, these two groups should be unwrapped with respect to each other. Group III has a smaller number of pixels than group I. Thus, 2π multiples of the difference between one pixel in group I and one pixel in group III can be calculated then added to or subtracted from all pixels in group III, which is then joined into group I. The quality-guided, path-following unwrapping is performed sequentially in the order of higher edge reliability until all pixels are unwrapped as shown in Figures 4.4d and 4.4e.

References

1. Takeda, M., Ina, H., and Kobayashi, S. (1982). Fourier-transform method of fringe-pattern analysis for computer-based topography and interferometry. *J. Opt. Soc. Am.* 72 (1): 156–160.
2. Wyant, J. and Creath, K. (1985). Recent advances in interferometric optical testing. *Laser Focus/Electro-Opt.* 21 (11): 118–132.
3. Ghiglia, D.C. and Pritt, M.D. (1998). *Two-Dimensional Phase Unwrapping: Theory, Algorithms and Software.* New York: Wiley.
4. Arevalillo-Herráez, M., Burton, D.R., Lalor, M.J., and Gdeisat, M.A. (2002). Fast two-dimensional phase-unwrapping algorithm based on sorting by reliability following a noncontinuous path. *Appl. Opt.* 41 (35): 7437–7444.
5. Pritt, M.D. and Shipman, J.S. (1994). Least-square two-dimensional phase unwrapping using FFTs. *IEEE Trans. Geosci. Remote Sens.* 32 (3): 706–708.
6. Ghiglia, D.C. and Romero, L.A. (1994). Robust two-dimensional weighted and unweighted phase unwrapping that uses fast transforms and iterative methods. *J. Opt. Soc. Am. A* 11 (1): 107–117.
7. Hung, K.M. and Yamada, T. (1998). Phase unwrapping by regions using least-squares approach. *Opt. Eng.* 37 (11): 2965–2970.
8. Baldi, A. (2001). Two-dimensional phase unwrapping by quad-tree decomposition. *Appl. Opt.* 40 (8): 1187–1194.
9. Strand, J., Taxt, T., and Jain, A. (1999). Two-dimensional phase unwrapping using a block least-squares method. *IEEE Trans. Image Process.* 8 (3): 375–386.
10. Fornaro, G., Franceschetti, G., Lanari, R., and Sansosti, E. (1996). Robust phase-unwrapping techniques: a comparison. *J. Opt. Soc. Am. A* 13 (12): 2355–2366.
11. Ghiglia, D.C. and Romero, L.A. (1996). Minimum L^p-norm two-dimensional phase unwrapping. *J. Opt. Soc. Am. A* 13 (10): 1999–2013.
12. Gierloff, J.J. (1987). Phase unwrapping by regions. In: *Current Developments in Optical Engineering II*, Proceedings of SPIE, vol. 818 (ed. R.E. Fischer and W.J. Smith), 2–9.

13 Charette, P.G. and Hunter, I.W. (1996). Robust phase-unwrapping method for phase images with high noise content. *Appl. Opt.* 35 (19): 3506–3513.
14 Fornaro, G. and Sansosti, E. (1999). A two-dimensional region growing least squares phase unwrapping algorithm for interferometric SAR processing. *IEEE Trans. Geosci. Remote Sens.* 37 (5): 2215–2226.
15 Arevalillo-Herráez, M., Gdeisat, M.A., Burton, D.R., and Lalor, M.J. (2002). Robust, fast, and effective two-dimensional automatic phase unwrapping algorithm based on image decomposition. *Appl. Opt.* 41 (35): 7445–7455.
16 Towers, D.P., Judge, T.R., and Bryanston-Cross, P.J. (1989). A quasi heterodyne holographic technique and automatic algorithms for phase unwrapping. In: *Fringe Pattern Analysis*, Proceedings of SPIE, vol. 1163 (ed. G.T. Reid), 95–119.
17 Stephenson, P., Burton, D.R., and Lalor, M.J. (1994). Data validation techniques in a tiled phase unwrapping algorithm. *Opt. Eng.* 33 (11): 3703–3708.
18 Arevalillo-Herráez, M., Burton, D.R., Lalor, M.J., and Clegg, D.B. (1996). Robust, simple, and fast algorithm for phase unwrapping. *Appl. Opt.* 35 (29): 5847–5852.
19 Itoh, K. (1982). Analysis of the phase unwrapping algorithm. *Appl. Opt.* 21 (14): 2470.
20 Goldstein, R.M., Zebker, H.A., and Werner, C.L. (1988). Satellite radar interferometry: two-dimensional phase unwrapping. *Radio Sci.* 23 (4): 713–720.
21 Cusack, R., Huntley, J.M., and Goldrein, H.T. (1995). Improved noise-immune phase-unwrapping algorithm. *Appl. Opt.* 34 (5): 781–789.
22 Gutmann, B. and Weber, H. (2000). Phase unwrapping with the branch-cut method: role of phase-field direction. *Appl. Opt.* 39 (26): 4802–4816.
23 Huntley, J.M. (1989). Noise-immune phase unwrapping algorithm. *Appl. Opt.* 28 (15): 3268–3270.
24 Quiroga, J.A. and Bernabeu, E. (1994). Phase unwrapping algorithm for noisy phase-map processing. *Appl. Opt.* 33 (29): 6725–6731.
25 Schafer, R.W. and Oppenheim, A.V. (1975). *Digital Signal Processing*. Englewood Cliffs, NJ: Prentice-Hall.
26 Karout, S.A., Gdeisat, M.A., Burton, D.R., and Lalor, M.J. (2007). Two-dimensional phase unwrapping using a hybrid genetic algorithm. *Appl. Opt.* 46 (5): 730–743.
27 Lu, Y., Wang, X., and He, G. (2005). Phase unwrapping based on branch cut placing and reliability ordering. *Opt. Eng.* 44 (5): 055601.
28 Souza, J.C., Oliveira, M.E., and Santos, P.A.M. (2015). Branch-cut algorithm for optical phase unwrapping. *Opt. Lett.* 40 (15): 3456–3459.
29 Zheng, D. and Da, F. (2011). A novel algorithm for branch cut phase unwrapping. *Opt. Lasers Eng.* 49 (5): 609–617.
30 Wang, J. and Yang, Y. (2019). Branch-cut algorithm with fast search ability for the shortest branch-cuts based on modified GA. *J. Mod. Opt.* 66 (5): 473–485.
31 Quiroga, J.A., González-Cano, A., and Bernabeu, E. (1995). Phase unwrapping algorithm based on an adaptive criterion. *Appl. Opt.* 34 (14): 2560–2563.

32 Arevalillo-Herráez, M., Burton, D.R., and Lalor, M.J. (2010). Clustering-based robust three-dimensional phase unwrapping algorithm. *Appl. Opt.* 49 (10): 1780–1788.
33 Su, X. and Chen, W. (2004). Reliability-guided phase unwrapping algorithm: a review. *Opt. Lasers Eng.* 42 (3): 245–261.
34 Abdul-Rahman, H.S., Gdeisat, M.A., Burton, D.R. et al. (2007). Fast and robust three-dimensional best path phase unwrapping algorithm. *Appl. Opt.* 46 (26): 6623–6635.
35 Cui, H., Liao, W., Dai, N., and Cheng, X. (2011). Reliability-guided phase-unwrapping algorithm for the measurement of discontinuous three-dimensional objects. *Opt. Eng.* 50 (6): 063602.
36 Ma, L., Li, Y., Wang, H., and Jin, H. (2012). Fast algorithm for reliability-guided phase unwrapping in digital holographic microscopy. *Appl. Opt.* 51 (36): 8800–8807.
37 Arevalillo- Herráez, M., Villatoro, F.R., and Gdeisat, M.A. (2016). A robust and simple measure for quality-guided 2D phase unwrapping algorithms. *IEEE Trans. Image Process.* 25 (6): 2601–2609.
38 Zhong, H., Tang, J., Tian, Z., and Wu, H. (2019). Hierarchical quality-guided phase unwrapping algorithm. *Appl. Opt.* 58 (19): 5273–5280.

5

Off-axis Digital Holographic Microscopy

5.1 Off-axis Digital Holographic Microscopy Designs

In this chapter, we will study an off-axis configuration of digital holographic microscopy (DHM) [1–5] for transmission imaging with transparent samples (e.g. biological cells) as shown in Figure 5.1. The basic architecture of DHM is based on the Mach–Zehnder interferometer: a beam expander produces a plane wave in which the coherent beam comes from a coherent laser source. The beam is then divided into object and reference beams using a beam splitter with a small tilt angle between them (see Figure 5.1). The object beam illuminates the specimen and creates the object wavefront. A microscope objective (MO) magnifies the object wavefront. Object and reference wavefronts are then combined by a beam collector at the exit of the interferometer to create a hologram.

At the exit of the interferometer, interference between the object beam $\mathbf{O}$ and the reference beam $\mathbf{R}$ creates the hologram intensity

$$I_H(x,y) = |\mathbf{R}|^2 + |\mathbf{O}|^2 + \mathbf{R}^*\mathbf{O} + \mathbf{O}^*\mathbf{R},$$

where $\mathbf{R}*$ and $\mathbf{O}*$ are complex conjugates of the reference beam and the object beam, respectively. A digital hologram is recorded by either a charge-coupled device (CCD) camera or a complementary metal oxide semiconductor (CMOS) camera and transferred to a personal computer for numerical reconstruction. The digital hologram $I_H(k, l)$ (CCD size: $L \times L$) is an array of $N \times N$ that results from a 2D sampling of $I_H(x, y)$ by the CCD or CMOS camera:

$$I_H(k,l) = I_H(x,y)\mathrm{rect}\left(\frac{x}{L}, \frac{y}{L}\right) \times \sum_{k=-\frac{N}{2}}^{\frac{N}{2}} \sum_{l=-\frac{N}{2}}^{\frac{N}{2}} \delta(x - k\Delta x, y - l\Delta y), \tag{5.1}$$

Artificial Intelligence in Digital Holographic Imaging: Technical Basis and Biomedical Applications, First Edition. Inkyu Moon.

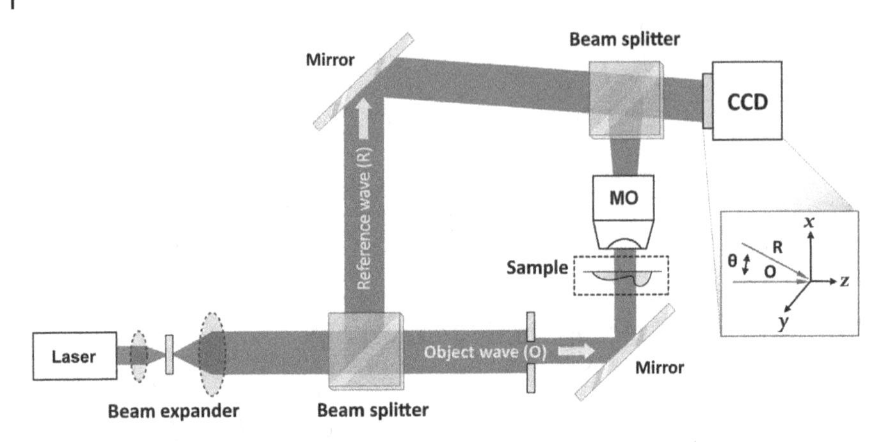

Figure 5.1 Schematic of the off-axis DHM setup.

where k and l are integers $\left(-\frac{N}{2} \leq k, l \leq \frac{N}{2}\right)$, rect() is the rectangular function, and Δx and Δy are sampling intervals in the CCD plane (pixel size: $\Delta x = \Delta y = \frac{L}{N}$).

DHM has been proposed in various configurations. Here, we will consider a geometry that includes an MO. The optical arrangement in the object arm of the off-axis DHM can be assumed to be an ordinary, single-lens imaging system (see Figure 5.2). The MO in the single-lens imaging system produces a magnified image of the object. The CCD plane x (the hologram plane) is placed between the MO and the image plane (x_i), at a distance d from the image plane. This situation can be considered equal to a holographic configuration with an object beam that

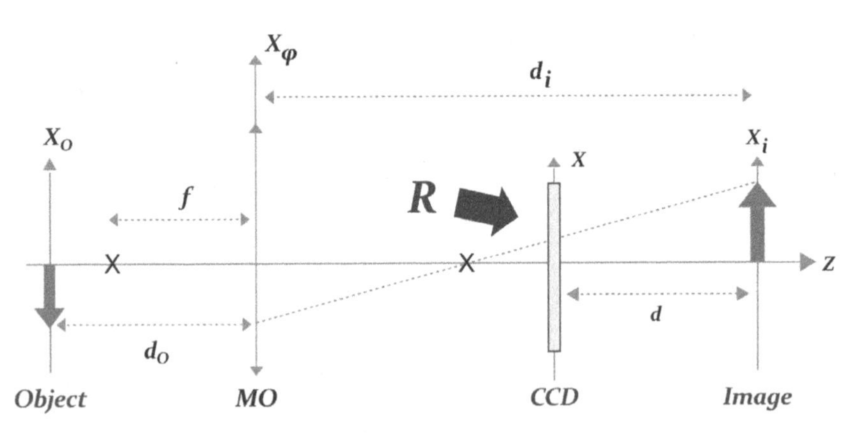

Figure 5.2 A single-lens imaging system for off-axis DHM.

emerges directly from the magnified image. Therefore, this procedure is called image holography. In classical microscopy, the focused image can be obtained by adjusting z-position of the object or the CCD plane. If the specimen is placed in the object focal plane of the MO, the object hologram is recorded with the Fourier transform of the object field since the distance between the image and the MO (d_i) is infinite. In this case, a numerical reconstruction can be performed by the Fourier transform of the digital hologram. Basically, in the off-axis DHM, the numerical reconstruction method consists of calculating the Fresnel diffraction pattern of the digital hologram.

5.2 Digital Hologram Reconstruction

In off-axis DHM, image focusing happens when the reconstruction distance is equal to the distance between the CCD plane and the image plane during the hologram recording (d in Figure 5.2). In classical off-axis holography, hologram reconstruction occurs by illuminating the hologram intensity with the reference beam **R**. The reconstructed wavefront can be defined as

$$\mathbf{\Psi} = \mathbf{R}I_H = \mathbf{R}|\mathbf{R}|^2 + \mathbf{R}|\mathbf{O}|^2 + |\mathbf{R}|^2\mathbf{O} + \mathbf{R}^2\mathbf{O}^*, \tag{5.2}$$

The first two terms of Eq. (5.2) are a zero order of diffraction. The third term produces a twin image. The fourth term produces a real image. Since the hologram is recorded in off-axis configuration, Fourier transform of the hologram can separately represent bandwidth of real image, virtual image, and zero-order noise (see Figure 8.2(c) in Chapter 8). Therefore, in off-axis DHM, a spatial filter with a properly defined size to cover only the bandwidth that corresponds to the real image is used (see Figure 23.1c). The resulting filtered hologram is reconstructed by Fresnel approximation. The reconstructed wavefront can be written as

$$\begin{aligned}\mathbf{\Psi}(\xi,\eta) = A exp\left[\frac{i\pi}{\lambda d}(\xi^2+\eta^2)\right] \times \iint I_H(x,y)\, exp\left[\frac{i\pi}{\lambda d}(x^2+y^2)\right] \\ \times\, exp\left[\frac{i2\pi}{\lambda d}(x\xi+y\eta)\right] dxdy,\end{aligned} \tag{5.3}$$

where λ is the wavelength, and $A = exp\left[\dfrac{\frac{i2\pi d}{\lambda}}{i\lambda d}\right]$ is a complex constant. As shown in Eq. (5.3), the Fresnel integral can be considered a Fourier transform in spatial frequencies $\frac{\xi}{\lambda d}$ and $\frac{\eta}{\lambda d}$ of the following function: $I_H(x,y)\, exp\left[\frac{i\pi}{\lambda d}(x^2+y^2)\right]$.

For fast numerical calculations, a discrete formulation of Eq. (5.3) using a two-dimensional fast Fourier transform can be expressed as

$$\Psi(m,n) = Aexp\left[\frac{i\pi}{\lambda d}(m^2\Delta\xi^2 + n\Delta\eta^2)\right] \times FFT\left\{I_H(k,l) \times exp\left[\frac{i\pi}{\lambda d}(k^2\Delta x^2 + l^2\Delta y^2)\right]\right\}_{m,n},$$

where k, l, m, and n are integers $-\frac{N}{2} \leq k,l,m,n \leq \frac{N}{2}$, and $I_H(k, l)$ is the digital hologram (Eq. (5.1)). Values of sampling intervals in the image plane ($\Delta\xi$ and $\Delta\eta$) can be derived from the relationship between sampling intervals in the CCD and Fourier domains in discrete Fourier-transform calculations $\left(\Delta\nu = \frac{1}{N\Delta x}\right)$. In our case, the Fourier transform must be calculated in the spatial frequencies $\frac{\xi}{\lambda d}$ and $\frac{\eta}{\lambda d}$. We have the following equations for $\Delta\xi$ and $\Delta\eta$ to define the transverse resolution in the image plane: $\Delta\xi = \Delta\eta = \frac{\lambda d}{N\Delta x} = \frac{\lambda d}{L}$.

For a typical reconstruction distance $d = 300$ mm with a wavelength $\lambda = 633$ nm and a CCD size $L = 5$ mm, the transverse resolution of the optical imaging system is limited to $\Delta\xi = 38$ μm. In an off-axis DHM, $\Delta\xi$ indicates the resolution for the reconstructed image of the object in a given magnification. Note that in image holography, a transverse resolution can be equal to the diffraction limit of the MO. For quantitative phase imaging, the digital hologram should be multiplied by a digital reference plane wave $\mathbf{R}_D$, which must be a replica of the experimental reference beam $\mathbf{R}$. In classical holography, the same operation is conducted optically when the hologram is illuminated with the reference beam. If we assume that a perfect plane wave is used as a reference beam for hologram recording, $\mathbf{R}_D$ is formulated as $\mathbf{R}_D(k,l) = A_R\, exp\left[i\left(\frac{2\pi}{\lambda}\right)(k_x k\Delta x + k_y l\Delta y)\right]$, where A_R is the amplitude, Δx and Δy are the sampling intervals in the CCD or hologram plane, and k_x, k_y are wave vectors that must be adjusted for the propagation direction of $\mathbf{R}_D$ to match the experimental reference beam as closely as possible.

As shown in Figure 5.3, the MO produces a curvature of the wavefront in the object arm of an off-axis DHM. This deformation does not affect the amplitude distribution of the object beam for amplitude-contrast imaging. However, for quantitative phase imaging in a DHM system, this phase aberration must be corrected. In our DHM, this phase aberration problem can be overcome experimentally by inserting the same MO in the reference arm of the DHM at the same distance from the exit of the interferometer. We can also design a numerical method that allows us to perform the correction by multiplying the reconstructed wavefront by the numerically computed, complex conjugate of phase aberration. If we assume an

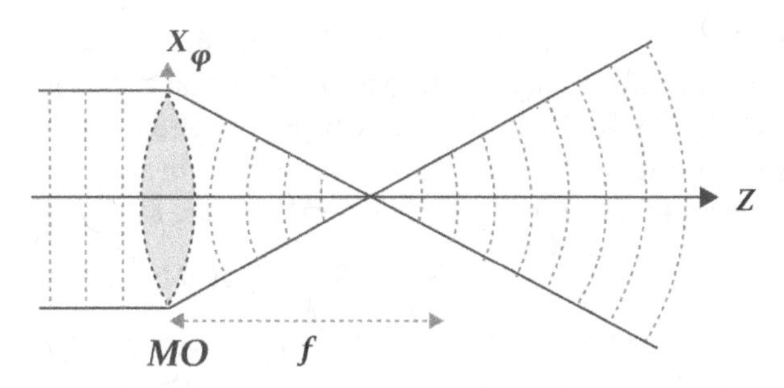

Figure 5.3 Schematic of the wavefront deformation by the microscope objective [1] / with permission of Optical Society of America.

optical imaging system with a monochromatic illumination, the relation between the optical fields $\mathbf{U}_o(x_o, y_o)$ in the object plane and $\mathbf{U}_i(x_i, y_i)$ in the image plane can be described as

$$\mathbf{U}_i(x_i, y_i) = \iint \mathbf{h}(x_i, y_i; x_o, y_o)\mathbf{U}_o(x_o, y_o)dx_o dy_o,$$

where $\mathbf{h}(x_i, y_i; x_o, y_o)$ is the point-spread function. If the object plane and the image plane form an object–image relation with the lens law $\frac{1}{d_i} + \frac{1}{d_o} = \frac{1}{f}$, $\mathbf{h}(x_i, y_i; x_o, y_o)$ can be written as

$$\mathbf{h}(x_i, x_o) = Cexp\left(\frac{i\pi}{\lambda d_i}x_i^2\right) exp\left(\frac{i\pi}{\lambda d_o}x_o^2\right) \times \int P(x_\varphi)\, exp\left[\left(\frac{-i2\pi}{\lambda}\left(\frac{x_o}{d_o} + \frac{x_i}{d_i}\right)x_\varphi\right)\right]dx_\varphi, \tag{5.4}$$

where x_φ is the coordinate of the MO plane, $P(x_\varphi)$ is the MO's pupil function, and C is a constant (we only considered the x coordinate for simplicity). Note that the integral in Eq. (5.4) is a Fourier transform of the pupil function $P(x_\varphi)$ (see Chapter 2 for more details). If we assume a perfect imaging system of magnification $M = \frac{d_i}{d_o}$ in which points (x_o, y_o) in the object plane becomes points $(x_i = -Mx_o, y_i = -My_o)$ in the image plane, the integral in Eq. (5.4) can be approximated by a Dirac function. We can obtain the following impulse response function by replacing x_o with $-\frac{x_i d_o}{d_i}$ in the quadratic phase term preceding the integral,

$$\mathbf{h}(x_i, y_i; x_o, y_o) \simeq Cexp\left[\frac{i\pi}{\lambda d_i}\left(1 + \frac{d_o}{d_i}\right)(x_i^2 + y_i^2)\right] \times \delta(x_i + Mx_o, y_i + My_o). \tag{5.5}$$

Equation (5.5) indicates that the image field is a magnified reproduction of the object field multiplied by the quadratic phase term. It also means that the phase aberration can be corrected by multiplying the reconstructed wavefront by the complex conjugate of the phase term that precedes the δ function in Eq. (5.5). Therefore, a digital phase mask $\mathbf{\Phi}(m, n)$ for the correction of the phase aberration can be calculated by $\mathbf{\Phi}(m,n) = exp\left[\frac{-i\pi}{\lambda D}\left(m^2\Delta\xi^2 + n^2\Delta\eta^2\right)\right]$, where D is a parameter that must be adjusted to compensate the wavefront curvature. According to Eq. (5.5), we can define $\frac{1}{D} = \frac{1}{d_i}\left(1 + \frac{d_o}{d_i}\right)$. Finally, the complete expression of the hologram reconstruction algorithm is

$$\mathbf{\Psi}(m,n) = A\mathbf{\Phi}(m,n)\, exp\left[\frac{i\pi}{\lambda d}\left(m^2\Delta\xi^2 + n^2\Delta\eta^2\right)\right]$$
$$\times\, FFT\left\{\mathbf{R}_D(k,l)I_H(k,l) \times exp\left[\frac{i\pi}{\lambda d}\left(k^2\Delta x^2 + l^2\Delta y^2\right)\right]\right\}_{m,n}, \tag{5.6}$$

Since $\mathbf{\Psi}(m, n)$ is an array of complex numbers, an amplitude-contrast image can be obtained by calculating the intensity: $I(x, y) = \mathrm{Re}[\mathbf{\Psi}(m, n)]^2 + \mathrm{Im}[\mathbf{\Psi}(m, n)]^2$.

We can obtain a quantitative phase image by calculating the argument

$$\varphi(x,y) = arctan\left\{\frac{\mathrm{Im}[\mathbf{\Psi}(m,n)]}{\mathrm{Re}\,[\mathbf{\Psi}(m,n)]}\right\}.$$

The numerical reconstruction algorithm in Eq. (5.6) contains four reconstruction parameters: the reconstruction distance (d), the two wave vectors in the digital reference beam (k_x, k_y), and the digital phase mask (D), which represents physical quantities (distances and angles). These parameters can be efficiently adjusted using the presented digital methods to produce high-quality phase-contrast images. An example of numerical reconstruction results obtained with red blood cells is shown in Figure 8.2.

References

1 Cuche, E., Marquet, P., and Depeursinge, C. (1999). Simultaneous amplitude and quantitative phase contrast microscopy by numerical reconstruction of Fresnel off-axis holograms. *Appl. Opt.* 38 (34): 6994–7001.

2 Marquet, P., Rappaz, B., Magistretti, P. et al. (2005). Digital holographic microscopy: a noninvasive contrast imaging technique allowing quantitative visualization of living cells with subwavelength axial accuracy. *Opt. Lett.* 30: 468–470.

3 Colomb, T., Cuche, E., Charrière, F. et al. (2006). Automatic procedure for aberration compensation in digital holographic microscopy and application to specimen shape compensation. *Appl. Opt.* 45: 851–863.
4 Cuche, E., Bevilacqua, F., and Depeursinge, C. (1999). Digital holography for quantitative phase-contrast imaging. *Opt. Lett.* 24 (5): 291–293.
5 Goodman, J. (1996). *Introduction to Fourier Optics*. New York: McGraw-Hill.

6

Gabor Digital Holographic Microscopy

6.1 Introduction

Direct observations of microscale biophysical processes such as kinematics and the dynamics of live cells require suitable tools to resolve both spatial and temporal scales at proper levels. An available candidate is optical microscopy in which, as the power increases and lateral resolution improves, the field of view and depth of field decrease nonlinearly. For example, increasing the power from 10× to 40× will decrease the theoretical depth of field from 12 to 2 μm, which greatly limits the size of the resolvable volume.

Basically, digital holographic microscopy (DHM) can record a 3D volumetric field on a charge-coupled device (CCD) plane and later reconstruct it. It can be used to investigate the spatial distribution and velocity of a dense particle cloud with an extended depth. Gabor DHM [1] can be implemented by combining Gabor holography and microscope objective (MO) using the same setup as an optical microscopy in which the light source is replaced with a collimated coherent beam and a sequence of magnified holograms is recorded on a CCD camera. Three-dimensional fields can be digitally reconstructed from these magnified holograms with a similar resolution to optical microscopy. Gabor DHM allows scientists to analyze particle dynamics by recording a time series of particle traces and the trajectory of biological specimens. This chapter introduces Gabor DHM as a promising tool for measuring particle motions and trajectories using time-lapse imaging.

6.2 Methodology

As shown in Figure 6.1, the optical configuration of Gabor DHM is similar to conventional bright-field microscopy except that it uses a laser beam as a light source. A hologram is a record of interference patterns between the light diffracted from

Artificial Intelligence in Digital Holographic Imaging: Technical Basis and Biomedical Applications, First Edition. Inkyu Moon.

samples and the non-scattered part of the light (reference beam). The optical field of samples can be represented at the hologram plane by

$$\mathbf{U}_H(x,y) = Ae^{jk_r n_H} + \sum_i \iint \mathbf{a}_i(x_o, y_o; z_i)\mathbf{h}_z(x - x_o, y - y_o; z_i)dx_o dy_o,$$

where k_r is the propagation vector of the reference beam and n_H is the norm vector of the hologram plane [1]. The first term represents the optical field of the reference beam in Gabor DHM. The angle of the reference beam is assumed to be zero. The second term is the superposition of light diffracted from discrete samples located at a distance z_i from the hologram plane, and produce fields with local distributions of $\mathbf{a}_i(x_o, y_o)$. Each particle is considered a superposition of point sources whose individual fields are $\mathbf{h}_z(x, y; z_i)$. Using a paraxial approximation for particles much smaller than z_i yields [1]

$$\mathbf{h}_z(x - x_o, y - y_o; z_i) = \frac{1}{j\lambda z_i} exp\left\{j\frac{k}{2z_i}\left[(x - x_o)^2 + (y - y_o)^2\right]\right\}. \tag{6.1}$$

Each particle can be thought of as a 2D aperture. Thus, diffraction from a single particle is the result of the convolution of a 2D aperture with the impulse response function in Eq. (6.1). The resulting interference intensity on the hologram plane can be presented as

$$I_z(x,y) = \mathbf{U}_H\mathbf{U}_H{}^* = A^2 - A\sum \mathbf{a}^*(x,y)\otimes\mathbf{h}_z^*(x,y) - A\sum \mathbf{a}(x,y)\otimes\mathbf{h}_z(x,y) + \sum\sum |\mathbf{a}(x,y)\otimes\mathbf{h}_z(x,y)|^2,$$

where $\otimes$ indicates a convolution integral. To determine the effect of the MO in Gabor DHM, we consider its compound lens system as a perfect thin lens. The optical field at distance d_i behind the lens that results from an optical field is $\mathbf{U}_o(x_o, y_o, d_o)$ where d_o, the distance before the lens (see Figure 6.1), can be represented by

$$\mathbf{U}_i(x_i, y_i; d_i) = \iint \mathbf{h}_l(x_i, y_i; x_o, y_o)\mathbf{U}_o(x_o, y_o)dx_o dy_o,$$

where

$$\mathbf{h}_l(x_i, y_i; x_o, y_o) = \frac{1}{M}\delta\left(\frac{x_i}{M} + x_o, \frac{y_i}{M} + y_o\right) \times exp\left[j\frac{k}{2M^2 d_o}(x_i^2 + y_i^2)\right] \times exp\left[j\frac{k}{2d_i}(x_o^2 + y_o^2)\right].$$

$M = \dfrac{d_i}{d_o}$ is the magnification. After integration, and replacing $\mathbf{U}_o(x_o, y_o)$ with $\mathbf{U}_H(x_o, y_o)$, the optical field generated by the hologram at the image plane is

$$\mathbf{U}_i(x_i, y_i; d_i) = \frac{1}{M}\mathbf{U}_H\left(-\frac{x_i}{M}, -\frac{y_i}{M}\right) \times exp\left[j\frac{k}{2M^2 d_o}(x_i^2 + y_i^2)\right] \times exp\left[j\frac{k}{2d_i}(x_o^2 + y_o^2)\right]. \tag{6.2}$$

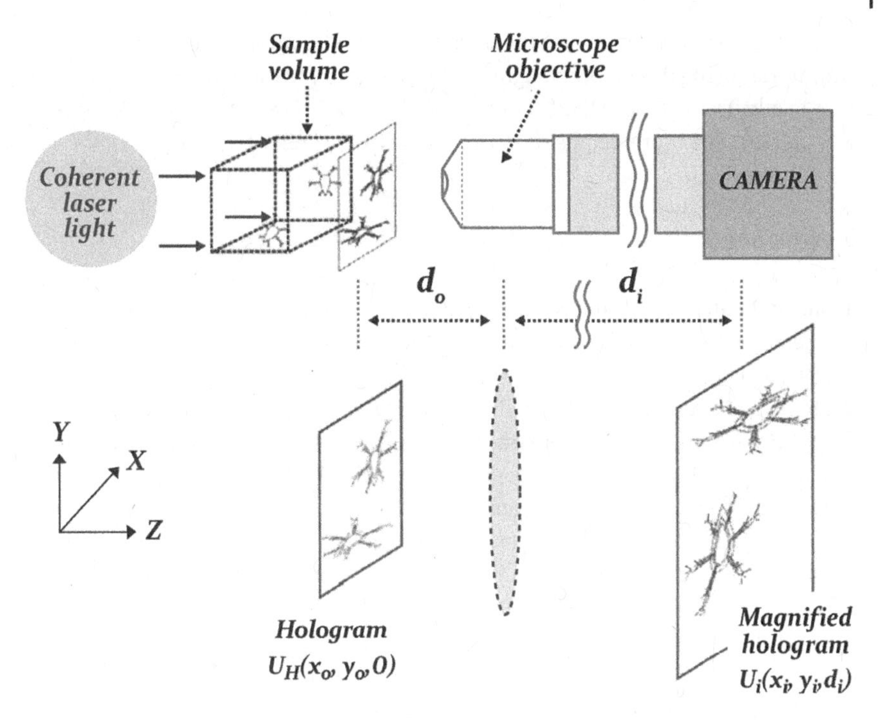

Figure 6.1 Optical setup of Gabor DHM [1] / with permission of Optical Society of America.

Equation (6.2) shows that if the image is highly magnified, the image plane contains a magnified holographic plane with a phase correction that becomes unity. The intensity distribution in the image plane simply becomes a magnified hologram:

$$I_i(x_i, y_i) = \frac{1}{M^2}\mathbf{U}_H\left(-\frac{x_i}{M}, -\frac{y_i}{M}\right)\mathbf{U}_H^*\left(-\frac{x_i}{M}, -\frac{y_i}{M}\right).$$

This true magnified hologram enables scientists to drastically relax the spatial resolution requirement of the recording medium such as the CCD camera. Furthermore, scientists can use magnification as a means to match the desired resolution with that of the recording medium. Finally, the magnified 3D particle field, $\mathbf{\Psi}_p(x, y; z)$, is numerically reconstructed using the Fresnel diffraction formula:

$$\mathbf{\Psi}_p(x, y; z) = I_i(x, y) \otimes \mathbf{h}_z(x, y; z), \qquad (6.3)$$

$$I_p(x, y; z) = \mathbf{\Psi}_p \mathbf{\Psi}_p^*.$$

Since Eq. (6.3) is a convolution of the magnified hologram with an impulse response function, both $\mathbf{\Psi}_p$ and $I_p(x, y; z)$ can be efficiently computed in the Fourier domain. This numerical reconstruction method allows the Gabor DHM to resolve the locations of several thousand particles and measure their motions and trajectories using time-lapse imaging. An example of the numerical reconstruction that results from Gabor DHM obtained with live cells is shown in Figure 10.10 in Chapter 10. For more details of Gabor DHM, please refer to [1–5].

References

1 Sheng, J., Malkiel, E., and Katz, J. (2006). Digital holographic microscope for measuring three-dimensional particle distributions and motions. *Appl. Opt.* 45 (16): 3893–3901.
2 Moon, I., Daneshpanah, M., Javidi, B., and Stern, A. (2009). Automated three-dimensional identification and tracking of micro/nanobiological organisms by computational holographic microscopy. *Proc. IEEE* 97: 990–1010.
3 Moon, I., Daneshpanah, M., Anand, A., and Javidi, B. (2011). Cell identification computational 3-D holographic microscopy. *Opt. Photon. News* 22: 18–23.
4 Moon, I. and Javidi, B. (2007). 3D identification of stem cells by computational holographic imaging. *J. R. Soc. Interface* 4 (13): 305–313.
5 Moon, I. and Javidi, B. (2008). 3-D visualization and identification of biological microorganisms using partially temporal incoherent light in-line computational holographic imaging. *IEEE Trans. Med. Imaging* 27 (12): 1782–1790.

Part II

Deep Learning in Digital Holographic Microscopy (DHM)

7

Introduction

Digital holographic microscopy (DHM) has recently emerged as a promising new technique well suited to explore cell structure and dynamics with a nanometric axial sensitivity without the need for labels, thus enabling the identification of new cellular biomarkers [1–6]. Recent advances in artificial intelligence (AI) and deep learning have opened up a new way for holographic image reconstruction with real-time performance. Specifically, in 2017, a deep-learning approach based on a convolutional neural network (CNN) began to be applied to DHM. Initially, simple problems using deep learning such as phase-image restoration and segmentation were investigated [7, 8]. Currently, more advanced deep-learning algorithms are applied to develop compact, low-cost, smart DHM systems. DHM using deep learning has outperformed conventional numerical methods in several applications such as depth estimation, phase unwrapping, and direct hologram reconstruction [9–14].

Part II provides an overview of previously published work on deep-learning techniques for digital, holographic image reconstruction of live cells for automated cell identification. Notable achievements include holographic image auto-focusing with deep learning, deep learning-based phase unwrapping in DHM, and noise-free Gabor holography by fusing deep learning and DHM [15–17]. In DHM, the numerical reconstruction algorithm allows the retrieval of both phase-contrast and amplitude-contrast cell images. This is possible when the exact distance between the charge-coupled device (CCD) plane and image plane is provided. Chapter 8 shows a deep CNN with a regression layer at the top to estimate the best focus distance. Experimental results with microsphere bead and red blood cells (RBC) showed that the deep-learning method could accurately estimate the propagation distance from the hologram. This method can significantly accelerate the numerical reconstruction time since the correct focus is provided by the CNN model without a need for digital propagation in different distances. Another advantage of this method is that since the distance is estimated at the single-object

Artificial Intelligence in Digital Holographic Imaging: Technical Basis and Biomedical Applications, First Edition. Inkyu Moon.

level, it can provide reconstruction distances with respect to the location of micro-sized objects.

Phase values of the reconstructed holographic image are limited between $-\pi$ and π. Therefore, discontinuity might occur due to the modulo 2π operation. Chapter 9 demonstrates that the deep-learning model based on generative adversarial network (GAN) can convert wrapped phase images to unwrapped ones. The deep-learning model is also free of phase jumping noise, which is typical for conventional unwrapping algorithms. In addition, we show that our model is twice as fast than quality-guided, path-following algorithms, which allows for the observation of morphology and cell movement in real-time. Chapter 10 shows that deep learning can eliminate a superimposed twin-image noise in phase images of Gabor holographic setup. This is achieved by using conditional generative adversarial network (C-GAN) trained by input–output pairs of noisy phase images obtained from synthetic Gabor holography and corresponding quantitative noise-free contrast-phase image obtained with off-axis digital holography. Two models are trained: a human red blood cell model and an elliptical cancer cell model.

References

1 Cuche, E., Bevilacqua, F., and Depeursinge, C. (1999). Digital holography for quantitative phase-contrast imaging. *Opt. Lett.* 24 (5): 291–293.

2 Marquet, P., Rappaz, B., Magistretti, P. et al. (2005). Digital holographic microscopy: a noninvasive contrast imaging technique allowing quantitative visualization of living cells with subwavelength axial accuracy. *Opt. Lett.* 30: 468–470.

3 Rappaz, B., Marquet, P., Cuche, E. et al. (2005). Measurement of the integral refractive index and dynamic cell morphometry of living cells with digital holographic microscopy. *Opt. Express* 13 (23): 9361–9373.

4 Colomb, T., Cuche, E., Charrière, F. et al. (2006). Automatic procedure for aberration compensation in digital holographic microscopy and application to specimen shape compensation. *Appl. Opt.* 45: 851–863.

5 Rappaz, B., Cano, E., Colomb, T. et al. (2009). Noninvasive characterization of the fission yeast cell cycle by monitoring dry mass with digital holographic microscopy. *J. Biomed. Opt.* 14 (3): 034049.

6 Moon, I., Daneshpanah, M., Javidi, B., and Stern, A. (2009). Automated three-dimensional identification and tracking of micro/nano biological organisms by computational holographic microscopy. *Proc. IEEE* 97 (6): 990–1010.

7 Rivenson, Y., Zhang, Y., Günaydın, H. et al. (2017). Phase recovery and holographic image reconstruction using deep learning in neural networks. *Light Sci. Appl.* 7 (2): 17141.

8 Yi, F., Moon, I., and Javidi, B. (2017). Automated red blood cells extraction from holographic images using fully convolutional neural networks. *Biomed. Opt. Express* 8 (10): 4466–4479.

9 Zhang, T., Jiang, S., Zhao, Z. et al. (2019). Rapid and robust two-dimensional phase unwrapping via deep learning. *Opt. Express* 27 (16): 23173–23185.

10 Rivenson, Y., Wu, Y., and Ozcan, A. (2019). Deep learning in holography and coherent imaging. *Light Sci. Appl.* 8: 85.

11 Pitkäaho, T., Manninen, A., and Naughton, T.J. (2019). Focus prediction in digital holographic microscopy using deep convolutional neural networks. *Appl. Opt.* 58: A202–A208.

12 Ren, Z., Xu, Z., and Lam, E. (2018). Learning-based nonparametric autofocusing for digital holography. *Optica* 5: 337–344.

13 Wu, Y., Rivenson, Y., Zhang, Y. et al. (2018). Extended depth-of-field in holographic imaging using deep-learning-based autofocusing and phase recovery. *Optica* 5 (6): 704–710.

14 Lin, X., Rivenson, Y., Yardimci, N. et al. (2018). All-optical machine learning using diffractive deep neural networks. *Science* 361 (6406): 1004–1008.

15 Jaferzadeh, K., Hwang, S., Moon, I., and Javidi, B. (2019). No-search focus prediction at the single cell level in digital holographic imaging with deep convolutional neural network. *Biomed. Opt. Express* 10 (8): 4276–4289.

16 Park, S., Kim, Y., and Moon, I. (2021). Automated phase unwrapping in digital holography with deep learning. *Biomed. Opt. Express* 12 (11): 7064–7081.

17 Moon, I., Jaferzadeh, K., Kim, Y., and Javidi, B. (2020). Noise-free quantitative phase imaging in Gabor holography with conditional generative adversarial network. *Opt. Express* 28 (18): 26284–26301.

8

No-search Focus Prediction in DHM with Deep Learning

8.1 Introduction

DHM allows for nondestructive investigations of biological samples as well as marker-free and time-resolved studies of cell biological processes. More specifically, interpretation of quantitative phase signals with DHM gives access to quantitative measurements of both cellular morphology and sample content with only a single shot. It is a real-time approach that can be used for time-lapse studies of biological samples in the absence of a mechanical focus adjustment. The propagation distance must be determined to obtain a quantitative phase image for phase objects. The distance between the hologram plane (CCD plane) and the observation plane (image plane) is defined by the reconstruction distance d. In digital holographic reconstruction, an in-focus image is reconstructed when the reconstruction distance is equal to the distance between the CCD plane and the image plane during the hologram recording (see Figure 8.1). An out-of-focus image appears if d is not precise (see Figure 8.2). Several automated approaches have been proposed to find the best focus plane in DHM [1–6]. Generally, multiple images at different focus planes are numerically reconstructed and a focus-evaluation function determines whether the image is focused by assessing the sharpness of either the amplitude-contrast or phase-contrast image. For example, pure phase objects have minimum visibility in amplitude-contrast images when in-focus reconstruction is estimated [3]. All these methods require several propagations at various distances, which are time-consuming since they require Fourier transforming. It becomes more apparent when time-lapse imaging is needed. Long-term biological studies require continuous focus readjustment to maintain optimum image quality.

In this chapter, we will introduce a deep learning convolutional neural network (CNN) for the estimation of propagation distance d [7]. This deep learning model can estimate the reconstruction distance from the recorded hologram for micro-sized objects without any focus-evaluation functions. This method has

Artificial Intelligence in Digital Holographic Imaging: Technical Basis and Biomedical Applications, First Edition. Inkyu Moon.

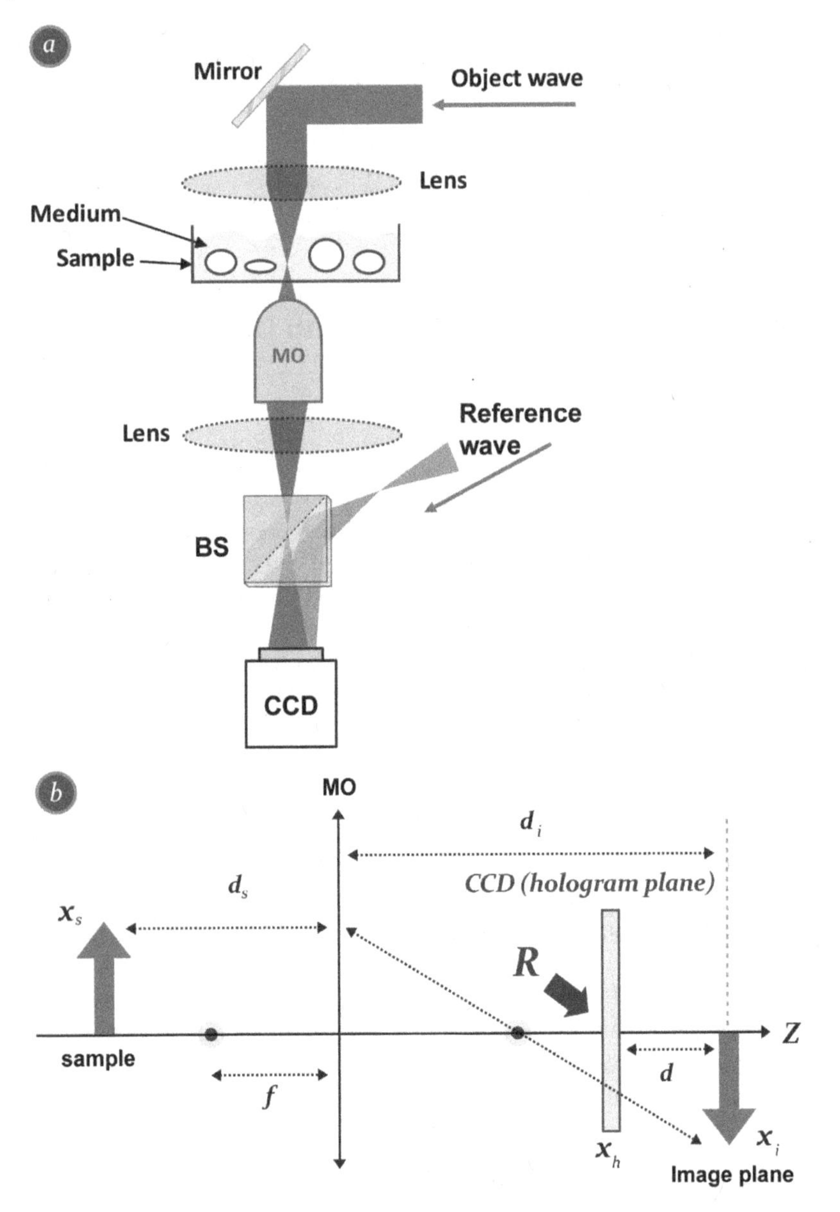

Figure 8.1 (a) Scheme of off-axis hologram recording and (b) configuration of image sensor (CCD camera), sample, and image plane.

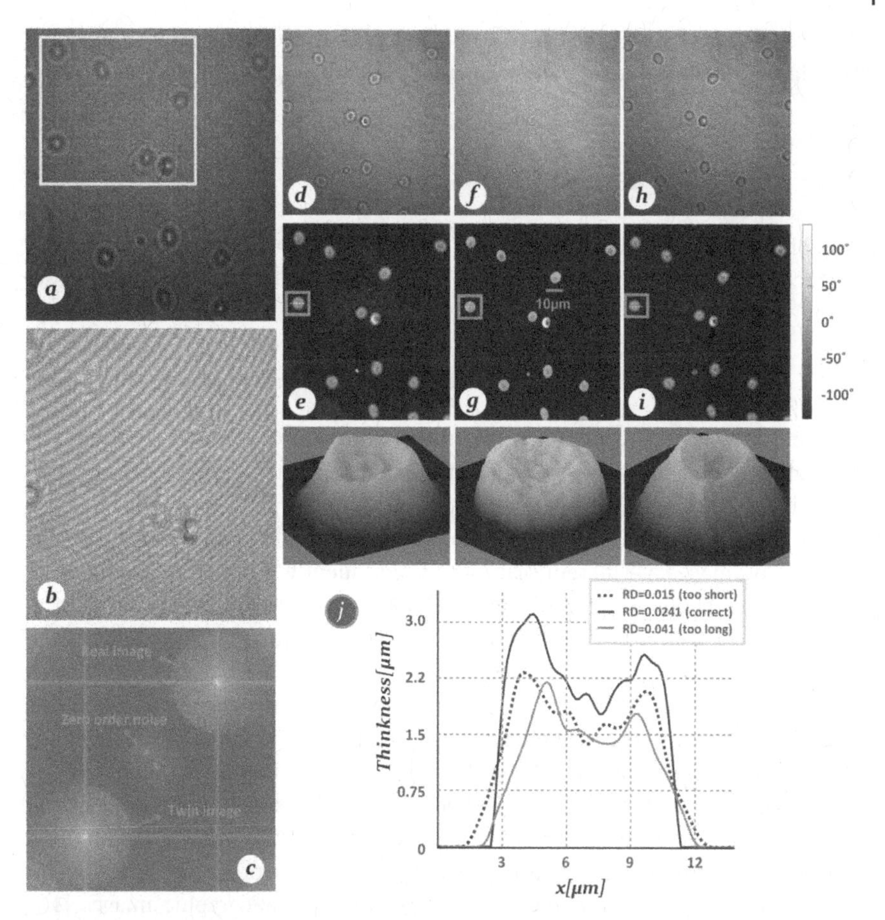

Figure 8.2 (a) A recorded red blood cell (RBC) hologram and (b) its magnification. Fringes are curved due to phase aberrations caused by MO. Fringes are not concentric with respect to the center of the image regarding the off-axis setup. (c) Fourier transformation of hologram shown in (a). Bandwidth of real image, twin image, and zero-order noise are separate due to the off-axis geometry. (d, e) Amplitude and phase of the numerically reconstructed signal when the reconstruction distance *d* is too short, (f, g) when the reconstruction distance *d* is correct, and (h, i) when reconstruction distance *d* is too long. The corresponding amplitude image for the in-focus phase contrast image has the lowest contrast. (j) Cross-section of the phase image of RBC in (e), (g), and (h). The phase is converted to thickness by Eq. (8.1) where $n_{RBC} = 1.39$ and $n_m = 1.3345$ [7] / with permission of Optical Publishing Group.

two main advantages. First, it is significantly faster because it does not require many digital propagations or a function to evaluate sharpness at each distance. Second, this model can provide a specific *d* regarding the location of cells with respect to other cells. This can be used to analyze the cell adherence on surfaces.

Focused images for training data are obtained with the focus-evaluation function, which uses the standard deviation of amplitude images.

8.2 Materials and Methods

The phase value in DHM is sometimes represented as the optical path length difference (OPD) between the reference beam and the object beam that passes through the sample:

$$OPD(x,y) = \phi(x,y) \times \frac{\lambda}{2\pi} = t(x,y) \times [n_s(x,y) - n_m], \tag{8.1}$$

where $n_s(x, y)$ is the integral refractive index of the sample at pixel (x, y) along the optical axis, n_m is the refractive index of the sample's surrounding medium, and $t(x, y)$ is the thickness of the sample at the $(x, y)_{th}$ pixel.

The deep architecture of the processing units in CNN allows the model to extract features of images at different scales [8]. Consequently, CNN models with these features can make accurate predictions thanks to the supervised learning of desired target values provided during the training process. The CNN has been applied to many applications including medical-image segmentation [9]. CNN has also been applied in several optical-related studies, such as depth estimation in inline holograms of natural images [10], the predication of the focus plane by AlexNet and VGG16 models [11], and non-parametric autofocusing with a regression CNN model [12]. Shimobaba, Kakue, and Itoonly used a portion of the intensity and spectrum from inline holograms to show that CNN with regression could predict depth [10]. In addition, deep-learning models have promising outcomes for phase recovery and eliminating twin-image noise in holographic images [13], which extends the depth of field in reconstructed images [14]. Furthermore, diffractive deep-neural networks for faster deep-neural networking have been proposed [15]. For a more comprehensive description of DHM using deep-learning models, please see a review paper [16].

Figure 8.3a shows a general scheme of our deep-learning convolutional neural network with a regression layer as the top layer (R-CNN) to estimate the best reconstruction distance. The linear regression layer in the R-CNN attempts to find the best fitting line between the feature map and continuous target values for the training. The feature map is obtained from the amplitude part of filtered holograms (see Figure 8.3a). Note that we used a spatial filter to extract only real images in the off-axis hologram (see Chapter 5).

Experiments demonstrated that non-filtered holograms reduced the accuracy of our estimation (data not shown). Target values are obtained by the focus-evaluation function, which will be explained in later parts of this section. Our

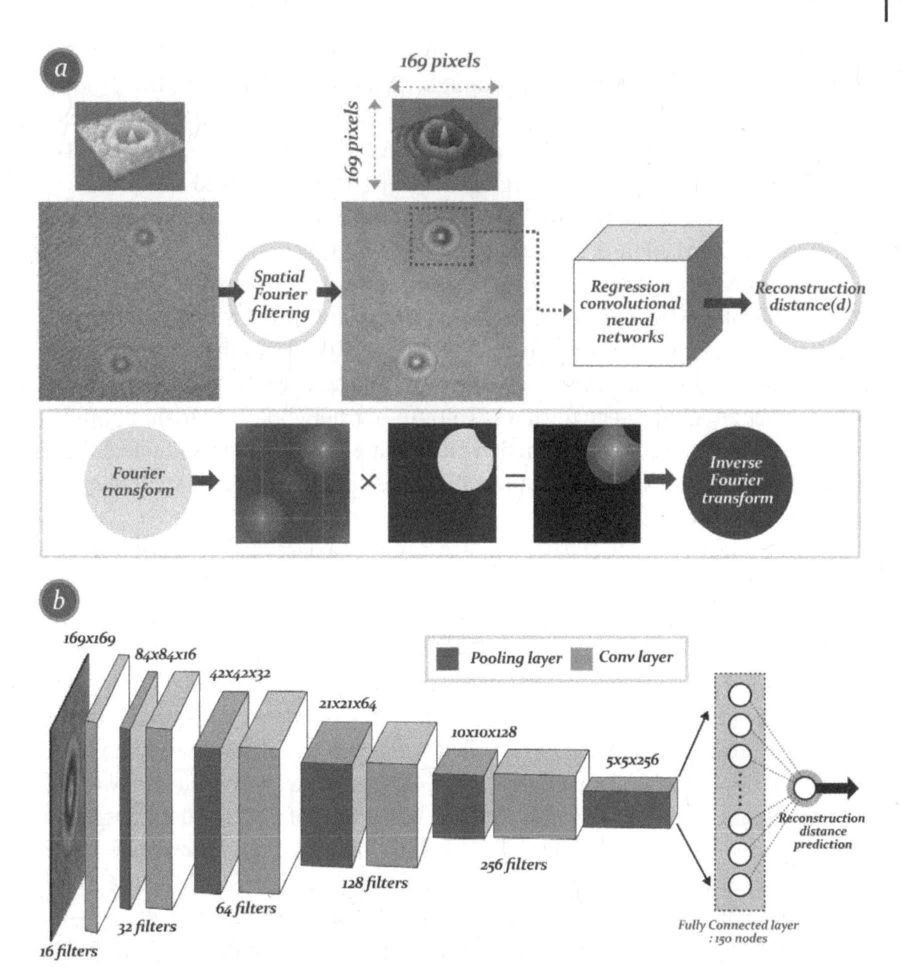

Figure 8.3 (a) General scheme of the proposed R-CNN for the estimation of reconstruction distance. Inset shows RBC hologram before and after filtering. (b) Details of R-CNN, various layers of the network, and configurations of R-CNN. Hologram after filtering has complex data. Thus, only the amplitude of the filtered hologram is fed into the model [7] / with permission of Optical Publishing Group.

R-CNN architecture consists of two main parts (see Figure 8.3b): a feature extraction part for learning features of images and a linear regression part to predict the focus distance. The feature extraction part consists of five stages. Each stage involves convolution layers, batch normalization layers, an activation function, and a pooling layer. All convolution layers have the same size filter of 5×5. A rectified linear unit (ReLU) activation function is applied to the output of every batch normalization since it has the advantage of reducing the vanishing gradient problem. A max-pooling layer is used for the pooling layer due to its good performance [8, 17]. After passing through each convolution layer, the depth of the feature map is doubled. Using the pooling layer allows scientists to greatly reduce the size of the input while obtaining translation invariant features. The fully connected layer for linear regression has 150 nodes linked to all units in the previous layer. Because our network model has to predict continuous focus values, a linear function is used as an output layer function.

In this study, two models with the same structure were developed: one for RBCs and one for microsphere beads. The main reason for designing these models was that the performance of a combined model was slightly worse than two independent models. We used the mean squared-error metric to minimize the difference between actual and predicted values. To train our network, the Adam optimizer was used to minimize the loss function and update all trainable parameters. Since the Adam optimizer is invariant to rescaling, it is suitable for a non-stationary loss function and automatic learning rate annealing [18].

The initial learning rate of 0.001 was decreased by a decay factor of 0.7 every five epochs. The momentum was set to be 0.9. The training process stopped when the validation loss did not change for 10 consecutive epochs. The mini-batch size for training was 128 for the RBCs model and 256 for the bead model. A data augmentation method (45° clockwise rotation at each mini batch, horizontal and vertical flip) was applied to our models to avoid over-fitting problems. All R-CNN simulations were done in Python. R-CNN models were built in Tensor flow (Keras, GPU only, NVIDIA Geforce GTX 690, version 2.2.4). The dataset is divided into two: 80% for training and 20% for validation. Several holograms of RBCs and microspheres were captured to train the two R-CNN models. Figure 8.4 shows variation in the reconstruction distance d regarding the distance between the sample and MO or d_s shown in Figure 8.1b. Several holograms are generated by adjusting d_s.

To generate a training set, the distance between the sample and MO was adjusted by a controllable stage with a resolution of 0.1 μm along the optical axis. For training the two R-CNN models, more than 3000 holograms containing 8 RBCs each and more than 2400 holograms containing 8 microspheres each were recorded (Figure 8.4). As previously mentioned, these holograms were filtered before they were fed into the R-CNN model. This filtering can reduce uninformative noise patterns stored in the hologram, thus providing more informative patterns for the R-CNN model. Figure 8.5 shows various examples of holograms at the object scale with corresponding reconstructed phase images at the single-object level. The focus image for the training data is obtained using the focus-evaluation function, which uses the standard deviation of amplitude images.

The best focus plane is determined by conducting 90 reconstructions at different focus planes for each hologram containing several micro-sized objects. The focus-evaluation function then estimates the best focus plane by computing the 2D

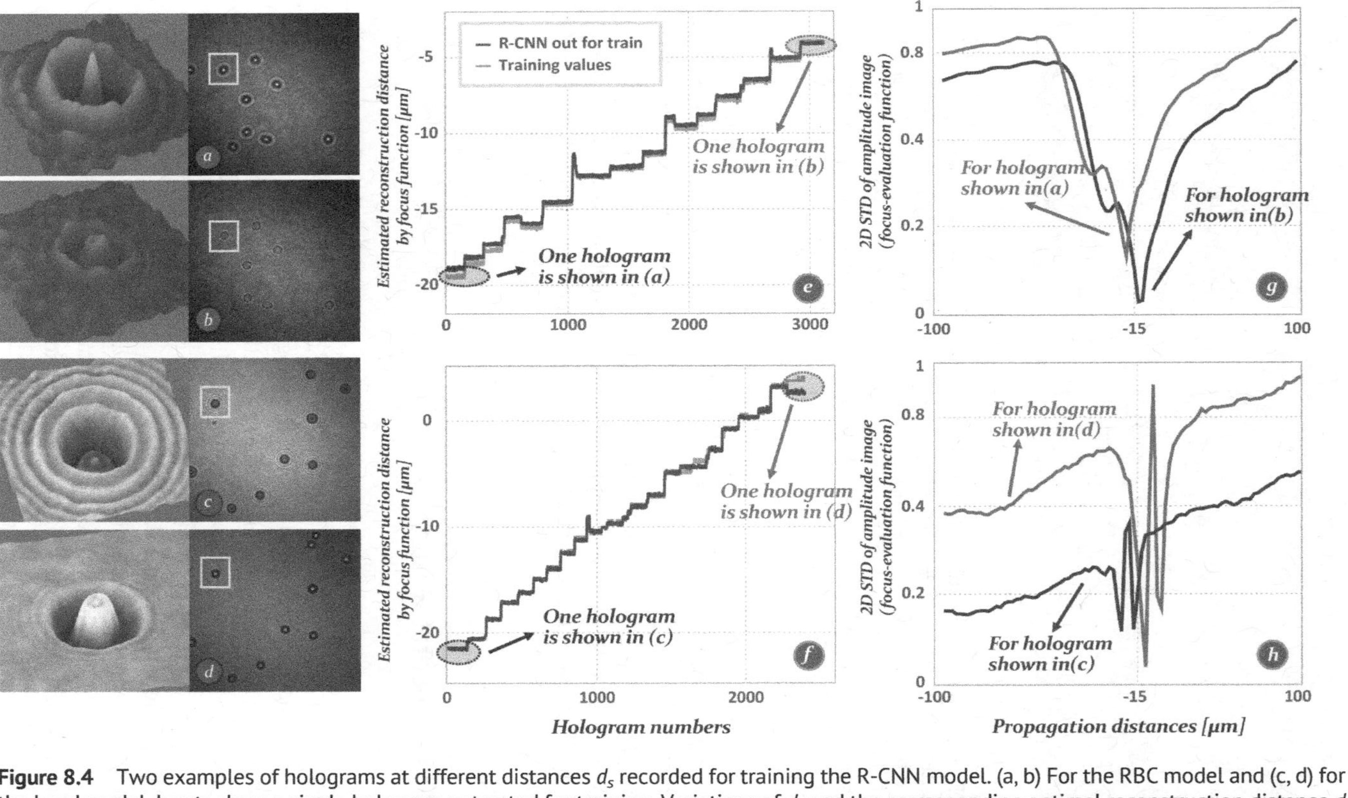

Figure 8.4 Two examples of holograms at different distances d_s recorded for training the R-CNN model. (a, b) For the RBC model and (c, d) for the bead model. Insets show a single hologram extracted for training. Variations of d_s and the corresponding optimal reconstruction distance d for the best reconstruction of the training set are illustrated for (e) the RBC model and (f) the bead model. The optimal distance for each hologram containing multiple cells or objects is evaluated by reconstructing 90 holograms at different d values and evaluating the amplitude image's 2D standard deviation for the 2D STD output for (g) one RBC hologram and (h) one bead hologram / with permission of Optical Publishing Group.

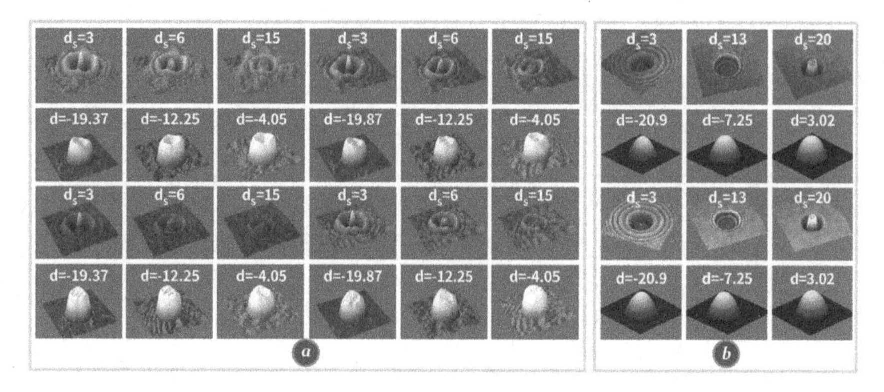

Figure 8.5 (a) Several single RBC and corresponding phase images by numerical reconstruction. Reconstruction distance *d* is found by the focus-evaluation function. *d* is considered the desired output for the regression layer of the CNN model. (b) Several bead holograms and corresponding phase images by numerical reconstruction. Hologram images at the single-object level are fed into the CNN regression model during the training stage. Units for d_S and d distance are 0.1 μm and 1.0 μm, respectively [7] / with permission of Optical Publishing Group.

standard deviation of the amplitude image (Figures 8.4g and 8.4h). This function evaluates the dispersion of the reconstructed amplitude image's pixel values by standard deviation measurements in x and y directions. Evaluating the amplitude image is a particularly useful approach for investigating transparent or semi-transparent objects, like biological samples, because these objects are nearly invisible in the amplitude image. Accordingly, the standard deviation of pixels is almost zero.

8.3 Experimental Results

The performance of the trained R-CNN model was analyzed by recording several new holograms at different d_S and comparing the R-CNN's estimation result with the output of a focus-evaluation function. The hologram test dataset was not used for the training. The R-CNN model provides estimation of the optimal reconstruction distance according to the spatial pattern of the input hologram. Figure 8.6 shows some examples of test holograms used to evaluate the performance of our model at different distances d_S.

Figure 8.7 presents the results of our R-CNN model to estimate the reconstruction distance of microsphere beads during hologram reconstruction. Since this method allows for the estimation of the reconstruction distance at the single object

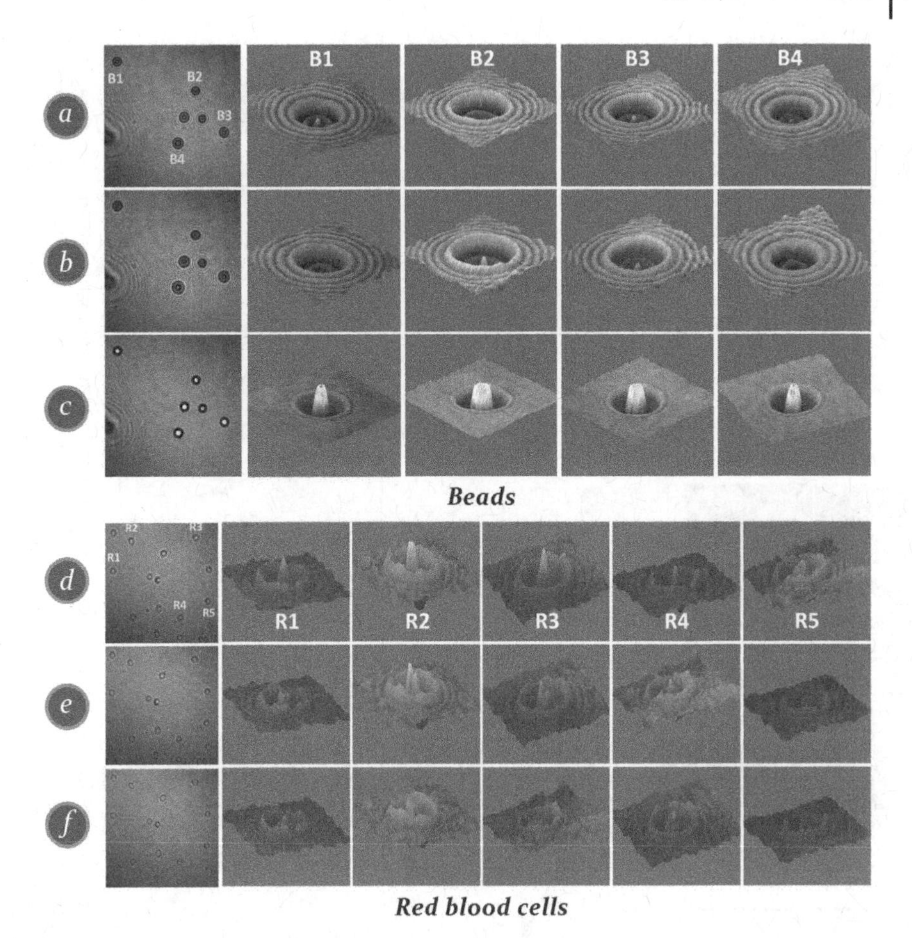

Figure 8.6 Holograms used for performance evaluation of the trained regression CNN model. (a)–(c) Three holograms for microbeads recorded at unknown distances d_S. Four beads were chosen from each sample for the bead R-CNN model test. All beads are located at the same distance along the z-axis. B1–B4 are 3D representations of the bead holograms. (d)–(f) Three holograms for RBCs recorded at unknown distances d_S. R1–R5 are 3D representations of the RBC holograms. R5 is located at a different depth with respect to R1–R4 [7] / with permission of Optical Publishing Group.

level in the hologram containing multiple objects, it is possible to automatically perform multiple reconstructions for each micro-object in the hologram. Accordingly, four beads were considered as shown in Figure 8.6. To validate the output of our model, the output of a single object's hologram was compared with the focus-evaluation function. A correlation analysis was then performed. Alongside the

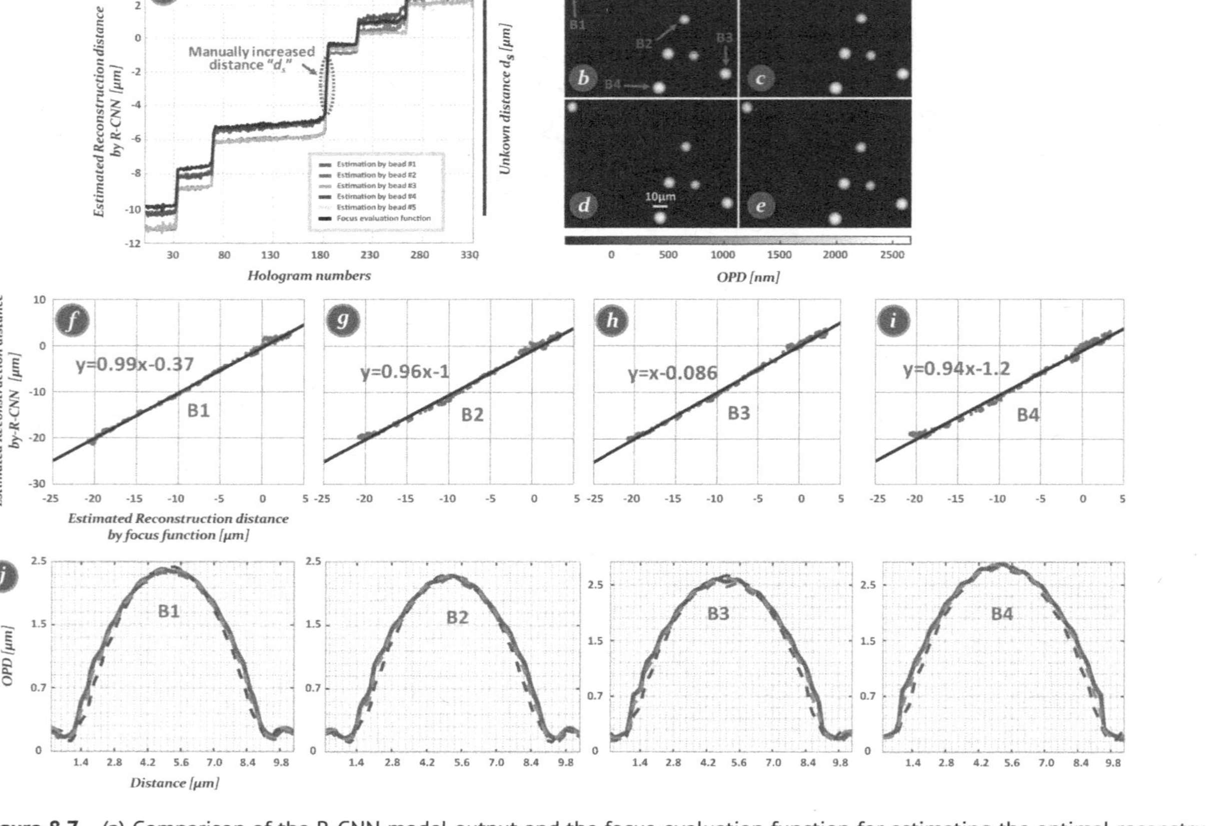

Figure 8.7 (a) Comparison of the R-CNN model output and the focus-evaluation function for estimating the optimal reconstruction distance. Holograms are recorded corresponding to different unknown d_s, which is manually increased (d_s is along the optical axis, by moving the stage d_s changes). A few bead holograms with multiple beads are shown in Figure. 8.6. (b–e) Reconstructed phase images by the value provided by R-CNN when the input to model respectively is B1–B4. The color map is similar for all figures. (f–i) Correlation analysis between CNN's output (input to the model is B1–B4, respectively) and focus-evaluation function. (j) Cross-section of the reconstructed phase images of B1–B4 when the R-CNN provides d for holograms of B1–B4. Reconstruction by the focus-evaluation function is also presented for comparison [7] / with permission of Optical Publishing Group

analysis, a cross-section of the 3D profile of each bead is also shown. Since beads are positioned at the same location along the z-axis, they are in the same focus distance from the CCD camera. We can see that the output of the R-CNN provides similar results for the four input holograms. The correlation between the focus-evaluation function's output and our R-CNN's output is significant.

The performance of our deep learning model was also analyzed using RBCs. Unlike microsphere beads, there is substantial cell-to-cell variation in RBCs. The variation is beneficial for testing a model's performance in various conditions. Five single RBCs (R1–R5) were selected to evaluate performance of the R-CNN model (Figure 8.6). As shown in Figure 8.6, R5 is placed at a different focus distance compared to other RBCs. Therefore, it might be difficult for conventional focus-evaluation functions to determine a focus distance at which the 3D profile of R5 is best resolved. We can see that the RBC-to-RBC single-hologram variation (Figures 8.6d–f) is considerable. Therefore, the performance of our model can be evaluated under conditions different from training. To validate the model's output, the output of a single RBC hologram was compared with that of the focus-evaluation function. A correlation analysis was then performed (Figure 8.8).

Figure 8.8 demonstrates that our model could correctly estimate the propagation distance with respect to the hologram of the micro-size sample. Moreover, our R-CNN model could predict the correct value for the R5, which is located at a different focus distance. This can be very helpful for studying samples where cells are placed at different levels along the optical axis. Figure 8.9 shows that the focus-evaluation function is unable to find a reconstruction distance at which the profile for R5 is well resolved. In contrast, when the input is the R5 hologram, our R-CNN model can find a perfect reconstruction distance at which R5 contrast is best resolved. Furthermore, a perfect profile of RBCs can be achieved by combining two reconstructions at two different distances (for example, R1 and R5).

In this work, 90 reconstructions were performed. The focus-evaluation function was then applied to examine standard deviations of pixel values within the whole amplitude image for multiple cells or objects. It is assumed that the plane that minimizes the standard deviations of amplitude image is the best focus plane. Our deep-learning models without reconstruction enabled us to estimate the focus plane of the corresponding object, which could be further used to determine the distance between micro-sized objects along the optical axis. This is critical for studying biological samples in 3D environment cultures because live cells move freely along the optical axis during the experiment. The R-CNN model is fast. It can instantly find the focus plane without reconstruction. This can significantly enhance the reconstruction time in DHM. One main challenge in designing CNN-based predication models is that the model needs a lot of samples to be well-trained.

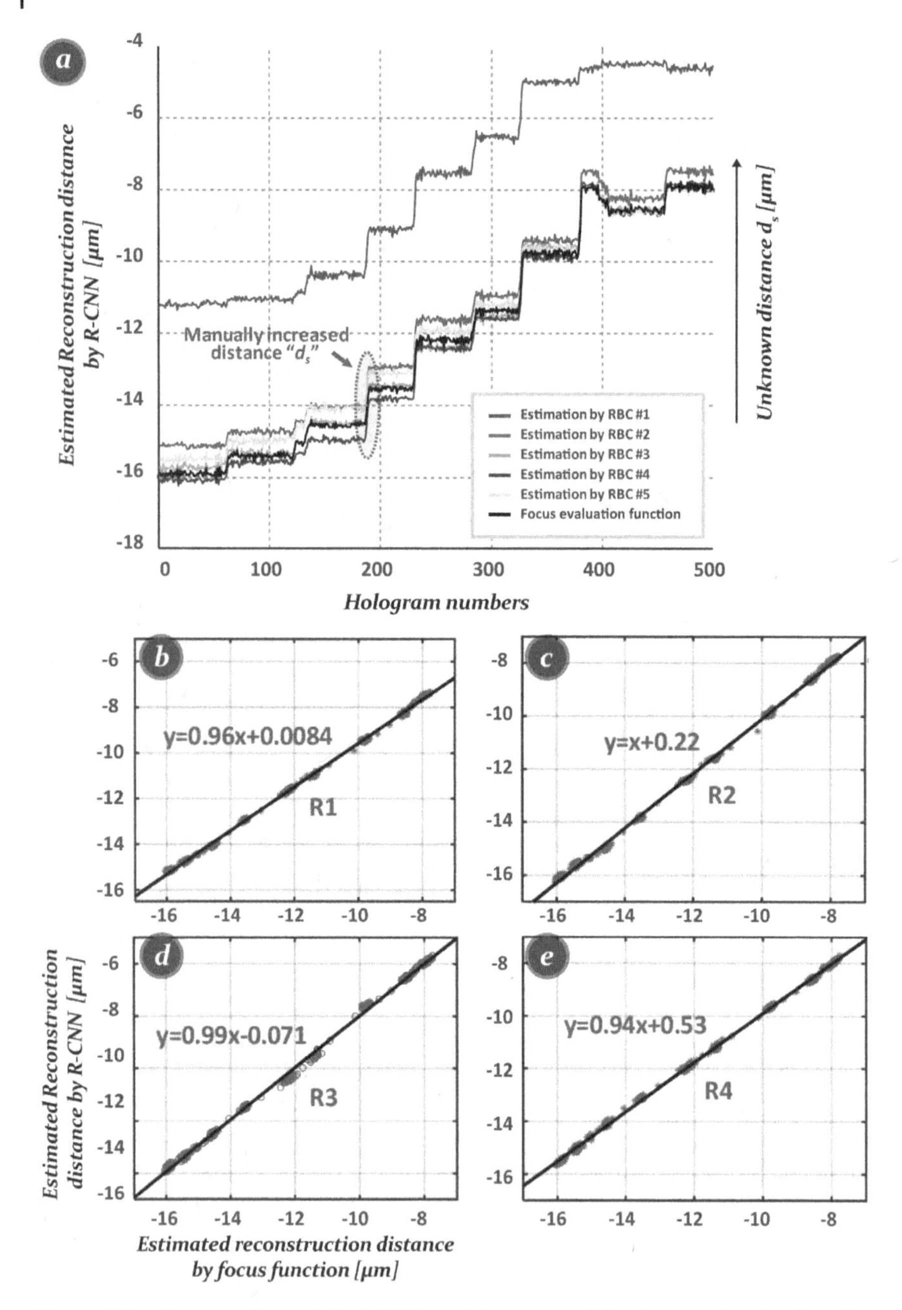

Figure 8.8 (a) Comparison of the R-CNN model output and the focus-evaluation function for estimating the optimal reconstruction distance. Holograms are recorded corresponding to different unknown d_s, which is manually increased (d_s is along the optical axis, by moving the stage d_s changes). A few RBC holograms recorded at various d_s are shown in Figure 8.6; R5 is located at a different distance from R1–R4. (b–e) Correlation analyses between the R-CNN model outputs and focus-evaluation function when the input to the model is R1–R4, respectively [7] / with permission of Optical Publishing Group.

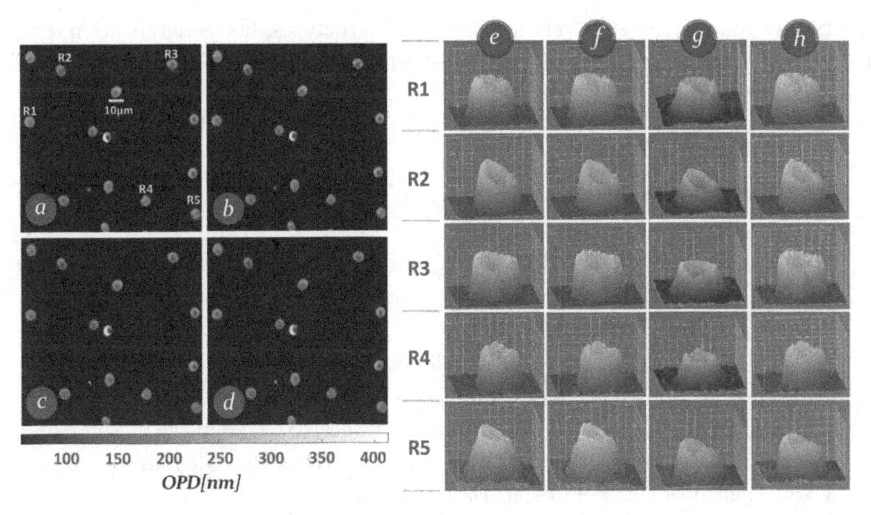

Figure 8.9 Reconstructed phase images when (a) *d* is estimated with a focus-evaluation function, (b) *d* is estimated with the R-CNN model and R1 input, and (c) *d* is estimated with the R-CNN model and R5 input. (d) Combination of (b) and (c); R5 is copied from (c) and inserted into (b). Color map is similar for all images. (e) 3D profile of R1–R5 when the focus is according to the focus-evaluation function. In this case, R5 is at a different focus level. Thus, the profile is not correct. (f) 3D profile of R1–R5 when the focus is according to CNN model with R1 as input. R5 is at a different focus level. Thus, the profile is not correct. (g) 3D profile of R1–R5 when the focus is according to CNN model with R5 as input. R5 profile is correctly reconstructed. (h) Profiles extracted from combination of (c) and (d) [7] / with permission of Optical Publishing Group.

8.4 Conclusions

In this chapter, we introduced a deep-learning CNN with a regression layer to estimate the best focus distance in the numerical reconstruction of micro-sized objects at the single-object level. Focused images and corresponding reconstruction distance for the training dataset were obtained using a conventional focus-evaluation function. Experimental results and comparison with the focus-evaluation function demonstrated that the presented models could properly estimate the reconstruction distance from a filtered hologram. We experimentally showed that the deep-learning method could significantly reduce the numerical reconstruction time to find the correct focus distance. The numerical focus-evaluation function requires many digital propagations at different distances making it computationally inefficient. Moreover, since the reconstruction distance of the object is numerically estimated at the single-cell level, our deep-learning model can offer reconstruction distances in accordance with the location of various micro-size objects. This method can be applied to biological cell studies, particularly cancer cells, since they are round cells with very similar visual structures. For generalization, it is required to train the model with plenty of cells and a wide range of reconstruction distances.

References

1 Dubois, F., Schockaert, C., Callens, N., and Yourassowsky, C. (2006). Focus plane detection criteria in digital holography microscopy by amplitude analysis. *Opt. Express* 14 (13): 5895–5908.

2 Langehanenberg, P., Kemper, B., and Bally, G.V. (2007). Autofocus algorithms for digital-holographic microscopy. In: *Biophotonics 2007: Optics in Life Science (Vol. 6633 of Proceedings of SPIE-OSA Biomedical Optics)* (ed. J. Popp and G. von Bally), Munich, Germany (17–21 June 2007). New York (NY): Optica Publishing Group.

3 Langehanenberg, P., Kemper, B., Dirksen, D., and Bally, G.V. (2008). Autofocusing in digital holographic phase contrast microscopy on pure phase objects for live cell imaging. *Appl. Opt.* 47 (19): 176–182.

4 Wang, H., Qin, A., and Huang, M. (2009). Autofocus method for digital holographic reconstruction of microscopic object. In: *The International Symposium on Photonics and Optoelectronics (SOPO 2009)*, Wuhan, China (14–16 August 2009). New York (NY): IEEE.

5 Li, W., Loomis, N.C., Hu, Q., and Davis, C.S. (2007). Focus detection from digital in-line holograms based on spectral l_1 norms. *J. Opt. Soc. Am. A* 24 (10): 3054–3062.

6 Fonseca, E.S., Fiadeiro, P.T., Pereira, M., and Pinheiro, A. (2016). Comparative analysis of autofocus functions in digital in-line phase-shifting holography. *Appl. Opt.* 55 (27): 7663–7674.

7 Jaferzadeh, K., Hwang, S., Moon, I., and Javidi, B. (2019). No-search focus prediction at the single cell level in digital holographic imaging with deep convolutional neural network. *Biomed. Opt. Express* 10 (8): 4276–4289.

8 Schmidhuber, J. (2015). Deep learning in neural networks: an overview. *Neural Networks* 61: 85–117.

9 Yong, Y.L., Tan, L.K., Mclaughlin, R.A. et al. (2017). Linear-regression convolutional neural network for fully automated coronary lumen segmentation in intravascular optical coherence tomography. *J. Biomed. Opt.* 22 (12): 126005.

10 Shimobaba, T., Kakue, T., and Ito, T. (2018). Convolutional neural network-based regression for depth prediction in digital holography. In: *2018 IEEE 27th International Symposium on Industrial Electronics (ISIE)*, Cairns, Australia (13–15 June 2018), 1323–1326. New York (NY): IEEE.

11 Pitkäaho, T., Manninen, A., and Naughton, T.J. (2019). Focus prediction in digital holographic microscopy using deep convolutional neural networks. *Appl. Opt.* 58: A202–A208.

12 Ren, Z., Xu, Z., and Lam, E. (2018). Learning-based nonparametric autofocusing for digital holography. *Optica* 5: 337–344.

13 Rivenson, Y., Zhang, Y., Günaydın, H. et al. (2017). Phase recovery and holographic image reconstruction using deep learning in neural networks. *Light Sci. Appl.* 7 (2): 17141.

14 Wu, Y., Rivenson, Y., Zhang, Y. et al. (2018). Extended depth-of-field in holographic imaging using deep-learning-based autofocusing and phase recovery. *Optica* 5 (6): 704–710.

15 Lin, X., Rivenson, Y., Yardimci, N. et al. (2018). All-optical machine learning using diffractive deep neural networks. *Science* 361 (6406): 1004–1008.

16 Zeng, T., Zhu, Y., and Lam, E.Y. (2021). Deep learning for digital holography: a review. *Opt. Express* 29 (24): 40572–40593.

17 LeCun, Y., Bengio, Y., and Hinton, G. (2015). Deep learning. *Nature* 521: 436.

18 Kingma, D. and Ba, J. (2014). Adam: A method for stochastic optimization. arXiv. https://doi.org/10.48550/ARXIV.1412.6980

9

Automated Phase Unwrapping in DHM with Deep Learning

9.1 Introduction

The digital holographic microscopy (DHM) can provide quantitative phase images related to the morphology and content of biological samples. Phase values in the reconstructed image are limited between $-\pi$ and π. Thus, discontinuity may occur due to modulo 2π operation. A phase unwrapping process must be carried out to remove 2π phase discontinuities in the phase image and estimate the true continuous phase image. Phase unwrapping consists of finding the location of the phase jump and connecting adjacent pixels by adding or subtracting multiples of 2π to remove phase discontinuities.

Many phase unwrapping algorithms have been studied to solve challenging problems such as phase discontinuities. Advanced phase unwrapping algorithms can be divided into three types: global, region, and path-following algorithms. Global algorithms can minimize differences between discrete gradients of wrapped and unwrapped phase images. Although these algorithms are robust, their computational requirements are large. Hence, they are unsuitable for real-time live-cell imaging applications. Region algorithms can split an image into smaller ones, unwrap regions with respect to each other, and merge them into larger regions. These algorithms have been regarded as a compromise between robustness and computational intensiveness. Region algorithms are further categorized into region-based algorithms and tile-based algorithms according to the procedure in defining a homogeneous region. Region-based algorithms can find homogeneous regions using phase gradients, while tile-based algorithms can split an image into a small local grid unwrapped by simpler algorithms. Path-following algorithms are classified into path-dependent, residue-compensation, and quality-guided algorithms. Path-dependent algorithms can perform unwrapping through a predetermined search path. However, they cannot remove noise well.

Artificial Intelligence in Digital Holographic Imaging: Technical Basis and Biomedical Applications, First Edition. Inkyu Moon.

Residue-compensation algorithms search for residues in a wrapped phase image and generate branch cuts to connect opposite orientation residues. These algorithms can determine the quality of an unwrapped phase image according to a cut selection strategy. For a detail description of the phase unwrapping algorithms, see Chapter 4.

Quality-guided algorithms are the most promising methods. They depend on the assumption that a good quality or phase map will lead to a reasonable unwrapping path while grouping pixels. According to the phase map, the highest-quality pixels are unwrapped first, while the lowest-quality pixels are unwrapped last to avoid error propagation. These methods are computationally efficient in real-time applications. There are cases where systematic phase unwrapping methods fail to obtain unwrapped phase values. Abrupt phase changes can occur at cell boundaries in phase images. When the phase continuously rises to π and exceeds it, the phase rapidly changes to $-\pi$ due to the modulo 2π operation on the phase. This results in 2π discontinuities which must be removed using phase unwrapping algorithms. However, if either the thickness of cells is out of the depth of focus or cells are reconstructed on a partially defocused image plane, the phase in the cell boundary may immediately shoot up above π. Thus, the phase difference in the local boundary in the wrapped phase image with these abrupt phase change problems can be much less than 2π (smaller than π). Consequently, the phase unwrapping algorithm will incorrectly interpret that the phases belonging to the two pixels are in the same range without requiring unwrapping. Hence, phase unwrapping is not correctly executed in the cell area. In this chapter, we will introduce a new deep learning model that can effectively resolve this incomplete phase unwrapping in real-time [1]. Moreover, this model can carry out an autofocusing that converts the out-of-focus wrapped phase image into an in-focus unwrapped phase image. Our model is a fusion of deep learning and off-axis DHM to recover the phase value of biological samples, which is essential for studying morphological and material changes in live cells at the single cell level.

Recently, deep learning models for phase unwrapping have been proposed using convolutional neural network (CNN) models, especially the U-net type model [2–7]. A CNN model can learn to minimize a certain loss function such as the Euclidean distance between actual and predicted image. Due to its squared characteristic of calculated distance, it can correct large errors well. However, it is tolerant of small errors, causing the CNN model to produce blurry images. Besides, CNN-based models face the problem of an abrupt phase shift that happens in the numerical phase unwrapping algorithms. To overcome these problems, we will introduce a generative adversarial network (GAN) to completely unwrap wrapped phase signals obtained using DHM which can automatically learn a proper adversarial loss function [8–10]. We will employ Pix2Pix GAN [11] which consists of a generator and a discriminator and learns image-to-image translation with label images to automatically reconstruct unwrapped focused-phase images.

In this study, our model is defined as UnwrapGAN which consists of a U-net generator and a discriminator [11, 12]. To train the UnwrapGAN model, we used three types of cancer cells and obtained wrapped defocused-phase images for each cell using DHM. Unwrapped focused images for true or label data were obtained from wrapped defocused-phase images using a quality-guided unwrapping algorithm (see Figure 9.1a). Wrapped defocused-phase images were used as input to a generator and the generator produced unwrapped focused-phase images. A discriminator determines whether the output image is well-formed and proceeds to train the generator to create a similar image to the true unwrapped phase image (see Figure 9.1b).

To test the trained model, we used defocused wrapped data not used in training as input to the generator. The trained model performed both unwrapping and focusing work on the untrained data. Results were compared with those of the numerical phase unwrapping algorithm through a single cell comparison and the entire image containing several sells. Compared with the U-net model, the trained model reconstructed more elaborate phase images. We also showed that it was possible to generalize our models since it performed phase reconstruction for other types of cells (liver cancer and colon cancer cells). Besides, we demonstrated that the presented model could overcome the problem of an abrupt phase change caused by a phase jump in which all numerical phase unwrapping methods failed to restore true phase images. Furthermore, the model is several times faster than the conventional quality-guided algorithm. The quality-guided method requires sorting. It must find the best unwrapping path, while our model uses fixed trained weights to unwrap the wrapped phase image, making it faster. Therefore, our deep learning model offers both phase unwrapping and autofocusing simultaneously in real time, which can greatly influence the process of imaging biological cells through DHM.

9.2 Deep-learning Model

Our deep learning model is based on a Pix2Pix GAN model with generators and discriminators. The generator conducts an image-to-image translation task. When a raw image is fed into the model, a modified output is generated. The discriminator is used to accurately train the generator. The generated or real image is fed as input into the discriminator which is trained to determine whether an input image is a generated image or a real image. The generator and discriminator have a

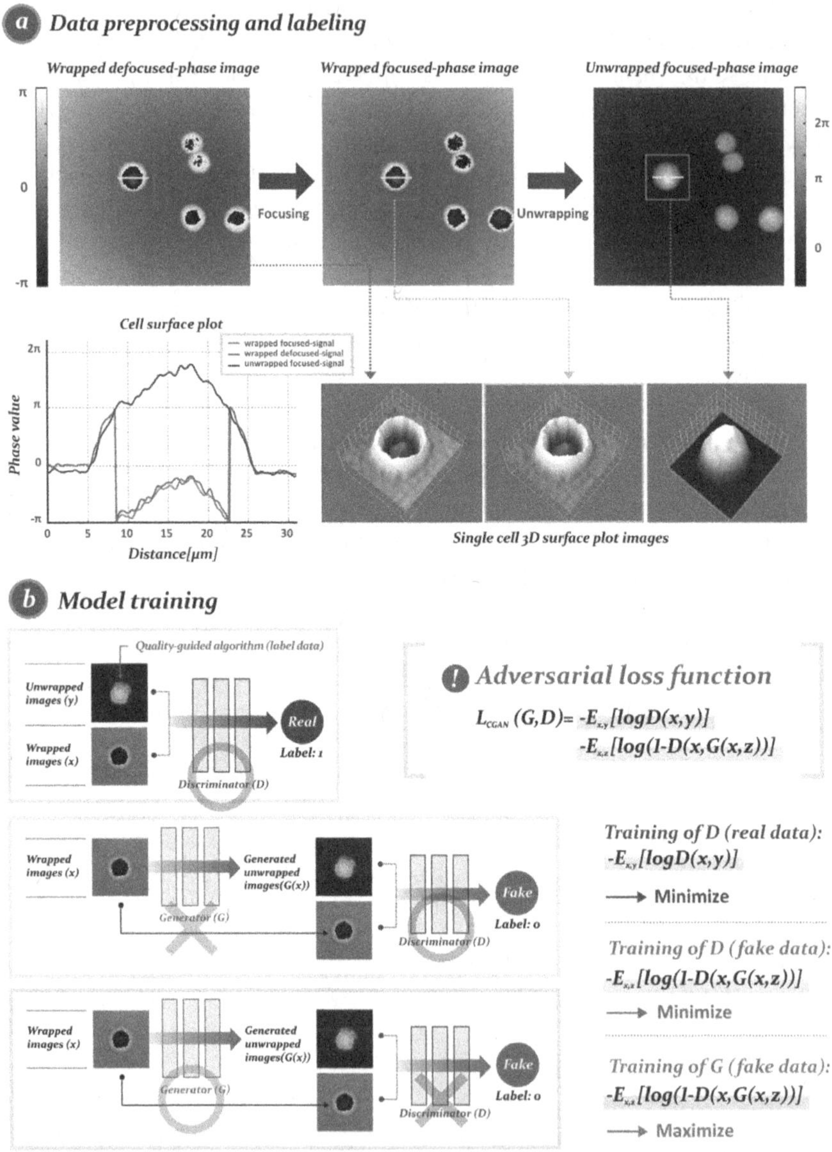

Figure 9.1 (a) Quantitative phase image of multiple lung cancer cells. These images are focused manually then unwrapped by the quality-guided unwrapping algorithm. The unwrapped focused-phase images are used for labeled training in the model. Cross-section and 3D representation of one cell with wrapped and unwrapped signals are shown. (b) Training of model where the UnwrapGAN model consists of a discriminator and a U-net generator. (c) Results for untrained cells. It tests whether the trained model can generate unwrapped focused-phase images from unseen images that have not been used for training and whether it is possible to recover phase values for other types of cells to evaluate model generalization. We found that the proposed model could enable the correction of problems on abrupt phase change. The proposed model's result is also compared with that of the U-net [1] / with permission of Optical Publishing Group.

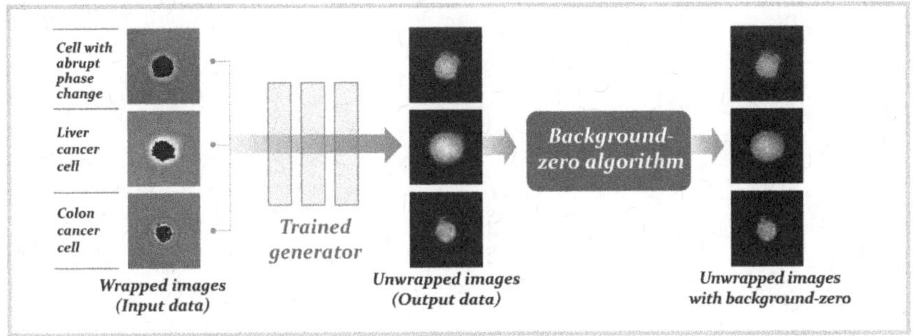

Figure 9.1 (Continued)

convolution-BatchNorm-Leaky ReLU with a 3×3 filter. The generator consists of down-sampling to extract features of the input image for translation and up-sampling to reconstruct the image based on extracted features (see Figure 9.2a). Both down-sampling and up-sampling have eight convolution layers. When down-sampling is conducted before up-sampling, much information of the original image is lost, resulting in a blurred output. Thus, a skip connection technique was used to share high-frequency information between the input and output. By doing this, blurry effects of the generated image can be reduced by connecting information in the ith layer of the down-sampling process to information about the $(n-i)$th layer of the up-sampling process with the general shape of a U-net [12].

The discriminator learns to distinguish between real and fake images (see Figure 9.2b). We used an Adam optimizer with adaptive momentum and parameters of $\beta 1 = 0.5$ and $\beta 2 = 0.999$. The number of epochs is 100 and the learning rate is 0.0002. Models are trained on a server with five NVIDIA RTX Quadro 6000 graphics cards.

An adversarial loss is used to train the Pix2Pix GAN model [11], which is defined as

$$L_{C-GAN}(G, D) = E_{x,y}[\log D(x, y)] + E_x[1 - \log D(x, G(x))],$$

where E is the expected value, x is the wrapped input image, y is the unwrapped label image, G is the generator, and D is the discriminator. The generator tries to minimize this objective against an adversarial D, which tries to maximize it. Additionally, the generator requires the following $L1$ loss function to produce the overall structure of the image and low-frequency information:

$$L_{L1}(G) = E_{x,y}\left[\|y - G(x)\|_1\right].$$

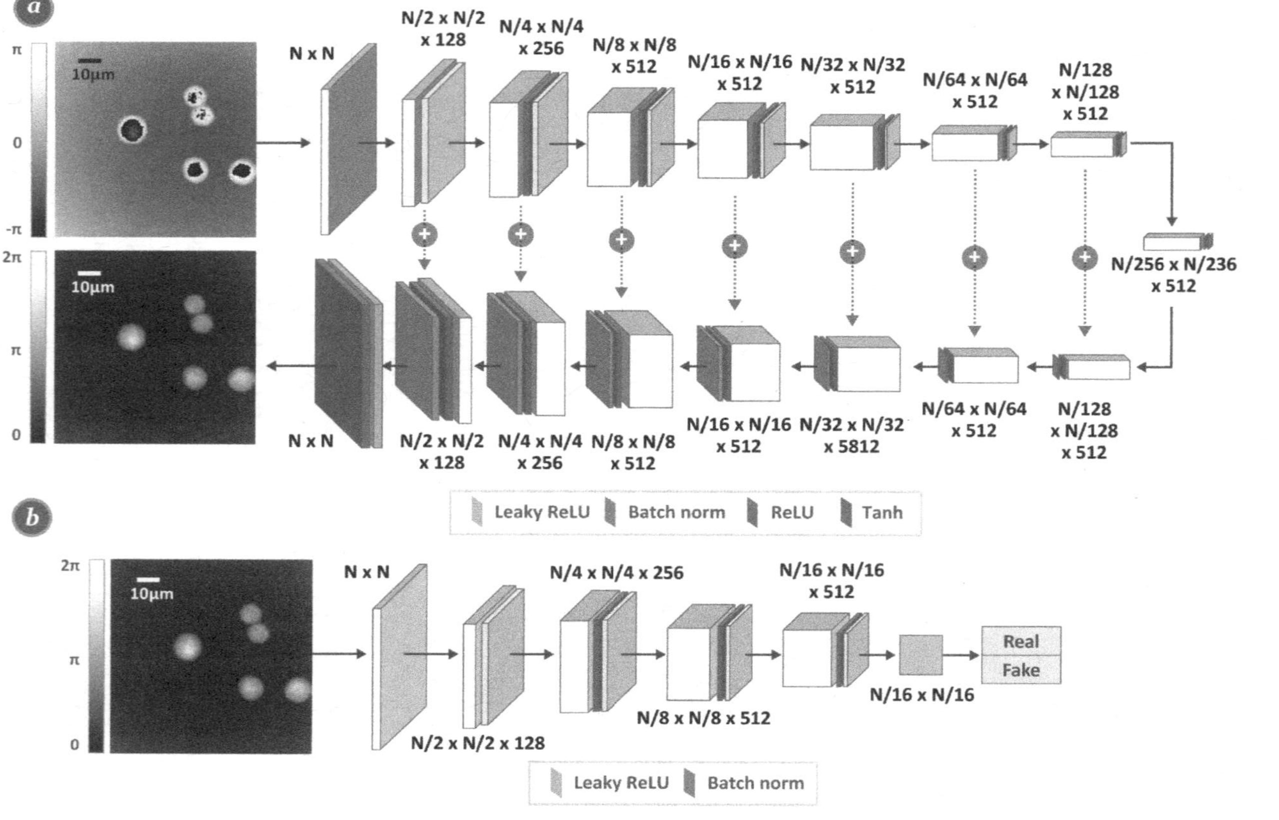

Figure 9.2 (a) Architecture of the generator similar to a U-net to recover an unwrapped phase image from wrapped phase images. The U-net has convolution layers, batch normalization, and various activation functions. (b) Discriminator to compare fake and real images with convolution layers. Tanh is the hyperbolic tangent function [1] / with permission of Optical Publishing Group.

Therefore, the objective to train our model is as follows:

$$L = \arg\min_{G} \max_{D} L_{C-GAN}(G, D) + \lambda L_{L1}(G).$$

We obtained wrapped phase images of three types of cancer cell lines: PC9 (lung cancer cells), SNU449 (liver cancer cells), and SW640 (colon cancer cells). The training dataset was generated using the numerical reconstruction algorithm in off-axis DHM (see Chapter 5). During numerical reconstruction, a quality-guided unwrapping algorithm was toggled on and off and the following two sets with the same area were stored for training: the wrapped phase and the corresponding unwrapped value. Reconstruction and unwrapping algorithms were run in MATLAB 2018. Cell segmentation was conducted using macro code in ImageJ [13].

We reconstructed cell images with a size of 900 × 900 pixels (a single cell covers an area of averagely 18 μm × 18 μm). The image size was changed to single cell level (256 × 256 pixels) or multiple-cell level (1024 × 1024 pixels) using an interpolation method to fit the model. The PC9 cell line was used as a training dataset. A total of 5200 pairs of defocused wrapped and unwrapped phase images were used. All unwrapped phase images for training were focused on through manual control. Figure 9.3 shows a gallery of images used for training. For training single cell level (256 × 256 pixels) with the proposed model, it took about 10 hours to train 100 epochs. It took about three days to train multiple cells (1024 × 1024 pixels) for the same epochs. Meanwhile, for comparison with our model, U-net was trained only on multiple cells. It took about 60 hours to train. The test dataset consisted of PC9, SNU449, and SW640 cell images.

9.3 Unwrapping with Deep-learning Model

9.3.1 Reconstruction of Unwrapped Phase Image at the Single Cell Level

We input wrapped phase images of lung cancer cells at the single cell level that were not used when training our model generator to validate the trained model. The model output was compared with that of the quality-guided unwrapping algorithm. Figure 9.4 shows results of the UnwrapGAN model for lung cancer cells. We could see that our deep learning model precisely removed 2π phase discontinuities in the wrapped phase image and restored correct unwrapped phase of cells.

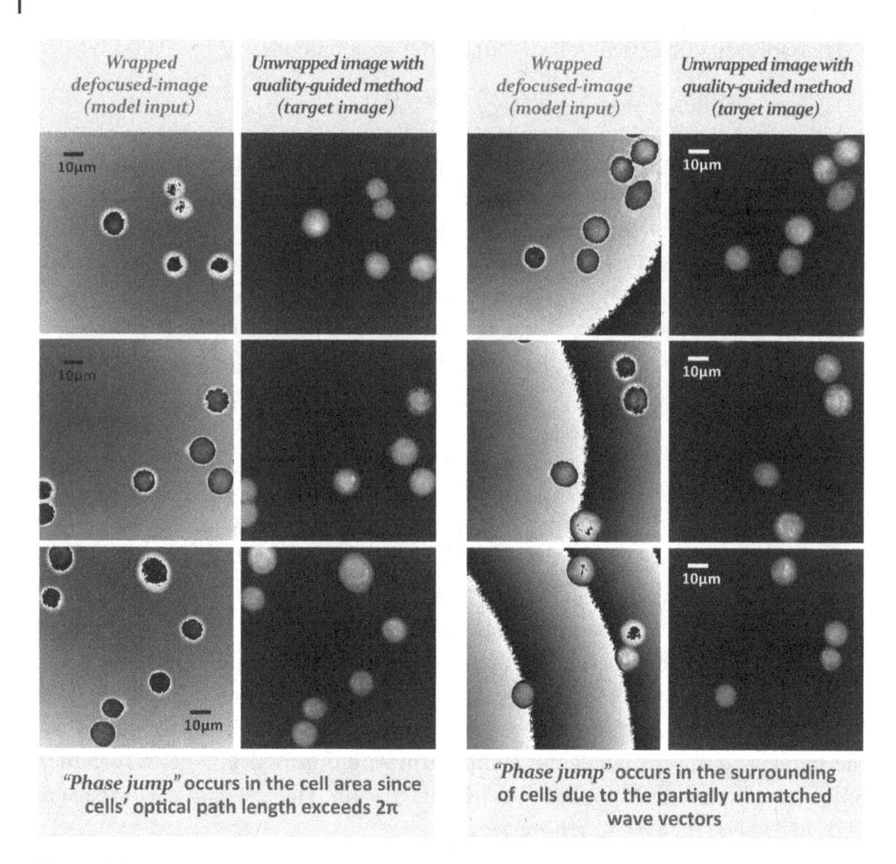

Figure 9.3 Gallery of lung cancer cell images used for training the model. Phase images were obtained using a numerical reconstruction algorithm from off-axis holograms. A quality-guided unwrapping method was switched off and on to provide input and target images. Image pairs were used for training [1] / with permission of Optical Publishing Group.

9.3.2 Reconstruction of In-focus Unwrapped Phase Image with Multiple Cells

In this section, we present an autofocus method based on our UnwrapGAN model. To train our model, we used phase images with a size of 1024 × 1024 pixels including multiple cells. For model training, wrapped defocused-phase images at random positions as input images to our model were generated and matched with unwrapped phase images in focus. Our model learned to accurately generate unwrapped phase images in focus from wrapped defocused-phase images, which were reconstructed at random positions deviated from the exact reconstruction distance.

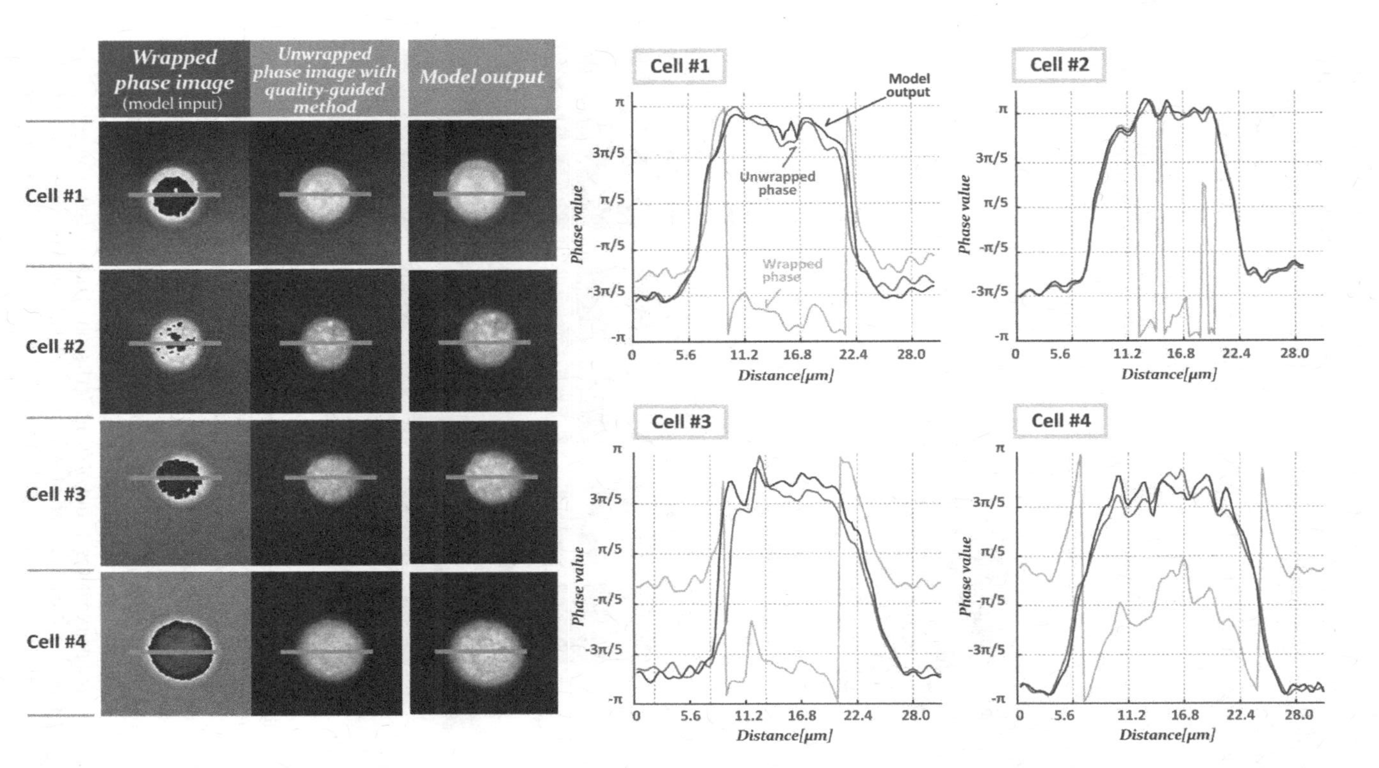

Figure 9.4 Results of the trained model compared with those of the quality-guided unwrapping method. The model was trained using PC9 cell line (lung cancer cells). Graphs presented on the right show phase profiles along the yellow line in cell images [1] / with permission of Optical Publishing Group.

To test how the deep learning model learned well to reconstruct unwrapped phase images in focus from wrapped defocused-phase images, we used wrapped defocused-phase images obtained at different reconstruction distances as test data and compared them with three cases (see Figure 9.5). In the first case, we used a quality-guided method for phase unwrapping. In the second case, the

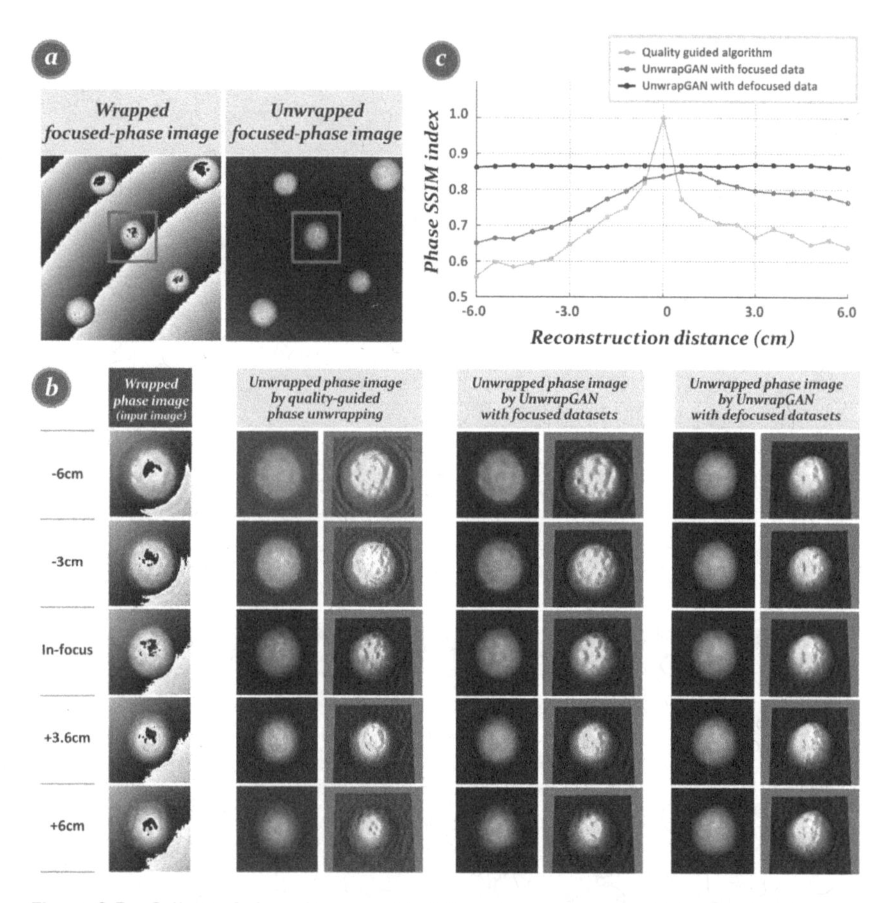

Figure 9.5 Gallery of phase images with different reconstruction distances. (a) Wrapped focused-phase image and corresponding unwrapped focused-phase image. (b) Images of the cell in the red box in (a) with specific reconstruction distance away from in-focus. The first column line of (b) indicates wrapped phase images, the second line indicates unwrapped phase images reconstructed using the quality-guided phase unwrapping method, the third line indicates unwrapped phase images reconstructed using the trained model with paired datasets in-focus, and the last line indicates unwrapped phase images reconstructed using our trained model with paired datasets out of focus. (c) Result of numerically computed SSIM indices between phase images obtained at different reconstruction distances and the unwrapped focused-phase image in (a) [1] / with permission of Optical Publishing Group.

UnwrapGAN model was trained using the input dataset in-focus. In the third case, we trained the UnwrapGAN model using the input dataset out of focus, where defocused images were reconstructed at random positions. Note that to create defocused images, reconstruction distances were deliberately moved away from the correct focus distance by specific values. In the case of phase image reconstruction with the quality-guided method (see Figure 9.5), the further away the reconstruction distance was from the in-focus, the bigger the difference between the correct phase and reconstructed phase image. In the case of the model trained using input datasets in-focus, there was less difference than that with the quality-guided method. However, the reconstruction accuracy decreased as the reconstruction distance deviated from the exact reconstruction distance. We observed that phase images produced from our model trained using the defocused input dataset were almost identical to the phase image in-focus, although wrapped defocused-phase images obtained at different reconstruction distances away from the focus position were fed into the UnwrapGAN model. This indicated that the deep learning model outperformed the numerical quality-guided path-following algorithm in phase unwrapping. It could even reconstruct a stable unwrapped phase image in focus from wrapped images restored at different reconstruction distances away from the focus position.

In addition, we quantified the structural similarity (SSIM) index [14] numerically to indicate how similar our results are to those of the original in-focus phase image. The SSIM is a framework for quality assessment based on the degradation of structural information. The SSIM index indicates the unity for two identical images. It drops below 1 if the similarity is low. SSIM indices for the unwrapped phase images at different reconstruction distances were calculated based on the in-focus unwrapped phase image, which was obtained using the quality-guided phase unwrapping algorithm. Circles on the solid lines in the graph of Figure 9.5 represent the average of SSIM indices for 30 different phase images as input data. The SSIM index of the in-focus unwrapped phase image from the quality-guided phase unwrapping method was exactly 1 since the two images were identical. When the reconstruction distance was away from the focus position, the SSIM index decreased rapidly. The trained model with in-focus input datasets showed the same trend as the quality-guided phase unwrapping method although the SSIM index was higher than that of the quality-guided method. However, the model trained with the defocused input datasets showed almost constant SSIM index value close to 0.9 regardless of the reconstruction distance value.

9.3.3 Model Comparison for Phase Unwrapping

In this section, we will compare the performance of the UnwrapGAN model with that of the U-net model based on CNN structure. Figure 9.6 shows that these two

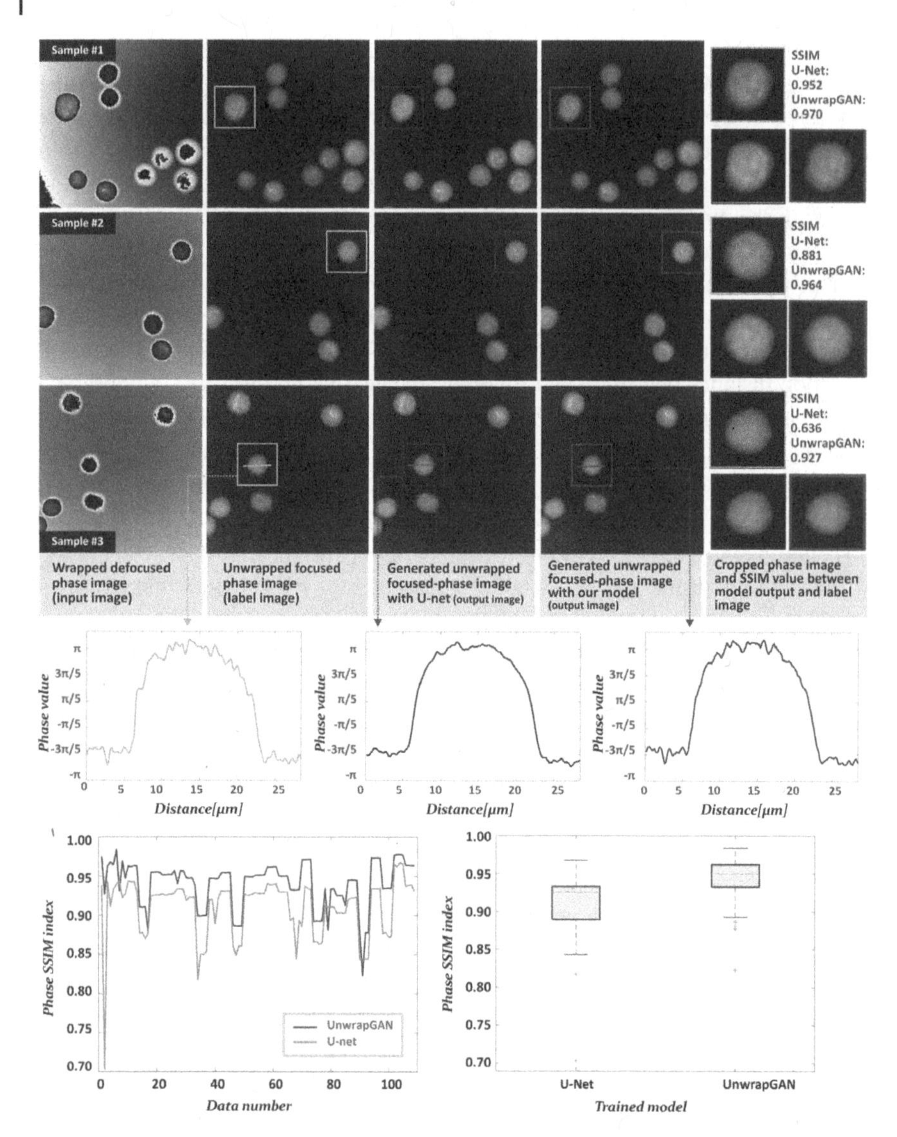

Figure 9.6 Gallery of phase recovery results generated by U-net and UnwrapGAN. Wrapped defocused-phase images were fed into the model as input. In-focus unwrapped phase images of corresponding inputs were obtained using a quality-guided path-following algorithm to make ground truth. The two output images indicate unwrapped focused-phase images reconstructed using trained U-net and UnwrapGAN models, respectively. Middle graphs show phase profiles along the straight line in sample #3. Bottom graphs show the calculated SSIM index between the label and output images of each model with 110 single cells' phase images, where the area marked with a square was considered as shown in the upper graph. The graph of the bottom left shows the SSIM index for each phase image. The graph of the bottom right shows the mean and standard deviation of SSIM indices for 110 phase images [1] / with permission of Optical Publishing Group.

deep learning models are remarkably different in the performance of phase unwrapping. The U-net model has a smoother phase distribution than the actual phase distribution (see Figure 9.6). However, we observed that the UnwrapGAN model tried restoring the phase values close to the actual phase distribution as much as possible. We also quantified the SSIM value numerically to indicate the similarity between each model and those of the label phase image. As shown in the bottom graph of Figure 9.6, SSIM values for the UnwrapGAN model were above 0.9 on average. However, for the U-net model, the SSIM was much smaller than that of the UnwrapGAN and deviation of SSIM values was very large.

9.3.4 Model-generalization with Different Cell Types

It is necessary to accurately unwrap wrapped phase images regardless of cell types. Liver and colon cancer cells were tested to verify the validity of our deep learning model, which was trained using lung cancer cells at the single cell level. This study evaluated whether the model well learned general phase reconstruction rather than phase reconstruction for specific cells. Figure 9.7 shows that our model can provide phase values correctly for other types of cells. This is because the training dataset of the same type of cells has different morphological features due to the heterogeneous cancer cell population. For example, the shape of colon cancer cells is partially different from that of lung cancer cells. However, the phase unwrapping was accurately performed according to ground truth phase values.

9.3.5 Abrupt Phase Change Problem

A phase image from DHM can be divided into two areas: cell area and background area. Phase distribution within the cell area is larger than that in the background area. Thus, phase distribution inside the cell is wrapped to values near or above $-\pi$ due to modulo 2π operation. When the optical path length at the cell boundary is smaller than that at the center of the cell as shown in the phase profile of the cell (Figure 9.8a), the cell boundary and background are grouped according to the grouping principle in quality-guided unwrapping. The phase unwrapping algorithm can unwrap the phase of the cell by adding or subtracting multiples of 2π to remove a phase discontinuity of the cell. To this end, the boundary between the interior of the cell and the background is crucial for phase unwrapping. However, if an abrupt phase change occurs due to either a partially defocused phase image or strong diffraction patterns on the cell boundary, the phase might have very big jumps above π at the boundary and wrap values much bigger than $-\pi$. Here, the phase difference between the cell and background becomes smaller (less than π). Thus, they are classified into the same group in quality-guided unwrapping. This results in phase unwrapping failure as shown in Figures 9.8a and 9.8b. On the other hand, our UnwrapGAN model can successfully perform phase

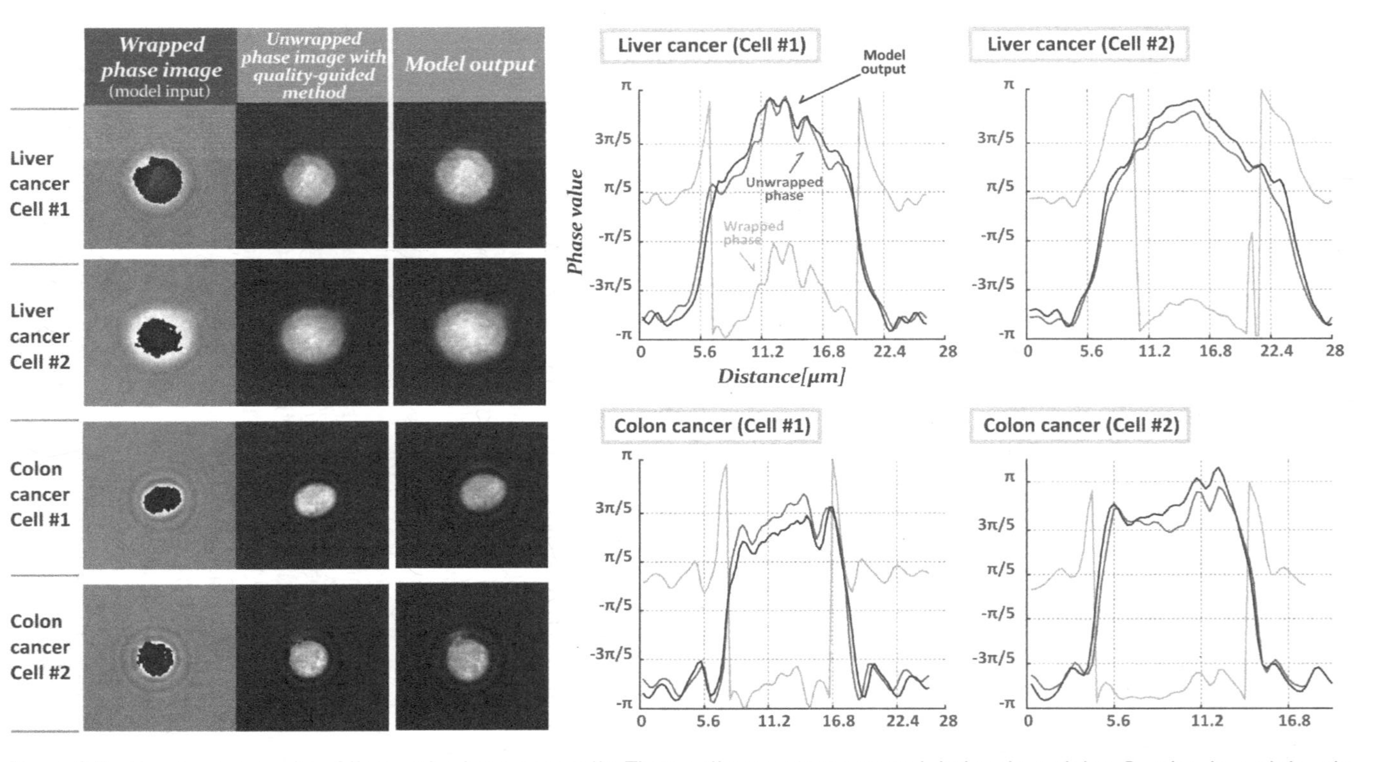

Figure 9.7 Unwrapping results of liver and colon cancer cells. These cell types were not used during the training. Results showed that the model was generalized. Right graphs show phase value of the yellow line in the cell image [1] / with permission of Optical Publishing Group.

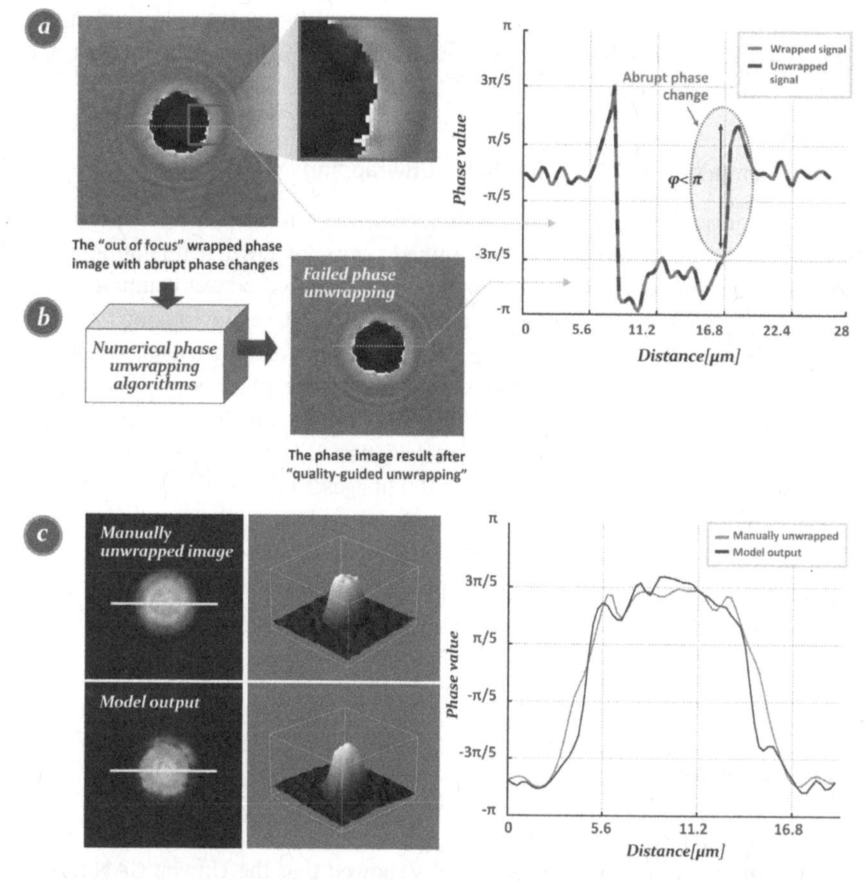

Figure 9.8 (a) Abrupt phase change noise at a cell boundary with one enlarged area and the cross-section of the phase value. (b) The quality-guided phase unwrapping algorithm fails to recover the image. (c) Manually unwrapped image and output of the proposed model. The 3D profile and cross section are also shown for visual comparison [1] / with permission of Optical Publishing Group.

unwrapping by removing the abrupt phase change (Figure 9.8c). To train our model, we obtained wrapped phase images partially out of focus which were reconstructed at the distance a little bit away from the exact focus distance. The trained model converted a wrapped phase image partially out of focus into a focused unwrapped phase image at the single cell level. The unwrapped phase image generated by the trained model was also compared with the unwrapped phase image in-focus obtained by manually removing abrupt phase changes. These results demonstrate that our model can remove phase jumps and successfully perform phase unwrapping. The quality-guided phase unwrapping algorithm performs phase unwrapping depending on edge reliability. However, our model can extract various features using a convolution layer and learn to use an accurate unwrapped image database to generate an unwrapped phase image.

9.3.6 Computational Time for Phase Unwrapping

One of the advantages of our model is that it can unwrap multiple phase images in a very short time. It only convolves trained filters for down-sampling and up-sampling an input image for phase unwrapping. Our method was compared with the quality-guided phase unwrapping algorithm. The phase unwrapping time was measured with 1000 images of the same size for the model trained using single cell level phase images (256×256 pixels). To calculate the phase unwrapping time taken regardless of the content of the image, the unwrapping time for 100 images was measured. The average and standard deviation were calculated by measuring the time taken for phase unwrapping of 100 images 10 times. The quality guided phase unwrapping algorithm is widely used for real-time phase unwrapping. It was confirmed that the deep learning model provided the output about twice as fast as the quality guided phase unwrapping algorithm. With our UnwrapGAN model, phase unwrapping can be done in real-time while solving various problems (autofocusing, generalization, and abrupt phase change).

9.4 Conclusions

It is essential to restore correct phase images of cells when studying live biological samples in real-time. Experimental results showed that the UnwrapGAN model could automatically reconstruct an unwrapped phase image in-focus from a wrapped phase image regardless of the reconstruction distance. It showed higher performance than recent U-net models. Moreover, we showed that the proposed model could be generalized to observe various cell types with DHM. The presented model outperformed numerical phase unwrapping methods since it could overcome problems related to abrupt phase changes and conduct phase unwrapping at a faster rate. Thus, the presented model can be used for analyzing morphology and movement of live cells in real-time applications.

References

1 Park, S., Kim, Y., and Moon, I. (2021). Automated phase unwrapping in digital holography with deep learning. *Biomed. Opt. Express* 12 (11): 7064–7081.

2 Schwartzkopf, W., Milner, T.E., Ghosh, J. et al. (2000). Two-dimensional phase unwrapping using neural networks. In: *4th IEEE Southwest Symposium on Image Analysis and Interpretation*, Austin, TX, USA (2–4 April 2000). New York (NY): IEEE.

3 Spoorthi, G.E., Gorthi, S., and Gorthi, R. (2019). PhaseNet: a deep convolutional neural network for two-dimensional phase unwrapping. *IEEE Signal Process. Lett.* 26 (1): 54–58.

4 Wang, K., Li, Y., Kemao, Q. et al. (2019). One-step robust deep learning phase unwrapping. *Opt. Express* 27 (10): 15100–15115.

5 Zhang, T., Jiang, S., Zhao, Z. et al. (2019). Rapid and robust two-dimensional phase unwrapping via deep learning. *Opt. Express* 27 (16): 23173–23185.

6 Spoorthi, G., Gorthi, R., and Gorthi, S. (2020). PhaseNet 2.0: phase unwrapping of noisy data based on deep learning approach. *IEEE Trans. Image Process.* 29: 4862–4872.

7 Dardikman-Yoffe, G., Roitshtain, D., Mirsky, S.K. et al. (2020). PhUn-Net: ready-to-use neural network for unwrapping quantitative phase images of biological cells. *Biomed. Opt. Express* 11 (2): 1107–1121.

8 Goodfellow, I., Pouget-Abadie, J., Mirza, M. et al. (2019). Generative adversarial nets. In: *Neural Information Processing Systems*, 2672–2680. Berlin, Germany: Springer.

9 Ma, J., Yu, W., Liang, P. et al. (2019). FusionGAN: a generative adversarial network for infrared and visible image fusion. *Inf. Fusion* 48: 11–26.

10 Khan, A., Zhijiang, Z., Yu, Y. et al. (2021). Gan-Holo: generative adversarial networks-based generated holography using deep learning. *Complex* 2021: 6662161.

11 Isola, P., Zhu, J., Zhou, T., and Efros, A.A. (2017). Image-to-image translation with conditional adversarial networks. In: *IEEE conference on Computer Vision and Pattern Recognition (CVPR)*, Honolulu, HI, USA (21–26 July 2017). New York (NY): IEEE.

12 Ronneberger, O., Fischer, P., and Brox, T. (2015). U-Net: convolutional networks for biomedical image segmentation. In: *18th International Conference on Medical Image Computing and Computer-Assisted Intervention, Munich, Germany (5–9 October 2015), 234–241. Berlin, Germany: Springer.*

13 Schneider, C.A., Rasband, W.S., and Eliceiri, K.W. (2012). NIH image to ImageJ: 25 years of image analysis. *Nat. Methods* 9 (7): 671–675.

14 Wang, Z., Bovik, A.C., Sheikh, H.R., and Simoncelli, E.P. (2004). Image quality assessment: from error visibility to structural similarity. *IEEE Trans. Image Process.* 13 (4): 600–612.

10

Noise-free Phase Imaging in Gabor DHM with Deep Learning

10.1 Introduction

Gabor holography [1] is particularly useful in conjunction with a digital reconstruction algorithm for particle image analysis, 3D tracking, cell identification, or swimming cells in a liquid flow [2–5]. The main advantage of Gabor holography is that it can be easily set up. The optical setup is very simple and compact. In addition, its building cost is lower than those of other popular configurations in optics since it requires only a few optical components. However, Gabor holography suffers a major limitation in that a focused real image and an unfocused twin-image are strongly superposed. To overcome this problem, several instrumental methods were proposed. However, they all required objects to stay immobile. This requirement makes it difficult to study live cells, especially in real-time flow cytometry applications. Iterative phase-recovery methods were also suggested for Gabor holography to remove the twin-image noise [6–9]. Their main drawback is that they need several back-and-forth propagations of light to obtain the phase value. Furthermore, they also require a convergence criterion, which is generally unknown. The determination of the criterion is particularly difficult for studying biological samples in real-time. Non-iterative methods and inverse problem solutions were also suggested for phase recovery in Gabor holography [8–14]. Another approach in DHM is off-axis recording, in which the object beam and reference beam are not on the same optical axis. A small tilt (a few degrees) is inserted between the two beams, thus allowing the separation of the real image and the twin image by spatial filtering in the spectrum domain. The off-axis setup requires several optical element adjustments for imaging sufficient to study biological samples. Optical path lengths of the reference beam and the object beam must be matched before recording the hologram of the biological sample. Any change made in the object arm needs readjustment on the reference arm. For example, if a microscope objective (MO) is inserted in the object arm, the same MO must be included in the

Artificial Intelligence in Digital Holographic Imaging: Technical Basis and Biomedical Applications, First Edition. Inkyu Moon.

reference arm. Otherwise, aberrations should be compensated for by adding extra equations in the numerical reconstruction algorithm. However, an off-axis configuration does not allow for use of an imaging sensor's entire bandwidth, like a charge-coupled device (CCD) camera, since real and twin images are recorded in separate, non-overlapped bandwidths of the sensor.

10.2 A Deep-learning Model for Gabor DHM

Convolutional neural network (CNN) and deep-learning approaches were proposed for several optical applications, including virtual staining of non-stained samples [15], increasing spatial resolution in a large field of view in optical microscopy [16, 17], color holographic microscopy with CNN [18], autofocusing and enhancing the depth-of-field in inline holography [19], lens-less computational imaging by deep learning [20], super-resolution fringe patterns by deep-learning holography [21], virtual refocusing in fluorescence microscopy to map 2D images to a 3D surface [22], and others [23–25]. Phase recovery based on a residual CNN model was also suggested [26]. However, its application is limited because the noise-free phase image necessary for the labeled data to train the deep learning model is generated by recording multiple holograms. For biological cells and particularly moving cells (e.g. cancer cells and blood cells in flow cytometry applications), it is problematic. Another drawback of this method is that blurriness may occur in the model output.

In this chapter, we will show that Gabor holograms can be digitally synthesized from an off-axis hologram, which allowed us to obtain the target or actual phase image for the Gabor hologram [27]. This unique method is very important for the purpose of generating the labeled datasets required for supervised learning. We also show that an image-to-image translation model can be trained to convert noisy Gabor phase images to noise-free Gabor phase images. To overcome the stationary condition of the biological cell during imaging, we suggest the use of off-axis holography and generation of corresponding Gabor holograms as shown in Figure 10.1.

Our approach uses CNNs (a conditional generative adversarial network [C-GAN] model) to eliminate twin image distortion in phase images generated from the numerical reconstruction of Gabor holograms. The model is trained with a pair of datasets that consist of phase images from Gabor holography (the model input) and corresponding quantitative phase images obtained from off-axis DHM (the desired model output). A novel method is used to replicate Gabor holograms from

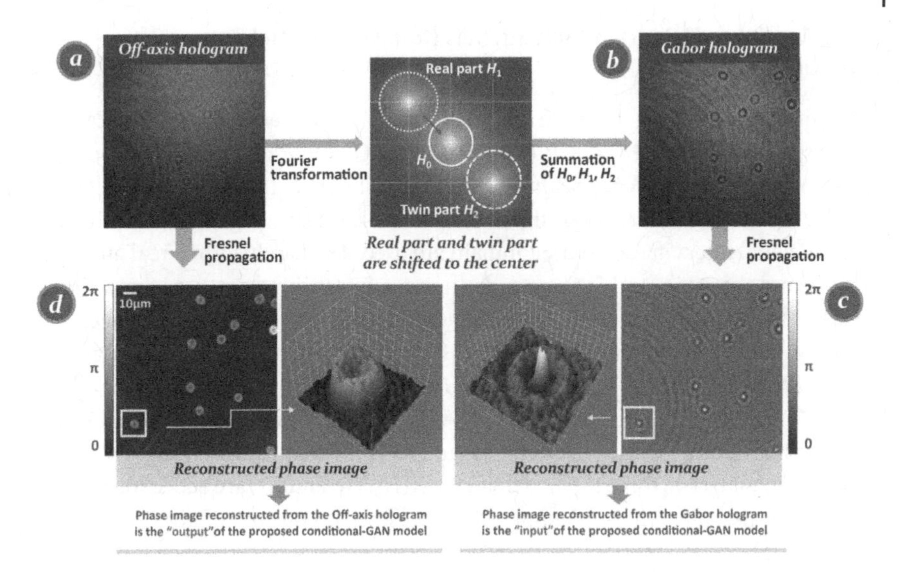

Figure 10.1 Scheme of the proposed method to generate superimposed noisy phase image and the corresponding noise-free high-contrast phase image. The original hologram is recorded in (a) an off-axis configuration with its spectrum obtained by a Fourier transformation. The bandwidth of the real image, twin image, and zero-order noise in the frequency domain are selected separately by spatial filtering. After filtering, the real-image spectrum and twin-image spectrum are shifted to the center frequency. By applying inverse Fourier transformation, three holograms are provided. (b) The intensities of these three holograms are added together, which is equivalent to a Gabor hologram. The Gabor hologram is numerically propagated at distance *d*. (c) The noisy superimposed phase image is the input for the C-GAN model. Fresnel propagation of H_1 at distance *d* can provide a noise-free phase image to be used for (d) the output of the C-GAN model [27] / with permission of Optical Publishing Group.

off-axis holograms as shown in Figure 10.1. This is an essential step because it offers a set of phase images for the exact same microscopic object. Two phase images (one obtained from a Gabor hologram and one from an off-axis hologram) of the exact same cell are fed into the model as input and output. The model is then trained for a few hundred epochs. Three bandwidths of the zero-order noise, real image, and twin image of the off-axis hologram are isolated separately with spatial filtering in the frequency domain. After frequency shifting the real image and twin image spectrum to the center, the three holograms (zero-order noise, real image, and twin image) are added together to generate a Gabor hologram from the off-axis hologram. Numerical propagation of the Gabor hologram provides a superimposed, noisy phase image. We will describe optical equations, optical configurations, and deep-learning models in the following subsections.

10.2.1 Gabor Hologram Construction from Off-axis Holograms and Optical Details

The off-axis hologram between object beam O and reference beam R can be expressed as $I_H = |O|^2 + |R|^2 + R^*O + O^*R$, where O^* and R^* denote complex conjugates of object and reference beams, respectively. The small tilt angle between O and R allows complete isolation of the zero-order noise, real image, and twin image. Three spatial filters in the Fourier domain are used. The bandwidth of real and twin images are the same, and for zero-order noise, a filter with a smaller bandwidth is used. Finally, three holograms corresponding to the real image, twin image, and zero-order data are obtained as shown in Figure 10.2 with the following equations: $H_1 = IFFT[FS[FFT(I_H) \times Filter_{real}]]$, $H_2 = IFFT[FS[FFT(I_H) \times Filter_{twin}]]$, $H_0 = IFFT[FFT(I_H) \times Filter_{noise}]$, and $GH = H_0 + H_1 + H_2$, where FFT and $IFFT$ are Fourier and inverse Fourier transforms, respectively, and FS is the frequency shifting. To obtain the Gabor hologram, three isolated holograms (H_0, H_1, and H_2) are added together in the spatial domain. Only the amplitude is preserved. For quantitative phase imaging, multiply the Gabor hologram or the off-axis hologram by the digital reference wave R_D during the reconstruction process (see Chapter 5). When a MO is inserted into the object wave arm, it introduces phase aberration in the off-axis configuration. This can be numerically resolved by multiplying the reconstructed wave front with the computed complex conjugate of the phase aberration (see Chapter 5). Reconstruction of each real-image complex field (Ψ_{H_1}) and Gabor hologram complex field (Ψ_G) can be expressed by the Fresnel approximation. Eventually, the phase image from the Gabor hologram and the noise-free quantitative phase image of the off-axis hologram are obtained from

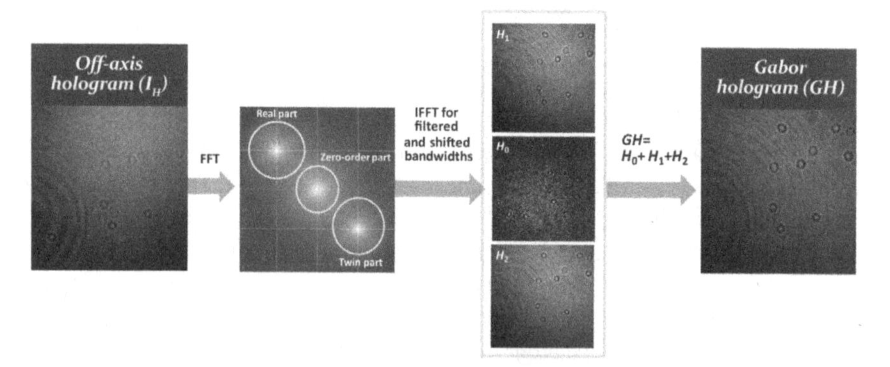

Figure 10.2 An off-axis hologram, shifting its bandwidth, filtering, and spectrum to the center generates three holograms: real image, twin image, and zero-order noise image. The intensity of the zero-order noise is adjusted for visualization [27] / with permission of Optical Publishing Group.

$$\phi_G(x,y) = arctan\left\{\frac{\text{Im}[\Psi_G(m,n)]}{\text{Re}[\Psi_G(m,n)]}\right\}, \phi_{real}(x,y) = arctan\left\{\frac{\text{Im}[\Psi_{H_1}(m,n)]}{\text{Re}[\Psi_{H_1}(m,n)]}\right\}.$$

During deep training of the model, ϕ_G is the input and ϕ_{real} is the desired output of the deep-learning model (C-GAN). A gallery of off-axis holograms, corresponding Gabor holograms, reconstructed phase images from the off-axis hologram, and reconstructed super-imposed phase images from the Gabor hologram is shown in Figure 10.3.

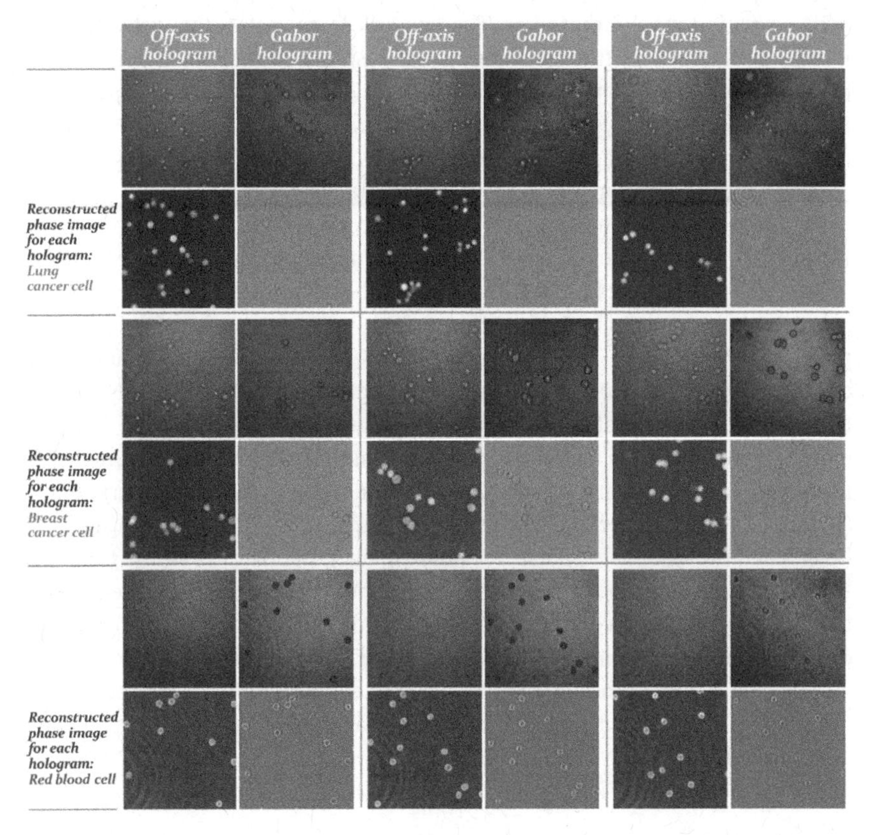

Figure 10.3 A gallery of phase images obtained with the proposed Gabor hologram construction approach. For each off-axis phase image, a corresponding Gabor hologram was made. Both holograms were numerically reconstructed. Phase images were used for training and testing of our proposed deep-learning model [27] / with permission of Optical Publishing Group.

10.2.2 A Deep-learning Model for Gabor DHM

CNN is a class of deep neural networks consisting of neurons. It was originally inspired by biological processes. A CNN generally includes an input layer, multiple hidden layers, and an output layer. The input to a CNN model is a matrix with size of (image number) × (image height) × (image width) × (image depth). Hidden layers of a CNN typically consist of a series of convolutional layers that convolve with a multiplication and pass its results to the next layer. The input image becomes abstracted to a feature map after passing through one convolutional layer. Layer parameters have a set of learnable kernels. During the training process, each filter is convolved across the width and height of the input volume. It computes the dot product between filter elements and input values and outputs a 2D activation map of that filter. The network learns filters that activate when it detects particular features at some spatial positions in the input. Neurons in a model produce an output value by applying a specific function to inputs from the previous layer's receptive field. The function applied to input values is determined by weights and a bias in a vector form. Learning in a CNN progresses by making iterative adjustments to these biases and weights [28].

For noise-free phase imaging in Gabor DHM, we will introduce deep learning based on a C-GAN used in several image-to-image translation studies [29–32] to solve the blurriness effect of CNN models. The advantage of this model is that it uses a structured loss function different from unstructured functions (pixel-to-pixel similarity). Thus, it can be generalized for image-to-image applications without changing the loss function or the structure of the model. The C-GAN consists of a U-net image generator and a PatchGAN classifier or discriminator. The generator (Figure 10.4a) is trained to produce images that cannot be distinguished from real images by an adversarial trained discriminator. The discriminator (Figure 10.4b) learns to classify between the generator's synthesized image (a fake image) and the real image (noise-free phase image). In other words, the generator attempts to produce a noise-free phase image with the same statistical features as the phase image obtained using an off-axis DHM, while the discriminator tries to distinguish if the input image is the actual off-axis phase image or the generator's output. The training procedure seeks a state of equilibrium in which the generator's output and the actual off-axis phase image share very similar statistical distributions.

10.2.3 Model Architecture

The presented image-to-image conversion is based on the concept of the conditional generative adversarial model (C-GAN) used in several applications [29–32]. The C-GAN model typically consists of a generator and a discriminator.

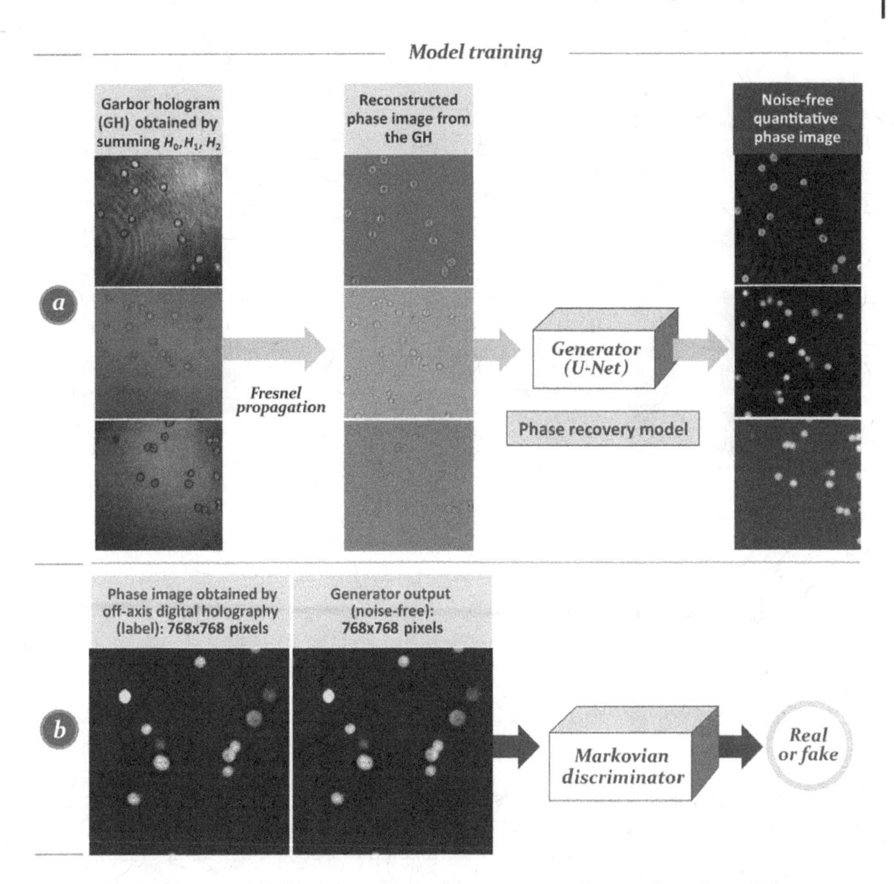

Figure 10.4 Structure of the proposed model to recover phase values from Gabor holograms. (a) After digital propagation of a Gabor hologram, the phase image is fed into a U-net shape generator. This generator tries to remove the noise. (b) The Markovian discriminator receives images (one is the output of the generator, and the other is the quantitative phase image). The discriminator outputs a probability value that the image is fake or real [27] / with permission of Optical Publishing Group.

In this study, the generator's input or output image size was 768 × 768 pixels to perform image-to image translation (see Figure 10.5a). The architecture of the C-GAN model allows for testing images with a bigger size if hardware resources are not limited. The discriminator receives a 1536 × 768 input image (two images are concatenated: the generator output and the real image) that passes through four convolution layers and derives a small patch (see Figure 10.5b). The discriminator learns to distinguish between real and fake patches. Evaluating images with patches allows the model to be trained faster with fewer parameters.

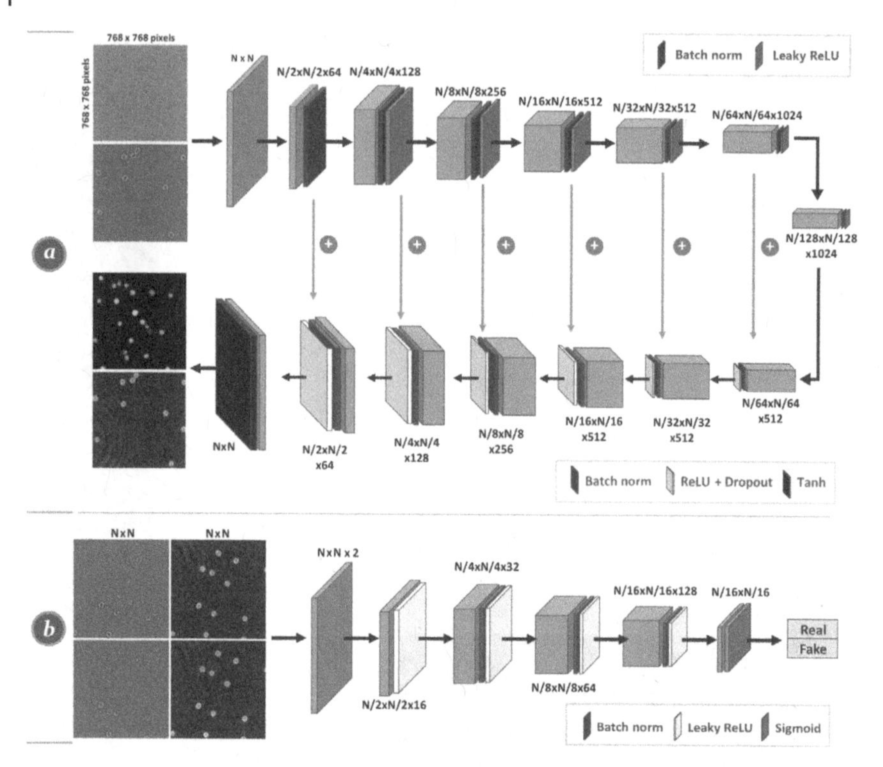

Figure 10.5 (a) Generator architecture similar to a U-Net to recover noiseless phase images in Gabor DHM, (b) Discriminator to compare fake and real images with convolution layers. Tanh is the hyperbolic tangent [27] / with permission of Optical Publishing Group.

10.3 Experimental Results

Two C-GAN models were trained with 500 RBCs for the RBC model and 1000 lung, skin, or breast cancer cells for the elliptical cell model. The accuracy of each model was evaluated with several unseen images. We used the Adam solver as the optimization process with an adaptive momentum and parameters of $\beta_1 = 0.5$ and $\beta_2 = 0.999$. The number of epochs was 450 and learning rate was 0.0002. We used an Nvidia Titan Xp graphical processing unit (GPU) for training. We selected 450 epochs because we did not observe any changes in the output images produced by our model's generator around 400 epochs. Batch training with a batch size of 10 was used. We implemented the model on a computer with an Intel(R) Xeon(R) Gold 6134 central processing unit (CPU) @3.20 GHz and 64 GB of RAM running Ubuntu 18.04.03. The network was implemented using Python version 3.6.8 and TensorFlow framework version 1.14.0. After the training, we tested the deep-learning generator with several images. Network testing was performed using an Nvidia Titan XP. The objective function of the C-GAN model is explained in Chapter 9 (see Section 9.1). A gallery of images used for the test is shown in Figure 10.6. In addition, for the RBC model, we tested the trained RBC model with real Gabor holograms recorded by blocking the reference beam in our off-axis DHM configuration. Figure 10.7a shows development of the training according to the generator output. At each epoch, the generator is updated with the help of the discriminator, resulting in a less noisy Gabor phase image. Figure 10.7b and 10.7c show normalized loss function for the generator and the discriminator at different epochs, respectively.

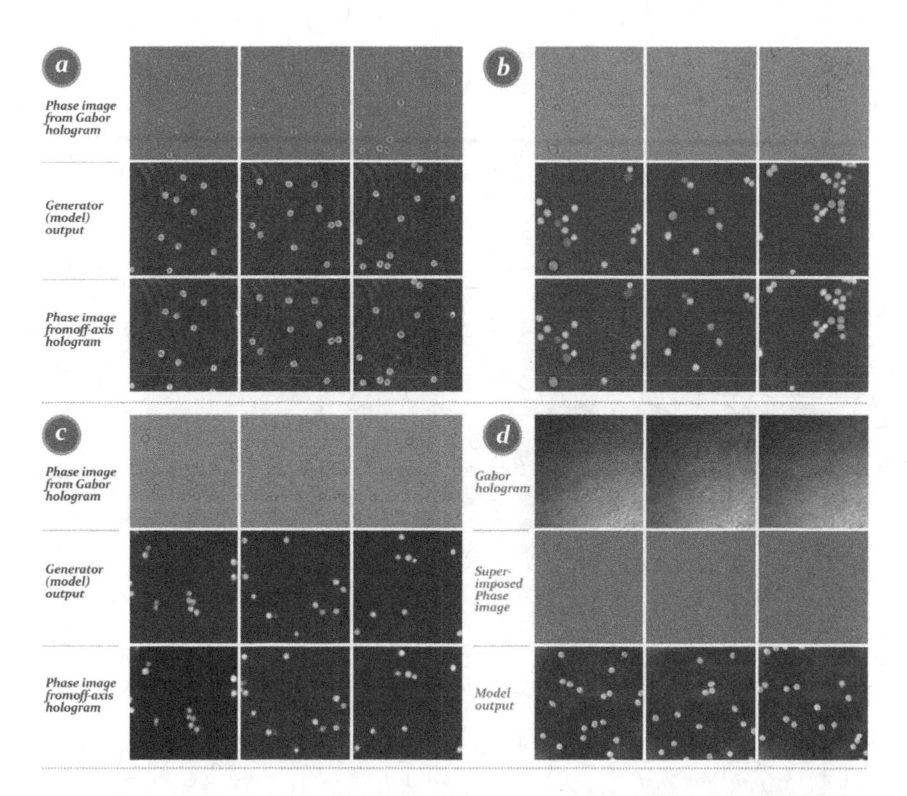

Figure 10.6 A gallery of test images. (a) RBC Gabor phase image, (b) breast cancer cell, and (c) lung cancer cell. For each sample, an off-axis quantitative phase image is also shown for visualization. (d) A gallery of test images with Gabor holograms recorded using only the object wave (the reference wave was blocked). A Gabor hologram alongside the reconstructed super-imposed phase image and noise-removed phase image is shown [27] / with permission of Optical Publishing Group.

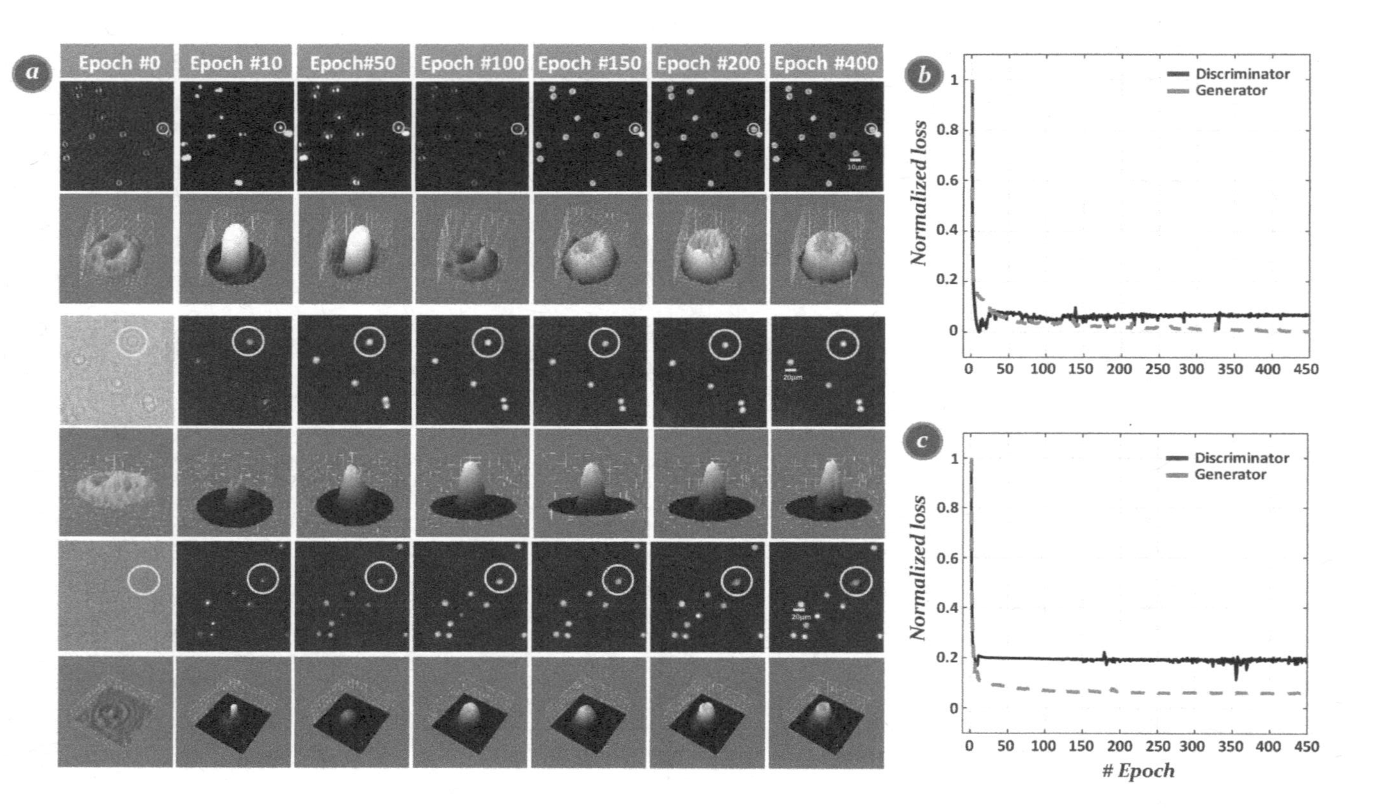

Figure 10.7 (a) Development of generator at different epochs. (b) Normalized loss for both generator and discriminator for the RBC model. (c) Normalized loss for both generator and discriminator for the cancer cell model [27] / with permission of Optical Publishing Group.

10.3.1 Quantitative Evaluations

To evaluate the performance of our deep-learning model, we calculated the mean square error (MSE) between the phase image obtained by the proposed method and the off-axis DHM. Histograms of MSE for 2000 RBC images and 1000 cancer cell images (all cell types pooled together) are shown in Figures 10.8a and 10.8b, respectively. Another criterion is the structural similarity index (SSIM), which can numerically evaluate the perceptual difference between two phase images. Figures 10.8c and 10.8d show histograms of the SSIM for 2000 pairs of RBC images (one is a quantitative phase image in off-axis DHM and the second is a noise-free phase image in Gabor DHM with our deep-learning model) and 1000 cancer cell images, respectively. One important advantage of quantitative phase imaging by DHM is that biophysical and morphological features can be analyzed at the single-cell level. To validate the output of our deep-learning model, the dry mass of 200 cells at the single-cell level was measured and compared with the exact same cells in the off-axis quantitative phase image. The dry mass is directly related to the mean corpuscular-hemoglobin (MCH) content since RBCs are mostly composed of hemoglobin. The dry mass can be computed by integrating phase values over the projected surface area of the cell, $DryMass = \frac{\phi\lambda\overline{S}}{2\pi\alpha}$, where $\overline{S}$ is the projected surface area of the cell, ϕ is the summation of all phase values within the projected surface area of the cell, and α is the specific refractive index increment factor (~0.193 ml/g for RBCs and 0.2 for most cancer cells) [33, 34]. Single-cell extraction is performed by binarization using the same binary mask for our model's phase-image output and the phase image obtained by off-axis DHM. A correlation analysis revealed that the MCH value obtained from the off-axis phase imaging strongly correlated with our method (see Figure 10.8e). The Pearson product–moment correlation coefficient was 0.78 with a 99% confidence level when t-test was applied. The MCH value was 30 ± 5 pg (average $\pm$ standard deviation) for the off-axis method and 28 ± 4 pg for our model. Additionally, there was a significant correlation between the dry mass of cancer cells obtained from off-axis DHM and our deep-learning model (these two values linearly increased [$y = 0.92x + 1.3$], see Figure 10.8f). The average dry mass values for off-axis DHM and our model output were 31 ± 6 pg and 29.6 ± 5 pg, respectively.

10.3.2 Evaluations for Gabor Holography Setup

To validate our method, the off-axis DHM setup was modified. One shutter was added to the off-axis configuration to block the reference beam as shown in Figure 10.9. In this case, the only object beam that illuminated the sample and reached the camera was a real Gabor hologram. After blocking the reference beam, several Gabor holograms were saved. After numerical propagation, superimposed

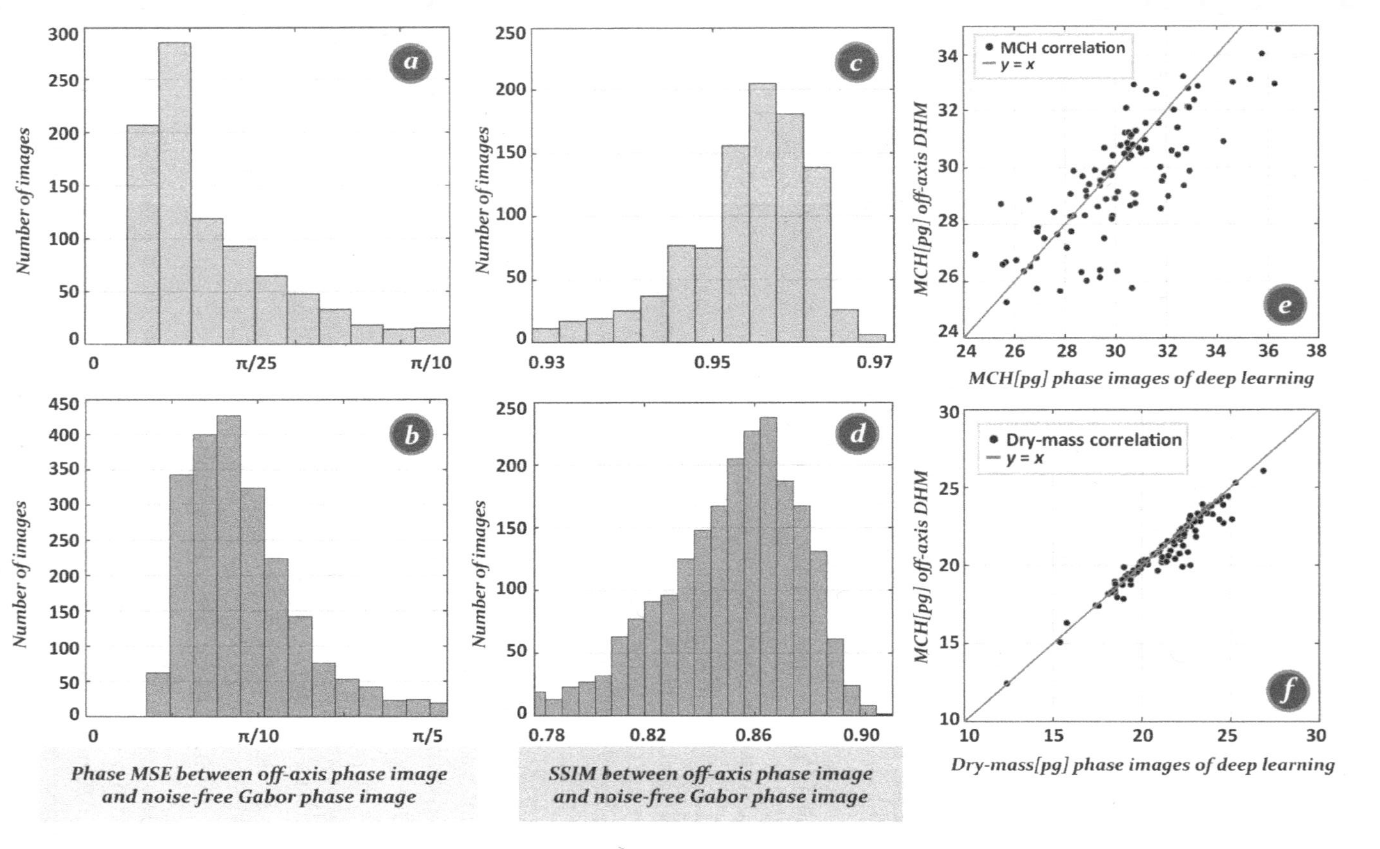

Figure 10.8 Histograms of phase MSE for (a) RBCs and (b) cancer cells. Histograms of phase SSID for (c) RBCs and (d) cancer cells between off-axis quantitative phase image and the proposed deep-learning phase recovery in Gabor holography. (e) Single-cell mean corpuscular hemoglobin analysis between off-axis phase image and our model. (f) Dry mass analysis between off-axis phase image and our model output.

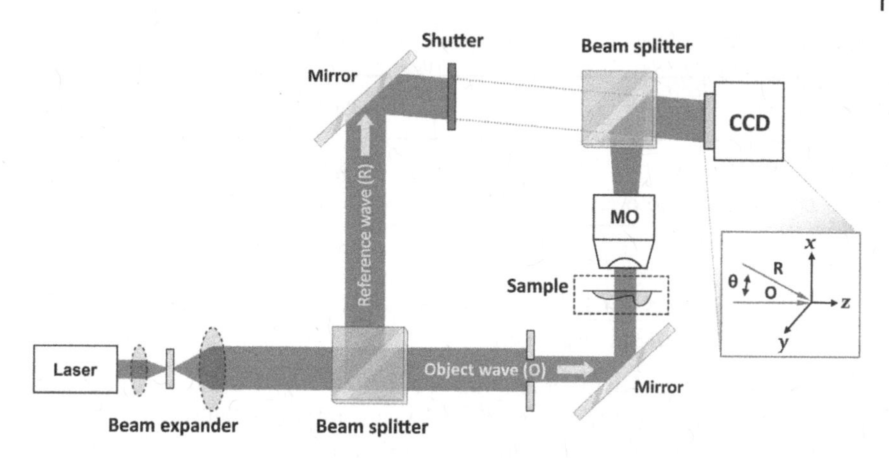

Figure 10.9 Gabor holography experimental setup. The Gabor holographic configuration was obtained using a shutter to block the reference beam from the off-axis DHM.

phase values were fed into the model and a noise-free phase image was obtained (one image is shown in Figure 10.10). A gallery of results is shown in Figure 10.6.

We also measured the volume and MCH value of RBCs obtained from the real Gabor holography. Gabor holograms were numerically reconstructed and noise in the phase image was removed with our trained model. The volume and MCH were compared with results obtained with off-axis DHM in Table 10.1. The volume was obtained using

$$V \simeq p^2 \sum_{(i,j) \in S_p} h(i,j),$$

where summation was achieved over all pixels (i, j) belonging to the RBCs' projected surface S_p, and p was the pixel size in the reconstruction plane; $h(i, j)$ was the thickness value at pixel (i, j), which was obtained using

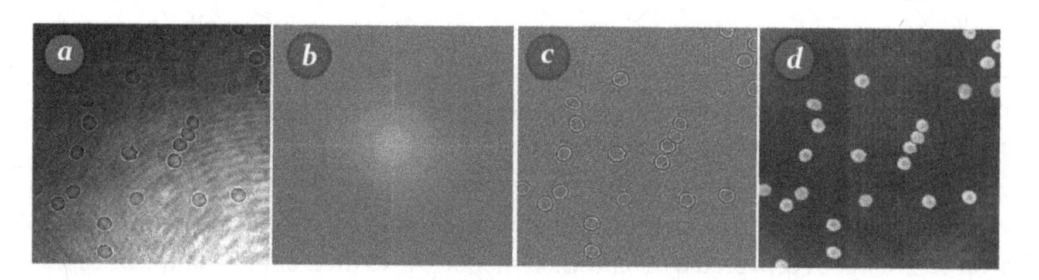

Figure 10.10 (a) A Gabor hologram recorded in Gabor configuration, (b) spectrum of the hologram, (c) reconstructed phase image from the Gabor hologram, and (d) noise-free result from our model [27] / with permission of Optical Publishing Group.

Table 10.1 Volume and MCH values for off-axis DHM and our model ($n > 70$).

Method	Volume (fl)	MCH (pg)
Off-axis DHM	90 ± 7	30.0 ± 5
The proposed phase recovery model	87 ± 6	28.5 ± 5

$$h(x,y) = \frac{\lambda \times \phi(x,y)}{2\pi(n_{RBC} - n_m)},$$

where λ was the wavelength of illumination light, $\phi(x, y)$ was the phase value at pixel (x, y), and $n_{RBC} = 1.42$ and $n_m = 1.3334$ were refractive indices of RBCs and HEPA buffer, respectively [35].

10.4 Discussion

Phase recovery in Gabor DHM can be obtained by deep-learning models that extract and separate spatial features of the desired data (the real image) from features of noise (the twin image and other undesired interference terms) in the phase image. Surprisingly, we also observed that the generator could artificially recover the shape of cells that were partially unfocused (see Figures 10.11a and 10.11b). This highlights that the deep-learning model can learn global and local features of the real phase image at the whole-image level and compensate for other errors at the single-cell level while imaging. A quantitative analysis of the un-focusing recovery is shown in Figure 10.12.

Note that if the correct focus is R, then the Gabor hologram is propagated at R + 0.25R, R + 0.50R, R + 0.75R, and R + 1.0R and the superimposed phase image is fed to our RBC model. If focus deviates around 50% of its correct value, then the model can roughly preserve the shape of an RBC. For 75 and 100% deviations, the biconcave shape is not preserved (see Figure 10.12). Additionally, our model can recover different cell lines not observed by the model during the training process. In this case, we fed our model with the Gabor phase image of bladder cancer cells (see Figure 10.11c). Noise-free phase images were then generated (see Figure 10.11d). This highlights that our model is applicable to other cell lines like elliptical cancer cells with spherical shapes. Generally speaking, all types of elliptical cells can be recovered by the trained model due to their similar shapes, despite the differences in organs from which they are produced. We also trained another model with Hela cancer cells cultured for several hours, and one hologram was recorded for 10 minutes. These cells have considerable shape variations

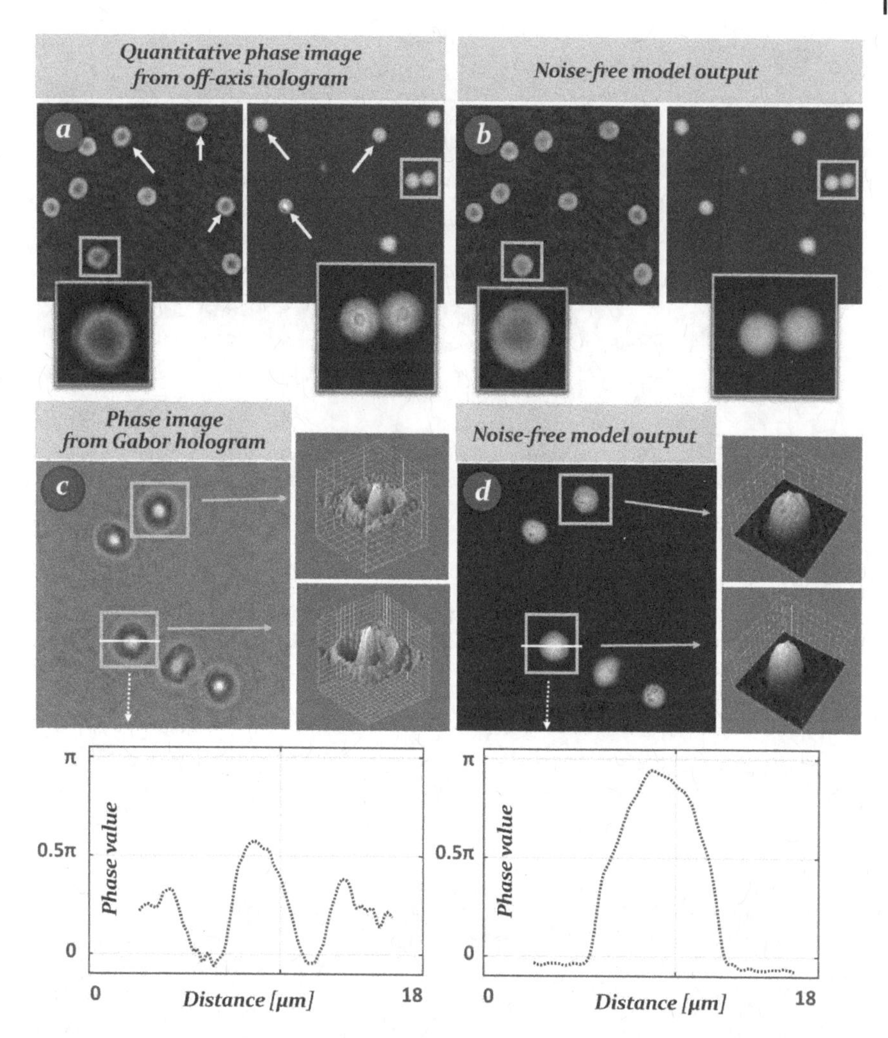

Figure 10.11 The proposed model can recover single cells that are not in focus compared to other cells. (a) Two off-axis quantitative phase images where some cells are unfocused (see arrows). (b) The exact same image obtained by the deep-learning method for noise-free Gabor holography. (c) Gabor phase image of bladder cancer cell not observed by the model while training. (d) Model output of cancer cell shown in (c). Two cells are magnified and plotted in 3D [27] / with permission of Optical Publishing Group.

throughout the culturing period since they can attach to the culture glass, detach, and divide. Some results are shown in Figure 10.13. Our deep-learning model is instant, making it suitable for high-throughput screening or high-content screening (around 0.02 seconds for each phase image; average of 3000 images). This can

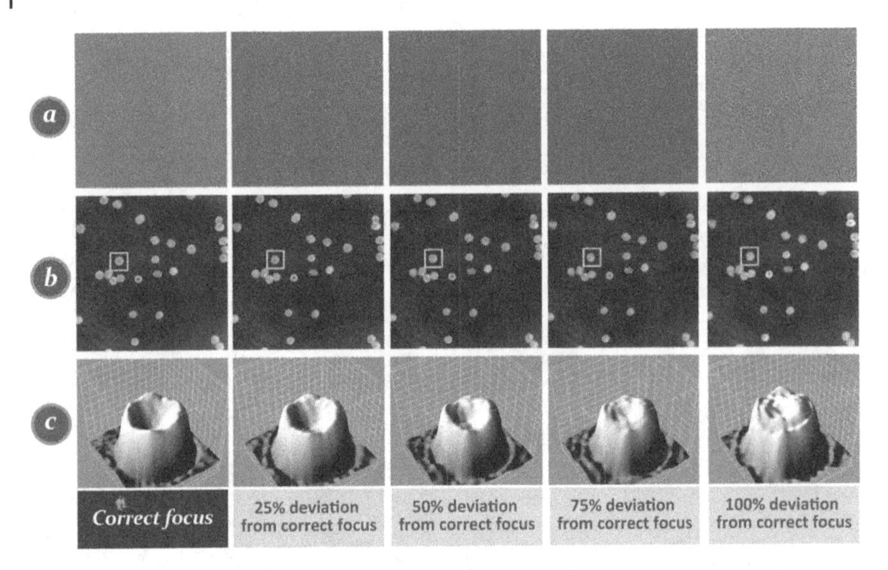

Figure 10.12 Representative model output when the input Gabor phase image is reconstructed with a deviated reconstruction distance. (a) Super-imposed phase images, (b) corresponding model outputs, and (c) a 3D mesh of one RBC [27] / with permission of Optical Publishing Group.

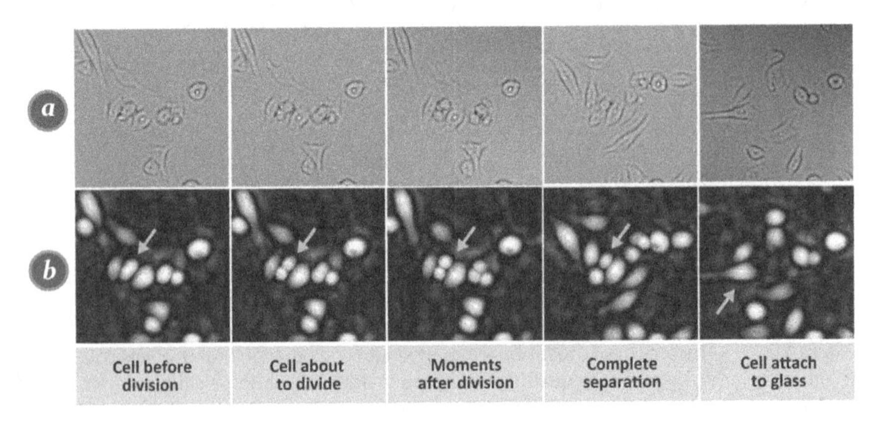

Figure 10.13 Gabor phase image of Hela cancer cells and the model's output. Cells go through several shape changes. (a) A noisy phase image and (b) a noise-removed phase image [27] / with permission of Optical Publishing Group.

be achieved using a PC with 16 GB of RAM, an Intel(R) Core-i7-9700K 3.60 GHz CPU, and NVIDIA GeForce RTX2080Ti GPU card.

Several points should be considered when applying deep learning to study live biological samples with optical setups since many variables can affect the results, like reconstruction distance, sample-to-sample variations, and so on. Based on our experiments, we believe that training separate models for similar biological organs is more efficient than using a single model with a mix of biological samples that have different morphologies. During the training process, deep learning tries to learn features of each morphology, which can differ from sample to sample and result in a poorly trained model. For example, biconcave RBCs basically consist of a dimple and a ring. They are different from elliptical cancer cells with a spherical geometry. It is also worth mentioning that plenty of sample images with different distances from the camera should be recorded for a well-trained model. This is essential for a very good training dataset. In Gabor holography, the superimposed phase image differs if the distance between the sample and camera varies. Additionally, it is very important to estimate the reconstruction distance for numerical propagation. Propagating the image into a plane (propagation of the hologram too short or too far) will result in a phase image that has an unknown feature for the trained model. In this case, the model may fail to remove the superimposed phase image. Training can be performed on a GPU server to speed up the training process. To perform the phase recovery for another unobserved sample, a conventional PC with a GPU card can be considered.

10.5 Conclusions

We presented a deep-learning model to obtain noise-free quantitative phase images in Gabor DHM. To achieve this, a C-GAN was trained with several RBCs and elliptical cancer cells. The model input consisted of separately isolating the spectrum of a real image, twin image, and zero-order noise of an off-axis digital hologram. These three images were added to construct the Gabor hologram. The Gabor hologram was digitally reconstructed, and the reconstructed phase image was fed into our C-GAN model. The desired output is a noise-free phase image of off-axis DHM. After training for several epochs, our model could remove the superimposed noise in actual Gabor holograms. Interestingly, our model was also able to solve depth-of-focus limitations at the single-cell level and restore unobserved elliptical cells during training.

References

1 Gabor, D. (1948). A new microscopic principle. *Nature* 161 (4098): 777–778.
2 Moon, I., Daneshpanah, M., Javidi, B., and Stern, A. (2009). Automated three-dimensional identification and tracking of micro/nano biological organisms by computational holographic microscopy. *Proc. IEEE* 97 (6): 990–1010.

3 Javidi, B., Moon, I., Yeom, S., and Carapezza, E. (2005). Three-dimensional imaging and recognition of microorganism using single-exposure on-line (SEOL) digital holography. *Opt. Express* 13 (12): 4492–4506.
4 Caprio, G., Mallahi, A., Ferraro, P. et al. (2014). 4D tracking of clinical seminal samples for quantitative characterization of motility parameters. *Biomed. Opt. Express* 5 (3): 690–700.
5 Memmolo, P., Miccio, L., Paturzo, M. et al. (2015). Recent advances in holographic 3D particle tracking. *Adv. Opt. Photon.* 7 (4): 713–755.
6 Denis, L., Fournier, C., Fournel, T., and Ducottet, C. (2008). Numerical suppression of the twin image in in-line holography of a volume of micro-objects. *Meas. Sci. Technol.* 19 (7): 074004.
7 Nakamura, T., Nitta, K., and Matoba, O. (2007). Iterative algorithm of phase determination in digital holography for real-time recording of real objects. *Appl. Opt.* 46 (28): 6849–6853.
8 Koren, G., Polack, F., and Joyeux, D. (1993). Iterative algorithms for twin-image elimination in in-line holography using finite-support constraints. *J. Opt. Soc. Am. A* 10 (3): 423–433.
9 Latychevskaia, T. and Fink, H. (2007). Solution to the twin image problem in holography. *Phys. Rev. Lett.* 98 (23): 233901.
10 Gire, J., Denis, L., Fournier, C. et al. (2008). Digital holography of particles: benefits of the 'inverse problem' approach. *Meas. Sci. Technol.* 19 (7): 074005.
11 Xu, L., Miao, J., and Asundi, A. (2000). Properties of digital holography based on in-line configuration. *Opt. Eng.* 39 (12): 3214–3219.
12 Cho, C., Choi, B., Kang, H., and Lee, S. (2012). Numerical twin image suppression by nonlinear segmentation mask in digital holography. *Opt. Express* 20 (20): 22454–22464.
13 Onural, L. and Scott, P. (1987). Digital decoding of in-line holograms. *Opt. Eng.* 26 (11): 261124.
14 Denis, L., Lorenz, D., Thiébaut, E. et al. (2009). Inline hologram reconstruction with sparsity constraints. *Opt. Lett.* 34 (22): 3475–3477.
15 Rivenson, Y., Liu, T., Wei, Z. et al. (2019). PhaseStain: the digital staining of label-free quantitative phase microscopy images using deep learning. *Light: Sci. & Appl.* 8 (1): 1–11.
16 Rivenson, Y., Göröcs, Z., Günaydin, H. et al. (2017). Deep learning microscopy. *Optica* 4 (11): 1437–1443.
17 Haan, K., Rivenson, Y., Wu, Y., and Ozcan, A. (2020). Deep-learning-based image reconstruction and enhancement in optical microscopy. *Proc. IEEE* 108 (1): 30–50.
18 Liu, T., Wei, Z., Rivenson, Y. et al. (2019). Deep learning-based color holographic microscopy. *J. Biophoton.* 12 (11): e201900107.
19 Wu, Y., Rivenson, Y., Zhang, Y. et al. (2018). Extended depth-of-field in holographic imaging using deep-learning-based autofocusing and phase recovery. *Optica* 5 (6): 704–710.

20 Sinha, A., Lee, J., Li, S., and Barbastathis, G. (2017). Lensless computational imaging through deep learning. *Optica* 4 (9): 1117–1125.

21 Ren, Z., So, H., and Lam, E. (2019). Fringe pattern improvement and super-resolution using deep learning in digital holography. *IEEE Trans. Ind. Inform.* 15 (11): 6179–6186.

22 Wu, Y., Rivenson, Y., Wang, H. et al. (2019). Three-dimensional virtual refocusing of fluorescence microscopy images using deep learning. *Nat. Methods* 16 (12): 1323–1331.

23 Barbastathis, G., Ozcan, A., and Situ, G. (2019). On the use of deep learning for computational imaging. *Optica* 6 (8): 921–943.

24 Rivenson, Y., Wu, Y., and Ozcan, A. (2019). Deep learning in holography and coherent imaging. *Light: Sci. & Appl.* 8 (1): 1–8.

25 Lin, X., Rivenson, Y., Yardimci, N. et al. (2018). All-optical machine learning using diffractive deep neural networks. *Science* 361 (6406): 1004–1008.

26 Rivenson, Y., Zhang, Y., Günaydın, H. et al. (2018). Phase recovery and holographic image reconstruction using deep learning in neural networks. *Light: Sci. & Appl.* 7 (2): 17141.

27 Moon, I., Jaferzadeh, K., Kim, Y., and Javidi, B. (2020). Noise-free quantitative phase imaging in Gabor holography with conditional generative adversarial network. *Opt. Express* 28 (18): 26284–26301.

28 LeCun, Y., Bengio, Y., Hinton, G., and G. (2015). Deep learning. *Nature* 521 (7553): 436–444.

29 Isola, P., Zhu, J., Zhou, T., and Efros, A. (2017)). Image-to-image translation with conditional adversarial networks. In: *IEEE conference on Computer Vision and Pattern Recognition (CVPR)*, Honolulu, HI (21–26 July 2017). New York (NY): IEEE.

30 Moeskops, P., Veta, M., Lafarge, M. et al. (2017)). Adversarial training and dilated convolutions for brain MRI segmentation. In: *Deep Learning in Medical Image Analysis and Multimodal Learning for Clinical Decision Support* (ed. et al.). Cham: Springer https://doi.org/10.1007/978-3-319-67558-9_7.

31 Mahmood, F., Borders, D., Chen, R. et al. (2020). Deep adversarial training for multi-organ nuclei segmentation in histopathology images. *IEEE Trans. Med. Imaging* 39 (11): 3257–3267.

32 Xue, Y., Xu, T., Zhang, H. et al. (2018). Segan: adversarial network with multi-scale L1 loss for medical image segmentation. *Neuroinformatics* 16 (3–4): 383–392.

33 Barer, R. (1953). Determination of dry mass, thickness, solid and water concentration in living cells. *Nature* 172 (4389): 1097–1098.

34 Rappaz, B., Cano, E., Colomb, T. et al. (2009). Noninvasive characterization of the fission yeast cell cycle by monitoring dry mass with digital holographic microscopy. *J. Biomed. Opt.* 14 (3): 034049.

35 Rappaz, B., Barbul, A., Emery, Y. et al. (2008). Comparative study of human erythrocytes by digital holographic microscopy, confocal microscopy, and impedance volume analyzer. *Cytometry A* 73 (10): 895–903.

Part III

Intelligent Digital Holographic Microscopy (DHM) for Biomedical Applications

11

Introduction

DHM is a new and highly promising approach to identify cellular biomarkers, particularly when it is combined with AI or deep-learning techniques for biomedical applications [1–10]. DHM is also a contact-less and label-free method. Thus, a sample can be studied without damaging it. Part III provides an overview of some of the recently published work on AI or deep-learning techniques in holographic image analysis as a tool to study the intracellular content and morphology of live cells [1, 11–30]. The overview focuses on the automated phenotypic analysis of live RBCs and cardiac cells via intelligent DHM.

Chapters 12 through 20 introduce an automated phenotyping platform based on DHM for the quantitative analysis of RBCs. Chapters 12 and 13 introduce RBC phase image-segmentation techniques essential for automated RBC analysis. Chapters 14–16 demonstrate that integrating DHM techniques integrated AI enable scientists to obtain rich, quantitative information about the structure of RBCs in non-invasive, real-time conditions for automatic phenotypic classification of RBCs. Chapter 17 shows that deep-learning DHM can also rapidly detect and count multiple cells in hologram images at the single-cell level, which is needed for high-throughput cell counting. Chapter 18 introduces a tracking algorithm to locate a single RBC through RBC image sequences obtained with time-lapse DHM and dynamically monitor its biophysical cell parameters. Chapter 19 introduces methods to quantitatively calculate the fluctuation rate of RBCs with nanometric axial and millisecond temporal sensitivity using time-lapse DHM, which are also important for quantitative RBC studies. Chapter 20 further demonstrates holographic, image-based methods to automatically analyze changes in membrane profile, mean corpuscular hemoglobin, and cell-membrane fluctuations of healthy RBCs at varying temperatures.

Chapters 21 through 25 describe an overview of DHM application in studying cardiomyocytes derived from induced pluripotent stem cells (iPSCs) under control and drug-treated conditions. DHM allows scientists to extract a set of parameters

Artificial Intelligence in Digital Holographic Imaging: Technical Basis and Biomedical Applications, First Edition. Inkyu Moon.

that characterize the beating patterns of cardiomyocytes from recorded quantitative phase images. This demonstrates that DHM can represent a promising label-free approach to identify new drug candidates by measuring their effects on iPSC-derived cardiomyocytes. Chapter 21 shows that integrating DHM with image-processing algorithms can provide automatic, dynamic, quantitative phase profiles of beating cardiomyocytes. We further demonstrate that multiple parameters of cardiomyocyte dynamics can be obtained from beating profiles of cardiomyocytes. In Chapter 22, we introduce an automated method to quantitatively investigate the rhythm strip and parameters of cardiomyocyte synchronization at the single-cell level. In addition, we present deep-learning models to characterize single cardiomyocytes by nucleus extraction from time-lapse holographic images using fully convolutional neural networking (FCN). Chapter 23 introduces an automated procedure for to assess the effects of drugs on cardiomyocyte beating patterns. We used DHM to image and quantify the beating movement of cardiomyocytes under control and drug-treated conditions. Chapter 24 further demonstrates holographic, image-based tracking algorithms for the automated measurement of drug-treated cardiomyocyte dynamics. These results clearly demonstrate that DHM is a promising label-free approach to identify new drug candidates by measuring their effects on human cardiomyocytes.

References

1 Moon, I., Javidi, B., Yi, F. et al. (2012). Automated statistical quantification of three-dimensional morphology and mean corpuscular hemoglobin of multiple red blood cells. *Opt. Express* 20: 10295–10309.

2 Rappaz, B., Moon, I., Yi, F. et al. (2015). Automated multi-parameter measurement of cardiomyocytes dynamics with digital holographic microscopy. *Opt. Express* 23 (10): 13333–13347.

3 Anand, A., Moon, I., and Javidi, B. (2017). Automated disease identification with 3-D optical imaging: a medical diagnostic tool. *Proc. IEEE* 105: 924–946.

4 Shaked, N., Satterwhite, L., Bursac, N., and Wax, A. (2010). Whole-cell-analysis of live cardiomyocytes using wide-field interferometric phase microscopy. *Biomed. Opt. Express* 1 (2): 706–719.

5 Carl, D., Kemper, B., Wernicke, G., and Bally, G. (2004). Parameter-optimized digital holographic microscope for high-resolution living-cell analysis. *Appl. Optics* 43 (36): 6536–6544.

6 Boss, D., Kohn, J., Jourdain, P. et al. (2013). Measurement of absolute cell volume, osmotic membrane water permeability, and refractive index of transmembrane water and solute flux by digital holographic microscopy. *J. Biomed. Opt.* 18 (3): 036007.

7 Merola, F., Miccio, L., Memmolo, P. et al. (2013). Digital holography as a method for 3D imaging and estimating the biovolume of motile cells. *Lab Chip* 13 (23): 4512–4516.

8 Mugnano, M., Memmolo, P., Miccio, L. et al. (2018). In vitro cytotoxicity evaluation of cadmium by label-free holographic microscopy. *J. Biophotonics* 11 (12).

9 Calabuig, A., Mugnano, M., Miccio, L. et al. (2017). Investigating fibroblast cells under "safe" and "injurious" blue-light exposure by holographic microscopy. *J. Biophotonics* 10 (6–7): 919–927.

10 Kühn, J., Shaffer, E., Mena, J. et al. (2013). Label-free cytotoxicity screening assay by digital holographic microscopy. *Assay Drug Dev. Technol.* 11 (2): 101–107.

11 Ahamadzadeh, E., Jaferzadeh, K., Park, S. et al. (2022). Automated analysis of human cardiomyocytes dynamics with holographic image-based tracking for cardiotoxicity screening. *Biosens. Bioelectron.* 195: 113570.

12 Kim, E., Park, S., Hwang, S. et al. (2022). Deep learning-based phenotypic assessment of red cell storage lesions for safe transfusions. *IEEE J. Biomed. Health Inform.* 26 (3): 1318–1328.

13 Ahmadzadeh, E., Jaferzadeh, K., Shin, S., and Moon, I. (2020). Automated single cardiomyocyte characterization by nucleus extraction from dynamic holographic images using a fully convolutional neural network. *Biomed. Opt. Express* 11: 1501–1516.

14 Jaferzadeh, K., Rappaz, B., Fabien, K. et al. (2020). Marker-free automatic quantification of drug-treated cardiomyocytes with digital holographic imaging. *ACS Photon.* 7: 105–113.

15 Jaferzadeh, K., Sim, M., Kim, N., and Moon, I. (2019). Quantitative analysis of three-dimensional morphology and membrane dynamics of red blood cells during temperature elevation. *Sci. Rep.* 9: 1–9.

16 Moon, I., Ahmadzadeh, E., Jaferzadeh, K., and Kim, N. (2019). Automated quantification study of human cardiomyocyte synchronization using holographic imaging. *Biomed. Opt. Express* 10: 610–621.

17 Moon, I., Jaferzadeh, K., Ahmadzadeh, E., and Javidi, B. (2018). Automated quantitative analysis of multiple cardiomyocytes at the single-cell level with three-dimensional holographic imaging informatics. *J. Biophotonics* 11: e201800116.

18 Jaferzadeh, K., Moon, I., Bardyn, M. et al. (2018). Quantification of stored red blood cell fluctuations by time-lapse holographic cell imaging. *Biomed. Opt. Express* 9: 4714–4729.

19 Yi, F., Moon, I., and Javidi, B. (2017). Automated red blood cells extraction from holographic images using fully convolutional neural networks. *Biomed. Opt. Express* 8: 4466–4479.

20 Moon, I., Yi, F., and Rappaz, B. (2016). Automated tracking of temporal displacements of a red blood cell obtained by time-lapse digital holographic microscopy. *Appl. Optics* 55: A86–A94.

21 Yi, F., Moon, I., and Javidi, B. (2016). Cell morphology-based classification of red blood cells using holographic imaging informatics. *Biomed. Opt. Express* 7: 2385–2399.

22 Jaferzadeh, K. and Moon, I. (2016). Human red blood cell recognition enhancement with three-dimensional morphological features obtained by digital holographic imaging. *J. Biomed. Opt.* 21: 126015.

23 Yi, F., Moon, I., and Lee, Y. (2015). Three-dimensional counting of morphologically normal human red blood cells via digital holographic microscopy. *J. Biomed. Opt.* 20: 016005.

24 Rappaz, B., Moon, I., Faliu, Y. et al. (2015). Automated multi-parameter measurement of cardiomyocytes dynamics with digital holographic microscopy. *Opt. Express* 23: 13333–13347.

25 Jaferzadeh, K. and Moon, I. (2015). Quantitative investigation of red blood cell three-dimensional geometric and chemical changes in the storage lesion using digital holographic microscopy. *J. Biomed. Opt.* 20.

26 Moon, I., Yi, F., Lee, Y. et al. (2013). Automated quantitative analysis of 3D morphology and mean corpuscular hemoglobin in human red blood cells stored in different periods. *Opt. Express* 21: 30947–30957.

27 Yi, F., Moon, I., and Lee, Y. (2013). Extraction of target specimens from bioholographic images using interactive graph cuts. *J. Biomed. Opt.* 18: 126015.

28 Moon, I., Anand, A., Cruz, M., and Javidi, B. (2013). Identification of malaria-infected red blood cells via digital shearing interferometry and statistical inference. *IEEE Photon. J.* 5: 6900207.

29 Yi, F., Moon, I., Javidi, B. et al. (2013). Automated segmentation of multiple red blood cells with digital holographic microscopy. *J. Biomed. Opt.* 18: 026006.

30 Yi, F., Park, S., and Moon, I. (2021). High-throughput label-free cell detection and counting from diffraction patterns with deep fully convolutional neural networks. *J. Biomed. Opt.* 26 (3): 036001.

12

Red Blood Cell Phase-image Segmentation

12.1 Introduction

With advances in DHM, 3D-imaging techniques have received increased attention. The 3D-image processing of holographic images is suggested for several tasks such as the segmentation, recognition, and tracking of objects [1–6]. However, due to the unnecessary background noise in the holographic image, a reliable holographic image segmentation is required [6].

Analyzing RBC morphological characteristics in the peripheral blood is important for studying hematological functions and detecting disease. However, a notable morphology variation in RBCs represents a challenge for automated analyzers. As a result, an important proportion of RBC samples still requires careful manual examination from an expert, which is time consuming. Within this framework, we will introduce an automated segmentation algorithm specifically designed to process quantitative phase images obtained from DHM. One significant advantage of the presented algorithm is that it provides relevant quantitative cell parameters including, but not limited to, cell size, shape, and volume. It is also possible to obtain the mean corpuscular hemoglobin (MCH) value and MCH concentration of RBCs [7]. Furthermore, segmented phase images of RBCs also benefit the tracking of single or multiple RBCs to study their dynamics (3D morphology and biomass changes).

Segmentation techniques can be classified into two main categories [8, 9]. One category is based on the intensity value. The other is related to the edge detection of objects. Some regions in a single RBC have phase values very close to the background value as shown in Figure 12.1a. Consequently, it is not easy to accurately obtain phase-image segmentation with only an intensity-based method like threshold algorithms. The other problem is that most RBCs have two gradients due to their discoid shape. One gradient is between the RBC boundary and the background. The other gradient lies inside the surface of a single RBC, as shown

Artificial Intelligence in Digital Holographic Imaging: Technical Basis and Biomedical Applications, First Edition. Inkyu Moon.

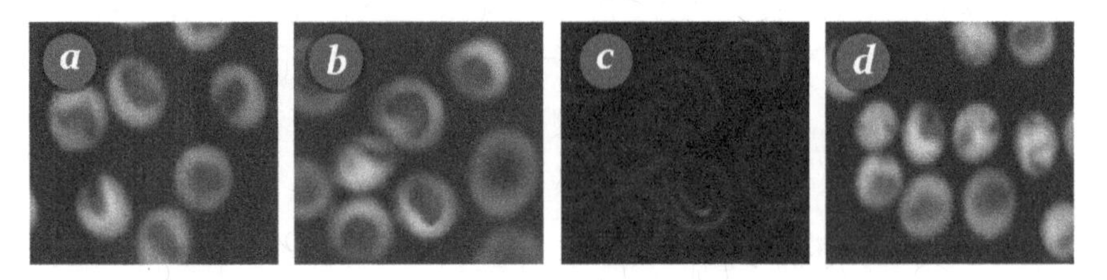

Figure 12.1 Some characteristics of red blood cell (RBC) quantitative phase images. (a) Some regions in a single RBC have phase values similar to the background value. (b) The original quantitative phase image of RBCs with two kinds of edges (inside and outside parts). (c) The gradient image of (b). (d) Some RBCs are connected to each other [10] / with permission of SPIE.

in Figure 12.1b and 12.1c. This makes it difficult to correctly segment RBC cells using an algorithm only based on edge detection. Furthermore, some individual RBCs are connected to each other, as shown in Figure 12.1d. This can also affect isolated RBC segmentation. All these problems complicate phase-image segmentation of RBCs.

This chapter presents a detailed explanation of the automatic algorithm to segment the RBC phase image to accurately calculate the RBC phase value in DHM. It also shows that the presented segmentation method can efficiently minimize over-segmentation and under-segmentation [10]. In particular, the automated RBC segmentation algorithm is suitable for the quantitative comparison of different RBC types because phase values in the backgrounds of RBC phase images can be set to zero. First, we used the Otsu algorithm to obtain a binary image. Next, a morphology operation was performed for the binary image. After conducting a series of morphology operations such as morphological opening, erosion, and reconstruction, we obtained internal markers of these RBCs, which allowed us to avoid effects of the internal gradient and connections between RBCs. With these internal markers, external markers were generated using the distance-transform algorithm combined with the watershed algorithm [9]. Finally, with these extracted internal and external markers, we applied the watershed algorithm to the modified gradient image obtained with a minima imposition technique [11]. We obtained good experimental results with these methods.

12.2 Marker-controlled Watershed Algorithm

The watershed transform algorithm is especially well-suited for generating the closed boundary of objects in question [12]. It also shows a good performance. However, it often leads to over-segmentation. To address this issue, the standard watershed transform algorithm can be improved with marker control. In this section, we will discuss both algorithms.

12.2.1 Watershed-transform Algorithm

The watershed-transform algorithm is based on flooding simulation. We consider intensity values like terrain elevations with valleys and peaks that represent regional minimal and maximal values, respectively. When water floods the terrain, dams are built between valleys. The watershed-transform algorithm finds the peak value between the two valleys as shown in Figure 12.2. These peaks form the watershed line. This can be implemented using Meyer's algorithm [13]. First, a set of markers are selected as valleys and labeled. All neighboring pixels of each labeled area are filled to the priority queue with a priority level determined by the label's intensity value. The label with the highest priority is then chosen from the priority queue. If neighbors of the selected value, which are already labeled, have the same label, these neighboring values are labeled the same. After that, all non-labeled neighbors are put into the priority queue. This step is repeated until the priority queue has no value. Finally, non-labeled intensity values are peak values (see Figure 12.2) that are included in watershed lines.

12.2.2 Marker-controlled Watershed Algorithm

In the watershed-transform algorithm, water fills the terrain starting from the regional minimal value. However, this goes according to markers. Maximum values between every two markers are determined when these maximal values become watershed lines. The marker-controlled watershed algorithm can further distinguish between internal and external markers. Internal markers represent objects that we are looking for. As a result, all objects must be marked as internal

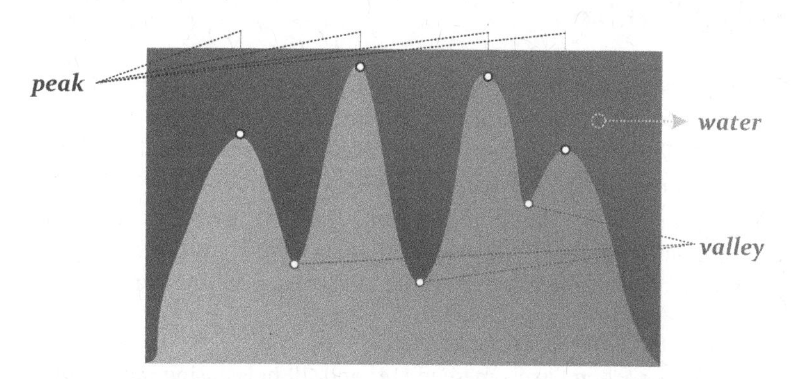

Figure 12.2 Flooding simulation model of the watershed algorithm.

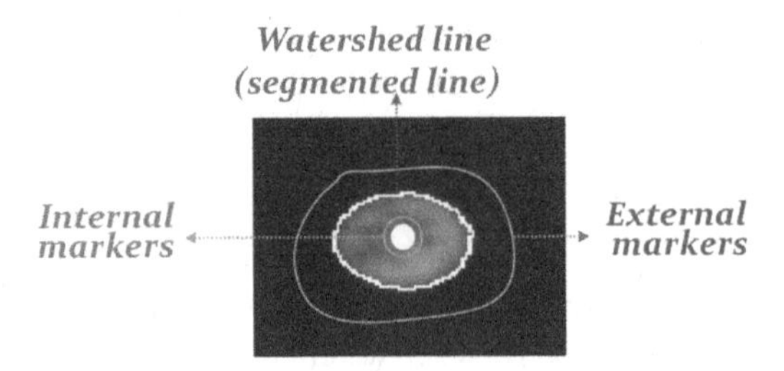

Figure 12.3 An example of experimental results for the marker-controlled watershed algorithm [10] / with permission of SPIE.

markers to obtain correct objects. Internal markers are usually obtained with a threshold algorithm. External markers, on the other hand, represent the background around objects. The distance-transform algorithm combined with the watershed algorithm with internal markers is a good way to obtain external markers. Distance transform measures the distance from every pixel to the nearest nonzero-valued pixel [9].

The watershed transform is usually applied to gradient images since objects and background both have low values while edges correspond to high values in the gradient image. However, the gradient image always contains a large number of regional minimal values due to noise and other local anomalies. For this reason, the segmentation result is not good enough. In the marker-controlled watershed algorithm, after internal and external markers are obtained, they are used to modify the gradient image. Using a minimal imposition method [11], only positions where marker values are located become regional minima. Thus, we can efficiently remove unnecessary regional minimal values and apply the watershed-transform algorithm to the updated gradient image. Figure 12.3 shows an example of such processing.

12.3 Segmentation Based on Marker-controlled Watershed Algorithm

Although the marker-controlled watershed algorithm described in [9] offers a good way to reduce over-segmentation, it cannot efficiently extract internal or external markers. Therefore, we present a method for efficient extraction of markers in the phase image of RBCs based on an enhanced marker-controlled watershed algorithm. Using this algorithm, we can overcome problems shown in Figure 12.1. We can also avoid under-segmentation. Below are the six steps of the enhanced marker-controlled watershed for RBCs phase-image segmentation:

Step 1: Normalize the RBC phase image (I_{norm}).

Step 2: Segment I_{norm} using Otsu's method [14] and fill holes using morphological reconstruction [9] (I_{bin}). The Otsu's method can be presented as

$\sigma^2(t) = \omega_1(t)\omega_2(t)[\mu_1(t) - \mu_2(t)]^2$ where $\sigma^2(t)$ is the variance of the inter-class, $\omega_i(t)$ is the class probability, and $\mu_i(t)$ is the class means. The variable t, which can maximize the inter-class variance $\sigma^2(t)$, is the required threshold.

Step 3: Obtain the gradient magnitude of the original phase image (I_{grad}). Here, we can use the Sobel operator to calculate the gradient in both vertical and horizontal directions:

$$g_x = \begin{bmatrix} -1 & 0 & 1 \\ -2 & 0 & 2 \\ -1 & 0 & 1 \end{bmatrix} I_{bin} \text{ and } g_y = \begin{bmatrix} -1 & -2 & -1 \\ 0 & 0 & 0 \\ 1 & 2 & 1 \end{bmatrix} * I_{bin}.$$

$$I_{grad} = \sqrt{\left(g_x^2 + g_y^2\right)},$$

where I_{grad} denotes the image of gradient magnitude, I_{bin} is the source image, and $*$ is the symbol for the convolution operation.

Step 4: Obtain internal markers.

a) Apply morphological opening [9] to I_{bin} with a disk structuring element of radius 9 (I_{open1}). Such an element is much smaller than the smallest RBC. Thus, all required objects will be preserved while additional noise is removed.

b) Apply morphological erosion to I_{open1} with a disk structuring element of radius 17. Get image I_{ero1}. Such an element is about the size of a medium RBC. As a result, connected objects will be separated.

c) Take image I_{ero1} as a marker and image I_{open1} as a mask. Apply the morphological reconstruction operation to them. Denote the obtained image asI_{rec1}.

d) Subtract $I_{rec1}$from I_{open1}. Denote the obtained image as I_{sma1}. When the separation of connected cells is necessary, as shown in Figure 12.4a, we can use a disk structuring element with radius 17 as described in Step 4a to erode the image. However, this can completely remove small RBCs, as

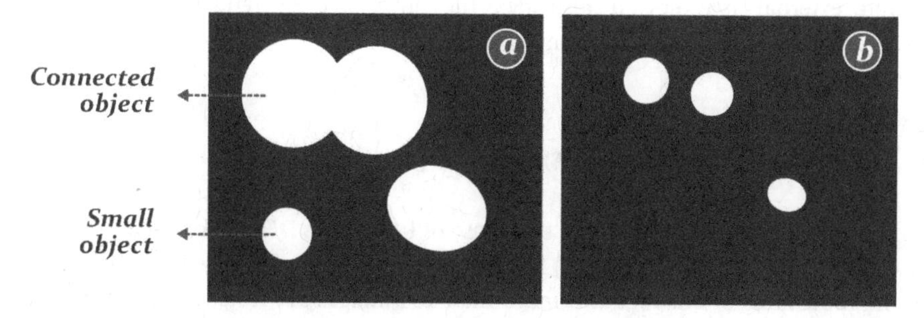

Figure 12.4 (a) Connected and small objects. (b) Erosion of the image in (a) where the disk structuring element is so large that small objects are lost [10] / with permission of SPIE.

shown in Figure 12.4b. If no precautions are taken, this will result in under-segmentation (small RBCs are lost). The goal of this step is to obtain small-sized RBCs.

e) Apply the morphological dilation to image I_{ero1} with a disk structuring element of radius 11. This step reduces the effect of internal gradients as described in Figure 12.1b by extending the center region to cover the internal gradient (I_{dila}).

f) Combine image I_{sma1} (Step 4d) with image I_{dila} (Step 4e). The result, which has most objects marked, can be used as internal markers (I_{inter}).

Step 5: Obtain external markers from image I_{inter} (Step 4e) using the distance-transform and watershed-transform algorithms (I_{exter}). To compute the distance from each pixel to the nearest nonzero-value pixel, use the equation

$$\begin{cases} D(x_i, y_i) = 0 \;\; if\; O(x_i, y_i) = 1 \\ D(x_i, y_i) = \sqrt{(x_i - x_j)^2 + (y_i - y_j)^2} \quad if\; O(x_i, y_i) = 0 \end{cases}$$

where $D(x, y)$ is the distance transform image, $O(x, y)$ is the source image, and $O(x_j, y_j)$ is the nearest non-zero value pixel of $O(x_i, y_i)$.

Step 6: Combine internal markers from I_{inter} with external markers from I_{exter} and obtain the final marker image I_{mark}. Now, modify the gradient magnitude image I_{grad} obtained in Step 3 using the minimal imposition algorithm [11, 15]: $I_{modify} = R^{\varepsilon}_{(I_{grad}+1)\wedge I_{mark}}(I_{mark})$ where $R^{\varepsilon}_{(I_{grad}+1)\wedge I_{mark}}(I_{mark})$ is the morphological erosion reconstruction of I_{mark} from $(I_{grad} + 1) \wedge I_{mark}$, and the symbol $\wedge$ stands for the pointwise minimum between $I_{grad} + 1$ and I_{mark}. Finally, we can apply the watershed-transform algorithm to the modified gradient image I_{modify} and obtain a reasonably segmented phase image I_{obj}.

Figure 12.5 shows a flow chart of the presented method. In the enhanced marker-controlled watershed, we can efficiently and properly extract internal and external markers. It also has the advantage of reducing both over-segmentation and under-segmentation issues.

12.4 Experimental Results

In this experiment, we obtained the phase of RBCs with DHM. Here, we used two classes of RBCs corresponding to two different durations of storage to demonstrate the robustness of our method (more than 100 images are tested). It is suggested that during storage, preserved RBCs will undergo progressive structural and functional changes that may affect RBC function and viability after transfusion [16].

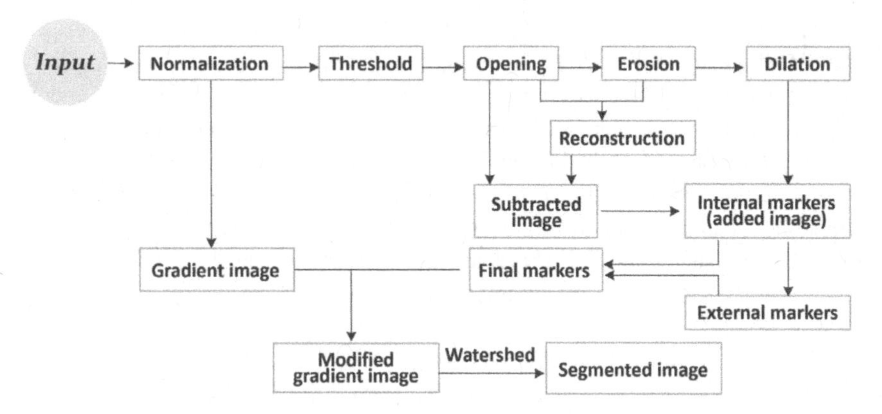

Figure 12.5 Flowchart of the presented phase-image segmentation method.

The first class of RBCs is newer, with a 14-day storage duration. The second group of RBCs is older, with a 38-day storage duration. Figure 12.6 shows respective segmented images using the classical watershed algorithm and the marker-controlled watershed algorithm described in [9].

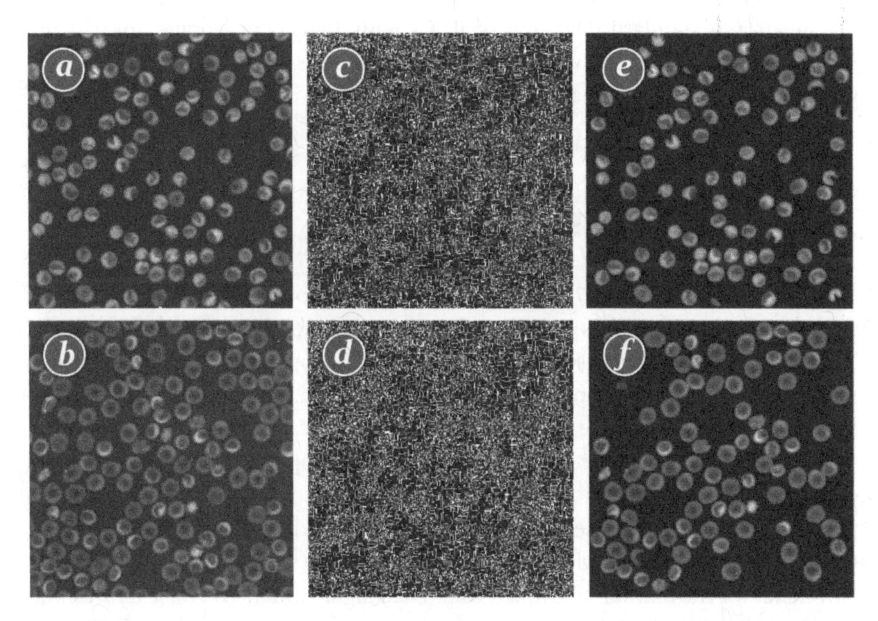

Figure 12.6 RBC phase images. (a) Newer RBCs. (b) Older RBCs. Segmentation results of (c) newer and (d) older RBCs using the standard watershed algorithm. (e),(f) Results using the marker-controlled watershed algorithm [10] / with permission of SPIE.

From Figure 12.6, we can see that without further processing, neither the standard nor the marker-controlled watershed method could provide sufficiently accurate segmentation results because they cannot properly handle over- or under-segmentation issues. Figure 12.7 shows experimental results obtained using the enhanced marker-controlled watershed algorithm described in 12.3; both internal and external markers are correctly extracted. Using our enhanced marker-controlled watershed algorithm, we obtained good experimental results for RBC phase-image segmentation. The algorithm is fairly efficient in reducing both over-segmentation and under-segmentation. It also can properly separate RBCs that touch each other.

After segmentation, the background phase value of the phase images can be set to zero. The average phase value in the background region can then be used to determine the average phase within a single RBC. It allows directly comparing phase images with each other, such as the phase images of two different RBC classes.

Figure 12.8 shows the statistical distribution of the average phase value of a single RBC between two different classes of RBCs (one stored for 14 days and the other for 38 days). The average phase value of a single RBC is calculated by subtracting the average phase value in the background of the corresponding phase image from the average phase value within each RBC. Figure 12.8 shows that the mean and standard deviation of RBCs with 14 days of storage are 97° and 9°, respectively. For RBCs with 38 days of storage, the mean and standard deviation are 74° and 15°, respectively.

12.5 Performance Evaluation

We used a scientific tool [17] for performance evaluation. This tool is based on the experimental design methodology and is independent of the output of systems. The performance comparison of two methods with biased segmentation results mainly depends on varied parameters in segmentation. The procedure of the performance evaluation can be briefly described as steps of data characterization, data sampling, primary parameter selection, parameter sampling, performance metrics definition, performance model calculation, and statistical analysis. The primary parameter that greatly affects the segmentation result in our procedure is the threshold value obtained with Otsu's method in step 2 (see section 12.3). The main parameter for the marker-controlled watershed algorithm in [9] is also a threshold used to find regional minimum values. For performance metrics, the segmentation accuracy is defined as the absolute value of correlation between segmented RBC images and manually segmented reference images. The closer the segmented image is to the reference image, the closer the segmentation accuracy will

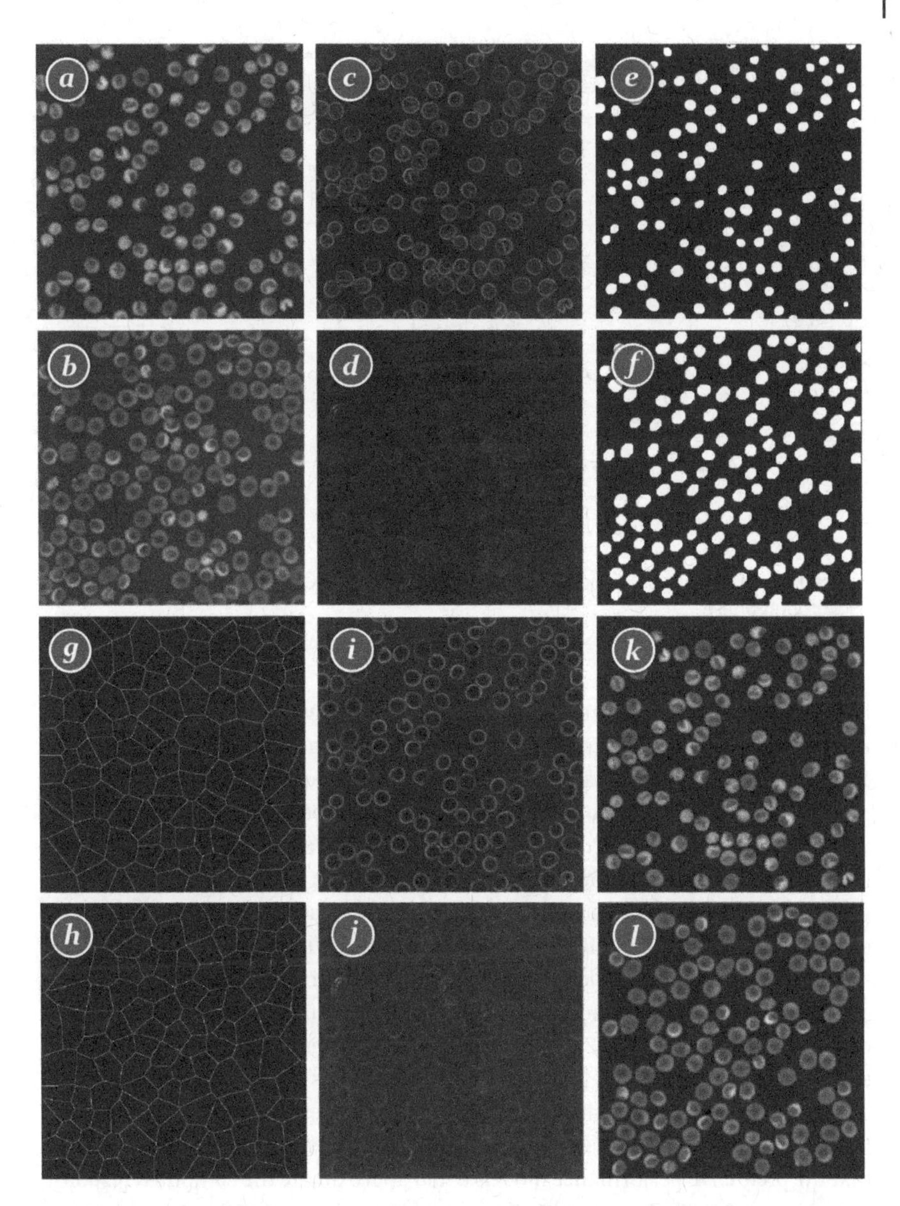

Figure 12.7 Key steps of the proposed phase image segmentation. (a) Newer and (b) older RBC phase images. Original gradient images of (c) newer and (d) older RBCs. Internal markers of (e) newer and (f) older RBCs, respectively. External markers of (g) newer and (h) older RBCs. Modified gradient images of (i) newer and (j) older RBCs. Segmented phase images of (k) newer and (l) older RBCs [10] / with permission of SPIE.

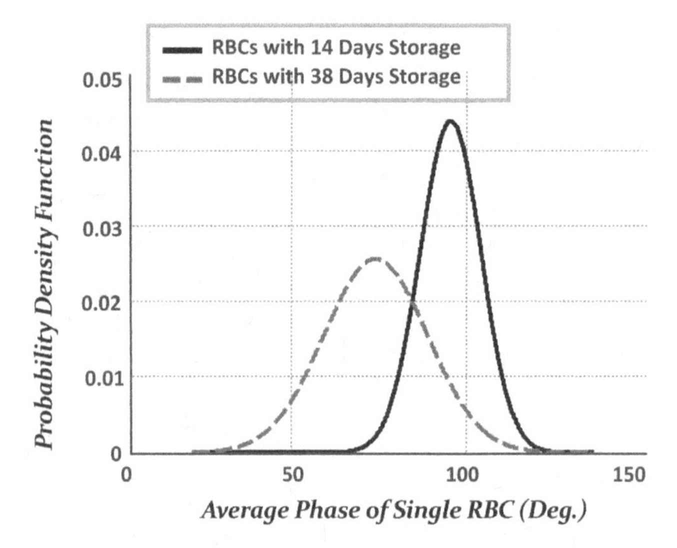

Figure 12.8 Distribution of the average phase of a single RBC after subtracting the average background phase.

approach one. The analyzed dataset consists of newer and older RBC phase images. The range of parameters and segmentation accuracy is between 0 and 1. Accordingly, sample data for 21 RBCs were extracted from the range of parameters and examined to construct a performance model for each segmentation method. The threshold used in [9] was sampled with an interval of 0.05 and the corresponding segmentation accuracy was calculated. Figures 12.9a,b show the performance model conducted between segmentation accuracy and varied threshold values for new and old RBC images, respectively. Threshold values in our segmentation method also varied from 0 to 1 with an interval of 0.05. Figures 12.9c,d show equivalent performance models between segmentation accuracy and thresholds for newer and older RBC images. For curve fitting, we used the least squares error estimation technique [17, 18]. Statistical analysis by a Chi-squared test was conducted to check the similarity between obtained results (segmented data and the fitted polynomial) [18, 19]. Consequently, *p*-values for the null hypothesis—that the predictive performance models could approximate the computed response curve—were 0.7578, 0.3571, 0.9135, and 0.1213 for Figures 12.9a–d, respectively. Therefore, the null hypothesis that the fitting curve is similar to the measured one should be accepted at the significance level of 0.05.

It should be noted that the maximum segmentation accuracy in our method outperformed that presented in [9]. Proper threshold values for newer and older RBC images were approximated at 0.27 and 0.30, respectively, similar to threshold

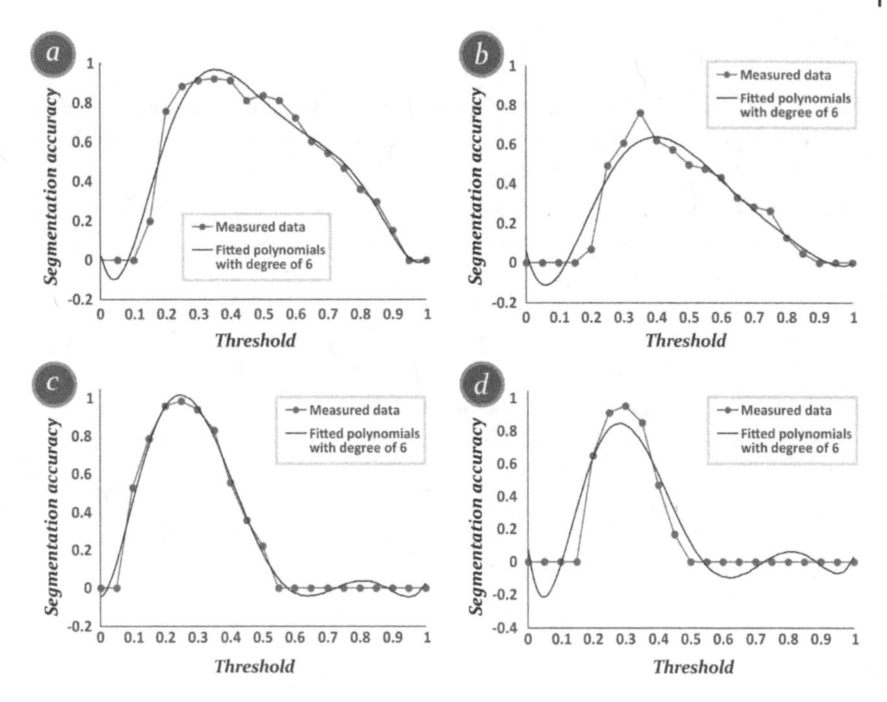

Figure 12.9 Performance models for (a) newer and (b) older RBC phase images in the method of [19] / John Wiley & Sons. Performance models for (c) newer and (d) older RBC phase images in our proposed procedure.

values obtained by Otsu's method used in step 2. It proves that Otsu's method can be used to automatically find the threshold, which allows us to reduce the uncertainty of segmentation results by setting a random value.

12.6 Conclusions

In this chapter, we presented a method to successfully segment phase images of RBCs to compute an accurate phase value of RBCs. Advantages of our method include reducing over- and under-segmentation events. Furthermore, we can obtain isolated RBCs without touching other cells. Our automated RBC segmentation algorithm enables an adequate comparison of different RBC types because phase values for the background of RBC images can be set to 0°. Classifying RBCs traditionally requires a time-consuming, manual inspection by a qualified person. Our segmentation algorithm can aid the automated classification of RBCs using phase images obtained by DHM.

References

1 Javidi, B., Moon, I., Yeom, S., and Carapezza, E. (2005). Three-dimensional imaging and recognition of microorganism using single-exposure on-line (SEOL) digital holography. *Opt. Express* 13: 4492–4506.

2 Dubois, F., Yourassowsky, C., Monnom, O. et al. (2006). Digital holographic microscopy for the three-dimensional dynamic analysis of in vitro cancer cell migration. *J. Biomed. Opt.* 11: 054032.

3 Rappaz, B., Cano, E., Colomb, T. et al. (2009). Noninvasive characterization of the fission yeast cell cycle by monitoring dry mass with digital holographic microscopy. *J. Biomed. Opt.* 14: 034049.

4 Moon, I. and Javidi, B. (2006). Volumetric 3D recognition of biological microorganisms using multivariate statistical method and digital holography. *J. Biomed. Opt.* 11: 064004.

5 Moon, I., Daneshpanah, M., Javidi, B., and Stern, A. (2009). Automated three-dimensional identification and tracking of micro/nanobiological organisms by computational holographic microscopy. *Proc. IEEE* 97: 990–1010.

6 DaneshPanah, M. and Javidi, B. (2006). Segmentation of 3D holographic images using bivariate jointly distributed region snake. *Opt. Express* 14: 5143–5153.

7 Rappaz, B., Barbul, A., Emery, Y. et al. (2008). Comparative study of human erythrocytes by digital holographic microscopy, confocal microscopy and impedance volume analyzer. *Cytometry A* 73: 895–903.

8 Chesnaud, C., Page, V., and Refregier, P. (1998). Improvement in robustness of the statistically independent region snake-based segmentation method of target-shape tracking. *Opt. Lett.* 23: 488–490.

9 Gonzalez, R.C. and Woods, R.E. (2002). *Digital Imaging Processing*. New York, NY: Prentice Hall.

10 Yi, F., Moon, I., Javidi, B. et al. (2013). Automated segmentation of multiple red blood cells with digital holographic microscopy. *J. Biomed. Opt.* 18: 026006.

11 Soille, P. (2003). *Morphologial Image Analysis: Principles and Application*, 2e. NY: Springer-Verlag.

12 Yang, X., Li, H., and Zhou, X. (2006). Nuclei segmentation using marker-controlled watershed, tracking using mean-shift, and kalman filter in time-lapse microscopy. *IEEE Trans. Circuits Syst.* 53: 2405–2414.

13 Meyer, F. (1994). Topographic distance and watershed lines. *Signal Process.* 38: 113–125.

14 Otsu, N. (1979). A threshold selection method from gray-level histograms. *IEEE Trans. Systems, Man, Cybernet. SMC-9* 1: 62–66.

15 Sun, C. and Wang, X. (2010). Spot segmentation and verification based on improve marker controlled watershed transform. In: *2010 3rd International Conference on*

Computer Science and Information Technology, Chengdu, China (09–11 July 2010), 63–66. New York (NY): IEEE.

16 Koch, C.G., Li, L., Sessler, D.I. et al. (2008). Duration of red-cell storage and complications after cardiac surgery. *N. Engl. J. Med.* 20: 1229–1239.

17 Sadjadi, F. (1990). Experimental design methodology: the scientific tool for performance evaluation. *Proc. SPIE* 1310: 100–107.

18 Press, W., Teukolsky, S., Vetterling, W., and Flannery, B. (1992). *Numerical Recipes in Fortran 77*. Cambridge University Press.

19 Rencher, A. (2002). *Methods of Multivariate Analysis*. Wiley-Interscience.

13

Red Blood Cell Phase-image Segmentation with Deep Learning

13.1 Introduction

In Chapter 12, we segmented RBC phase images obtained by DHM using the marker-controlled watershed algorithm combined with morphological operations. However, this method cannot properly segment heavily overlapped RBCs or those touching multiple cells. Therefore, is essential to develop a more robust algorithm for RBC phase-image segmentation.

Deep learning is a promising technique that can offer results superior to those obtained by traditional methods. Consequently, it is extensively studied in the computer vision community [1–8]. CNNs are used for image classification with great success. Recurrent neural networks provide reasonably good performance for text classification and translation. Fully convolutional neural networks (FCNs) are proposed for semantic segmentation with surprising outcomes. FCNs have the advantage of end-to-end training with pixel-wise prediction. Moreover, the size of the image provided to an FCN algorithm can be arbitrary. Other FCN algorithms such as U-net and SegNet are also suggested for semantic segmentation and are applied to biological images. In this chapter, we will apply the FCN method to RBC phase images for RBC segmentation [9]. We will introduce two RBC segmentation schemes. In the first scheme, FCN-1, RBC phase images and manually segmented RBCs were used as a true label to train the FCN model. The trained FCN model was then applied to predict foreground (RBC) or background phase-image pixels for RBC segmentation. In the second scheme, FCN-2, we combined the FCN model with the marker-controlled watershed algorithm to segment RBCs. In FCN-2, we used the FCN to predict the inner part of each RBC then used the predicted results as internal markers of the marker-controlled watershed algorithm to further segment RBCs. In the second scheme, the training label image was not the mask of all segmented RBCs. It was an erosion result of that mask representing the inner portion of each RBC.

Artificial Intelligence in Digital Holographic Imaging: Technical Basis and Biomedical Applications, First Edition. Inkyu Moon.

We prepared two kinds of training images from RBC phase images and trained FCNs for two different schemes. One FCN was used to predict all RBCs whereas the other was only used to predict the inside of each RBC. Predicted results were further combined with the marker-controlled watershed algorithm to fully segment RBCs. We then compared segmentation results from the two methods with those obtained by other methods in terms of the segmentation accuracy (*SA*) and cell separation ability. Experimental results showed that our methods achieved overall good segmentation results, with the FCN-2 model offering the best performance in separating overlapped RBCs.

13.2 Fully Convolutional Neural Networks

FCNs, an extension of CNNs [3], are the mainstream algorithms in the field of semantic segmentation due to their amazing performance [6]. FCNs are successfully applied to biomedical images such as cardiac segmentation in magnetic resonance imaging (MRI) and liver or lesion segmentation in computed tomography (CT) [10, 11].

Unlike CNNs, FCNs do not have fully connected layers [6]. A general network architecture of an FCN is shown in Figure 13.1 (see row A). It has some basic layers consisting of convolution (conv), pooling (pool), activation, and deconvolution (deConv) [3, 6]. The convolution layer performs convolution between image or feature map and a kernel for feature extraction. Pooling mainly refers to max pooling in the FCN algorithm that shrinks feature maps in spatial dimension. Max pooling has the advantage of leading to a faster convergence rate by selecting superior invariant features that enhance the performance of generalization. The activation layer in FCN refers to rectified linear units (Relu) [6] that can add nonlinearity in a network. Because an FCN is an end-to-end and pixel-to-pixel training or prediction technique, the FCN output must have the same size as the input image. Consequently, the deconvolutional layer is used to map feature resolution to the same size of the input image. The deconvolutional operation is achieved by upsampling previous coarse output maps followed by convolutional manipulation. As a result, the FCN can consume an image of arbitrary size and output a dense prediction map of the same size. FCN has a translation-invariant feature due to the local connectivity properties of convolutional, pooling, Relu, and deconvolutional layers [6]. A loss layer is included in the FCN training phase so that network parameters can be learned by minimizing the cost value [6]. Some other layers such as batch normalization, dropout, and softmax are also commonly used in FCNs [1, 2, 6]. Specifically, each layer of data in the FCN are contained in a 3D array of $h \times w \times d$, where h and w are spatial dimensions, and d is the dimension

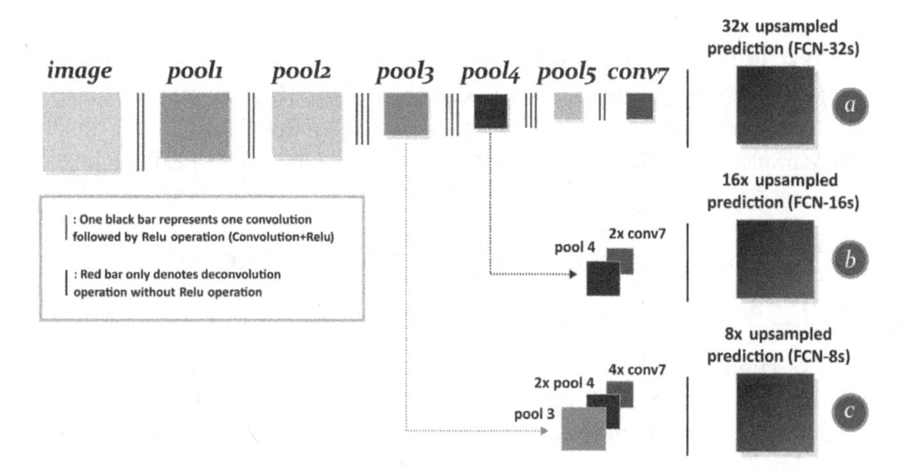

Figure 13.1 Fully convolutional neural networks [6] / with permission of IEEE. Row A: Single-stream net, upsamples stride 32 predictions back to pixels in a single step (FCN-32s); Row B: Fusing predictions from both the final convolutional layer and the pool4 layer for additional prediction (FCN-16s); Row C: Fusing predictions from the final convolutional layer, pool4, and pool3 for additional prediction (FCN-8s).

of feature. The core units in FCN (convolution, pooling, and activation functions) can only operate on local input regions. They depend on relative spatial coordinates.

Since parameters of an FCN model only exist in the kernel used in the convolutional and deconvolutional layers, the total number of parameters required for an FCN model is much smaller than that for a fully connected deep neural network when the same number of hidden units is used. The number of parameters is even smaller than that in CNNs. The relatively small number of parameters required by an FCN is beneficial for network training. In an FCN, the feed-forward passing through the network provides a dense prediction map, and the backpropagation algorithm minimizes the loss function—a sum over spatial dimensions of the final layer combined with information from the ground truth label image—to learn the network [12]. In other words, in an FCN, the forward path is for inference and the backward direction is for learning.

Following a series of successful FCN model applications to semantic segmentation, many new algorithms based on the FCN method were proposed. They are widely studied in fields of image segmentation, classification, and tracking [6, 13, 14]. Long et al. [6] presented two other FCN architectures (FCN-16s and FCN-8s) with different upsampling scales to compensate for a shortcoming of the main FCN architecture, which requires 32×upsampling. They provided better semantic segmentation results than the original one. Network architectures of

FCN-16s and FCN-8s are also shown in Figure 13.1 (row B is FCN-16s and row C is FCN-8s). In FCN-8s, for example, the coarse output from the FCN model is first 4×upsampled and the pool4 image is 2×upsampled. These upsampled images are then fused with the image at the pool3 layer. Fused images are finally 8×upsampled to obtain the prediction image with the same size as the input image.

13.3 RBC Phase-image Segmentation via Deep Learning

In this section, we introduce the RBCs phase-image segmentation procedure based on deep learning. The RBC hologram was first recorded using DHM. The corresponding RBC phase image was numerically reconstructed using the numerical algorithm described in Chapter 5. We prepared two training datasets for RBC segmentation using the FCN model. In the first scheme (FCN-1), we manually segmented RBCs in the RBC phase image and used the mask of the segmented RBC phase image as the ground-truth labeled image (the RBC target). We then zeroed the background. Figure 13.2 shows one of the RBC phase images obtained by DHM along with the corresponding prepared ground-truth labeled images. The FCN was trained by minimizing the error between the ground-truth label image and the FCN prediction image. The trained FCN was then used to predict the class of each pixel in the RBC phase image (0: background, 1: RBC target). In this approach, the predicted results are viewed as final RBC segmentation results since

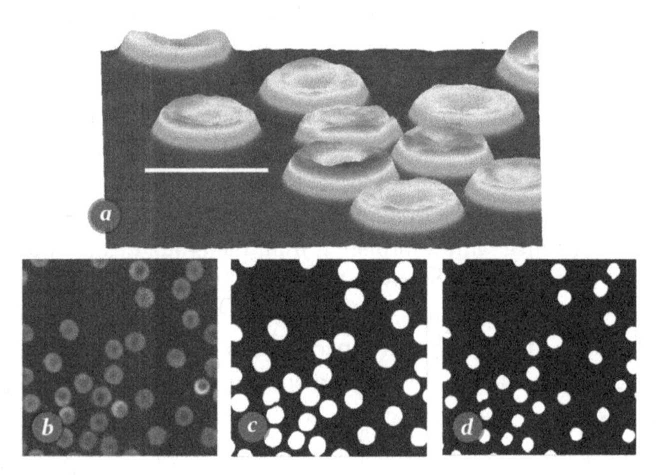

Figure 13.2 RBC phase images and ground-truth label images. (a) RBC 3D profile obtained by off-axis DHM, (b) a reconstructed RBC phase image, (c) a ground-truth label image for the FCN-1 model, and (d) a ground-truth label image for the FCN-2 model. Bar is 10 μm [9] / with permission of Optical Publishing Group.

the training dataset expresses the segmented RBCs. In the second scheme (FCN-2), the ground-truth label image only represented the middle part of each RBC in the RBC phase image. These ground-truth label images were obtained by conducting morphological erosion on the ground-truth label image in the first scheme (FCN-1) with structuring elements of size seven. Figure 13.2d shows one of the ground-truth label images used for the FCN-2 model. Consequently, the FCN-2 scheme was trained and used to predict the center area of each RBC. Because the second FCN method could not segment RBCs directly, we combined this model with the marker-controlled watershed algorithm for RBC phase-image segmentation. The predicted center part of RBCs from the FCN could be efficiently used as internal markers of the marker-controlled watershed algorithm. RBC phase images were finally segmented using the marker-controlled watershed segmentation algorithm. Figure 13.3 shows flowcharts of these two FCN schemes.

The original FCN model in [6], which performs max pooling layer five times, is not very robust for small-object segmentation [8] because of its large upsampling scale value. Figure 13.4 shows our FCN structure for RBC phase-image segmentation. The presented model only uses the max pooling layer four times. There is no max pooling operation at the second layer. The image size in the pool2 layer is the same as that of the previous layer (see Figure 13.4). Further, the image in the pool5 layer is 4×upsampled and fused with the 2×upsampling image at the pool4

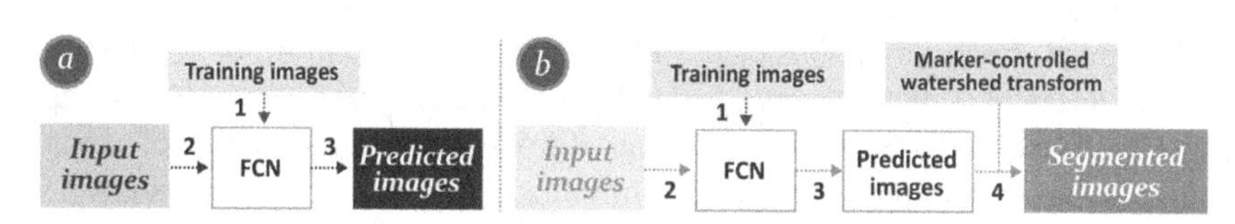

Figure 13.3 Flowcharts for the (a) FCN-1 and (b) FCN-2 models.

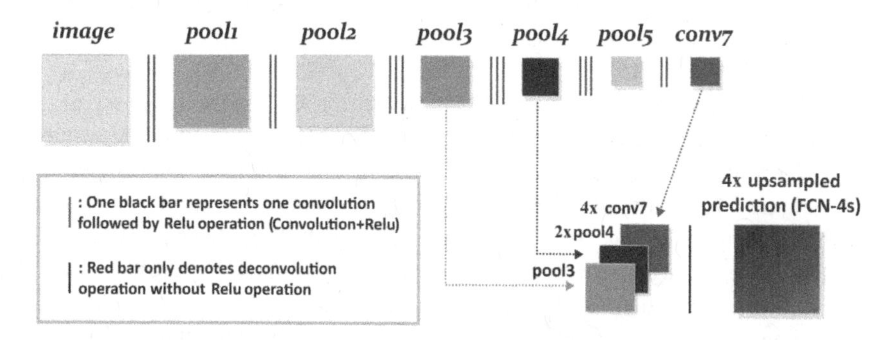

Figure 13.4 The proposed FCN structure for RBC phase-image segmentation.

layer and the image at the pool3 layer. The final layer is 4×upsampled from the fused image. The relatively small upsampling scale value in the last layer can help obtain fine segmentation results. For FCN training, the pre-trained VGG-16 Caffe model [15] was used to initialize the parameters in the two schemes. Here, the parameters within layers that also existed in the VGG-16 network were initialized with their corresponding weight values in a pre-trained VGG-16 Caffe model [15] while other parameters were randomly initialized [1–3]. Training a deep-learning model with a pre-trained model can help the network converge faster and improve accuracy while training a network from scratch usually requires more training images and time [6].

13.4 Experimental Results

We manually segmented 50 RBC phase images for training and testing datasets. Each RBC phase image was 700 × 700 pixels in size. We randomly cropped five images with a size of 384 × 384 from each 700 × 700 RBC phase image to increase the size of training and testing images. The ratio of the training dataset to the testing dataset was set at 7 : 3. Figure 13.2 shows one example of the two ground-truth label images for the RBC phase image in FCN-1 and FCN-2 models. For FCN training, the stochastic gradient-descent algorithm [16] was used to optimize the loss function of the FCN model. A momentum of 0.99 and a weight decay of 0.0005 were used to regularize the loss function. The learning rate was initially set at 0.01 and decreased by a factor of 10 every 1000 iterations. Weights for shared layers were initialized with the pre-trained VGG-16 neural network. Those for varied layers were initialized with values randomly extracted from a normal distribution. The iteration number was set at 4000. Our FCN models were trained on a personal computer equipped with an NVIDIA Tesla K20 GPU running Ubuntu 15.04. The training time for our FCN models with the given specification was approximately 58 minutes on the Caffe deep-learning framework [15].

To show the feasibility of the two FCN schemes for RBC phase-image segmentation, they were compared to two other methods: the marker-controlled watershed algorithm described in Chapter 12 and a method described by Yang et al. [17]. Figure 13.5 shows three RBC phase-image segmentation results from these methods. Visually, it was clear from Figure 13.5 that all methods worked well for RBC phase images. There were no overlapped RBCs. On the other hand, our methods, especially the FCN-2 model, appeared to perform better on RBC phase images for touching and overlapped RBCs. For the quantitative analysis of segmentation results, *SA* was used. The *SA* is defined as

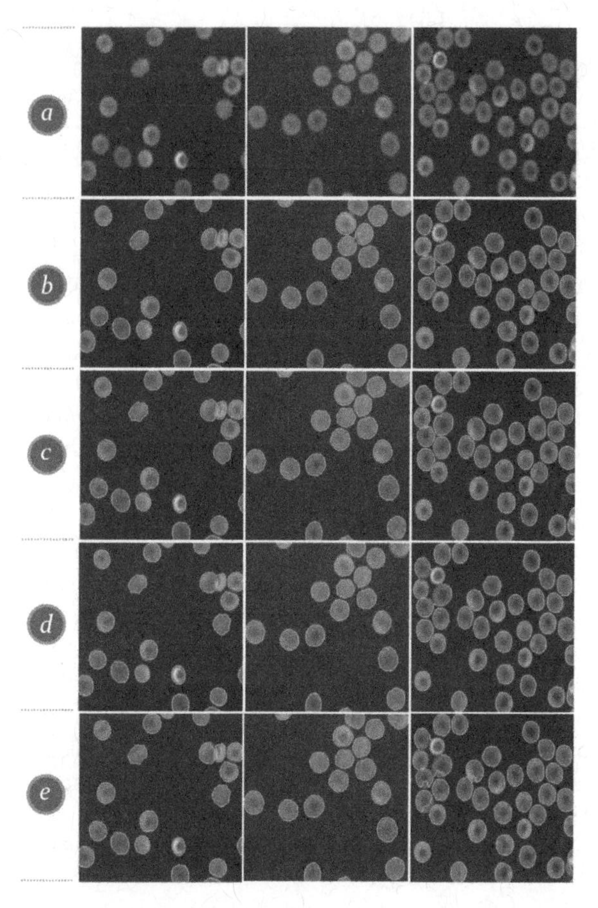

Figure 13.5 Segmentation results for the four segmentation algorithms. (a) Original RBC phase images, (b) segmentation results using FCN-1, (c) segmentation results using FCN-2, (d) segmentation results using marker-controlled watershed algorithm (see Chapter 12), and (e) segmentation results using the Yang et al. method [9] / with permission of Optical Publishing Group.

$$SA\left(S_{seg}, S_{\text{gt}}\right) = 2\frac{\left|S_{seg} \cap S_{\text{gt}}\right|}{\left|S_{seg}\right| + \left|S_{\text{gt}}\right|},$$

where S_{seg} and S_{gt} are the segmented region and the ground-truth region, which are manually extracted as the gold standard; $|\bullet|$ signifies the number of pixel points in a certain region, S_{seg} or S_{gt}. The SA tends toward one when the segmentation results are very similar to the ground truth. The higher the SA, the better the segmentation algorithm performs. In this study, 20 RBC phase images, each consisting of approximately 70 RBCs, were used to compute the segmentation accuracy of

Table 13.1 Segmentation accuracy of the RBC phase image.

Visualization	FCN-1	FCN-2	Method in Ch 12	Yang et al.
SA (average/std)	0.9503 (0.0085)	0.9557 (0.0168)	0.9440 (0.0126)	0.9283 (0.0306)

each method. Table 13.1 provides quantitative evaluation of segmentation results with our deep learning methods, marker-controlled watershed algorithm, and Yang et al.

Our FCN-2 method achieved the best segmentation results in terms of *SA*. This is because it can properly handle RBCs with touching and overlapping problems in RBCs phase images, whereas marker-controlled watershed algorithm and Yang et al. were unable to separate multiple connected RBCs or heavily overlapped RBCs. To highlight the separation ability for connected or overlapped RBCs of these segmentation methods, some segmentation results for regions with connected or overlapped RBCs are shown in Figure 13.6. Our FCN-2 scheme clearly separated RBCs well in RBC phase images. The Yang et al. method used two structuring elements with different sizes to isolate the connected target. However, this method can only divide two connected cells and, it is difficult to define the size of the structuring element. The marker-controlled watershed algorithm separates connected RBCs using morphological operations. However, it also has difficulty determining the size of the structuring element because each RBC and connected area have different sizes. The FCN-1 model could achieve a better performance in terms of RBC separation if more data containing connected or overlapped RBCs were used for training.

Metrics of under-separating, over-separating, and encroachment errors were employed to quantitatively measure the RBC-separation abilities of these segmentation methods. Under-separating is defined as the number of non-separated, connected, or overlapped RBCs. Over-separating refers to the number of RBC divisions within a single, non-touching RBC. The encroachment error refers to the number of incorrect RBC separations. Table 13.2 shows the measured values for under-separating, over-separating, and encroachment errors for 33 RBC phase images with 150 overlapped RBC regions and approximately 1000 RBCs. RBC-separation evaluation curves for the four methods are also shown in Figure 13.7. It was clear that our methods had improved separation above the marker-controlled watershed algorithm and Yang et al. method. Moreover, the FCN-2 method offered the best segmentation result in terms of RBC separation ability. This means that integrating FCN with the marker-controlled watershed algorithm can further boost segmentation performance.

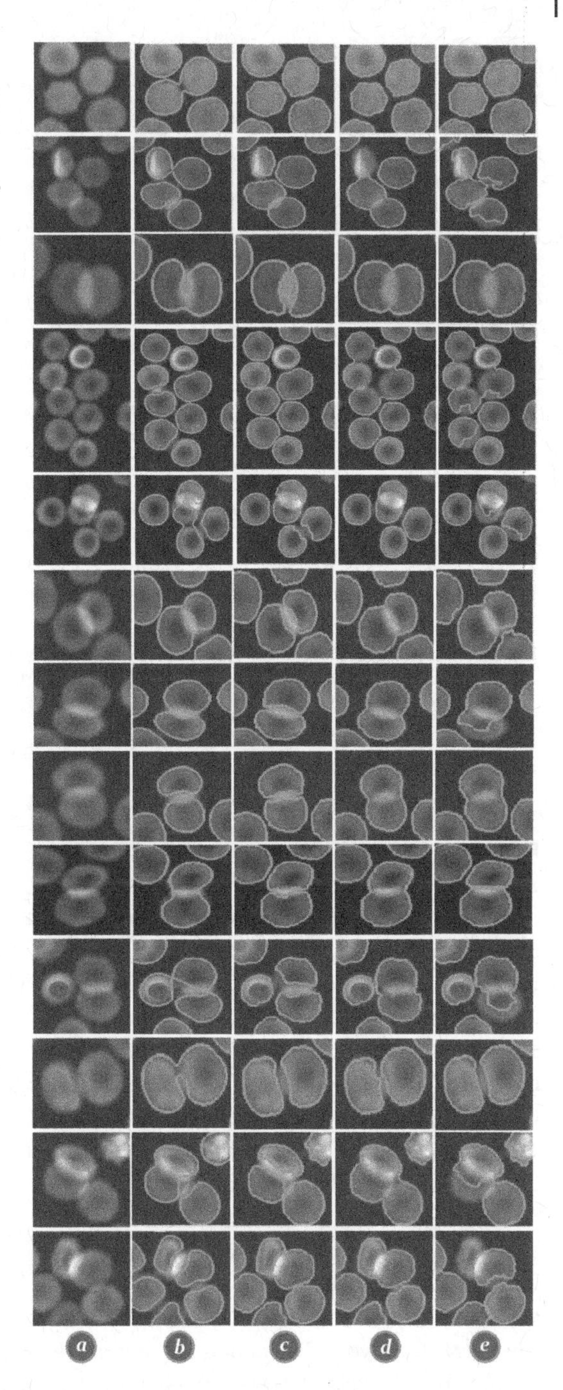

Figure 13.6 RBC separation. (a) Connected RBC region in original RBC phase images. RBC separation results using (b) FCN-1, (c) FCN-2, (d) the marker-controlled watershed algorithm, and (e) the Yang et al. method [9] / with permission of Optical Publishing Group.

Table 13.2 RBC separation evaluation results.

	Under split	Over split	Encroachment error
FCN-1	32	2	2
FCN-2	9	1	1
Method in Ch 12	56	2	4
Yang et al. [17]	34	11	15

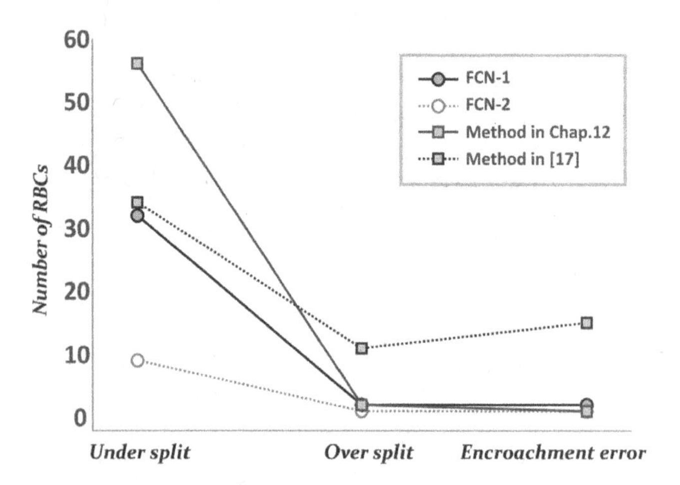

Figure 13.7 RBC separation evaluation results.

Our FCN fine-tuning process, based on the Caffe deep-learning framework, was 58 minutes. The average computing time for RBC phase-image prediction or segmentation was 11.36 seconds for the FCN-1 model and 12.96 seconds for the FCN-2 model for 20 RBC phase images with a size of 700 × 700 pixels. In contrast, the average computing time on 700 × 700 images was 4.67 seconds using the marker-controlled watershed algorithm and 7.83 seconds using the Yang et al. method. Our methods achieved outstanding *SA* and RBCs separation performance but sacrificed efficiency in terms of computation time. Since computing power continues to improve, this is not a major problem.

13.5 Conclusions

We presented two FCN models for automated RBC extraction of RBC phase images obtained using DHM. In the first FCN model, only fully convolutional networks were used for semantic segmentation of RBC phase images, whereas the second FCN model integrated fully convolutional networks with the marker-controlled watershed algorithm for RBC segmentation. FCN parameters were initialized using a VGG 16-layer net, then fine-tuned by manually labeling RBC phase images in the two models separately. Experimental results showed that our two methods could automatically segment RBCs in phase images. However, connected and overlapped RBCs in RBC phase images were better handled by our second model than by others. Comparison results indicated that our methods achieved better performance than the other two algorithms in terms of RBC *SA* and separation ability for overlapped or connected RBCs. This is the first work to apply deep-learning algorithms to holographic images for RBC segmentation. The proposed methods can be useful for quantitatively analyzing RBC morphology and other features to allow the detection of RBC-related disorders.

References

1 Ronneberger, O., Fischer, P., and Brox, T. (2015). UNet: Convolutional networks for biomedical image segmentation. In: *Medical Image Computing and Computer-Assisted Intervention – MICCAI 2015 (Lecture Notes in Computer Science)*, Munich, Germany (5–9 October 2015) (ed. N. Navab, J. Hornegger, W. Wells and A. Frangi). New York (NY): Springer, Cham.

2 Badrinarayanan, V., Kendall, A., and Cipolla, R. (2017). SegNet: A deep convolutional encoder-decoder architecture for image segmentation. *IEEE Trans. Patt. Anyl. Mach. Intell.* 39: 2481–2495.

3 Krizhevsky, A., Sutskever, I., and Hinton, G. (2017). ImageNet classification with deep convolutional neural networks. *Comm. ACM.* 60: 84–90.

4 Mikolov, T., Karafiát, M., Burget, L. et al. (2010). Recurrent neural network based language model. In: *INTERSPEECH 2010, 11th Annual Conference of the International Speech Communication Association (ISCA)*, Makuhari, Chiba, Japan (26-30 September 2010) (ed. T. Kobayashi, K. Hirose and S. Nakamura). Baixas (France): ISCA.

5 Liu, S., Yang, N., Li, M., and Zhou, M. (2015). A Recursive Recurrent Neural Network for Statistical Machine Translation. In: *Proceedings of the 52nd Annual Meeting of the Association for Computational Linguistics (Volume 1: Long Papers)*, Baltimore, MD, USA (22–27 June 2014), 1491–1500. Stroudsburg (PA): Association for Computational Linguistics.

6 Long, J., Shelhamer, E., and Darrell, T. (2015). Fully convolutional networks for semantic segmentation. In: *2015 IEEE Conference on Computer Vision and Pattern Recognition (CVPR)*, Boston, MA, USA (7–12 June 2015), 3431–3440. New York (NY): IEEE.

7 Gu, B., Sun, X., and Sheng, V. (2016). Structural minimax probability machine. *IEEE Trans. Neural Netw. Learn. Syst.* 99: 1–11.

8 Takeki, A., Trinh, T., Yoshihashi, R. et al. (2016). Combining deep features for object detection at various scales: finding small birds in landscape images. *IPSJ Trans. Comput. Vision Appl.* 8 (1): 5.

9 Yi, F., Moon, I., Javidi, B. et al. (2017). Automated segmentation of multiple red blood cells with digital holographic images using fully convolutional neural networks. *Biomed. Opt. Express.* 8: 4466–4479.

10 P. Tran. (2016). A Fully Convolutional Neural Network for Cardiac Segmentation in Short-Axis MRI. arXiv. https://doi.org/10.48550/ARXIV.1604.00494

11 Christ, P., Elshaer, M., Ettlinger, F. et al. (2016). Automatic Liver and Lesion Segmentation in CT Using Cascaded Fully Convolutional Neural Networks and 3D Conditional Random Fields. In: *Medical Image Computing and Computer-Assisted Intervention – MICCAI 2016* (ed. S. Ourselin, L. Joskowicz, M. Sabuncu, et al.), Athens, Greece (17–21 October 2016), 415–423. New York (NY): Springer, Cham.

12 Schmidhuber, J. (2015). Deep learning in neural networks: an overview. *Neural Netw.* 61: 85–117.

13 Wang, L., Ouyang, W., Wang, X., and Lu, H. (2015). Visual tracking with fully convolutional networks. In: *2015 IEEE International Conference on Computer Vision (ICCV)*, Santiago, Chile (7–13 December 2015), 3119–3127. New York (NY): IEEE.

14 Maggiori, E., Tarabalka, Y., Charpiat, G., and Alliez, P. (2016). Fully convolutional neural networks for remote sensing image classification. In: *2016 IEEE International Geoscience and Remote Sensing Symposium (IGARSS)*, Beijing, China (10–15 July 2016), 5071–5074. New York (NY): IEEE.

15 Jia, Y., Shelhamer, E., Donahue, J. et al. (2014). Caffe: Convolutional architecture for fast feature embedding. In: *MM '14: Proceedings of the 22nd ACM International Conference on Multimedia*, Orlando, FL, USA (3–7 November 2014), 675–678. New York (NY): Association for Computing Machinery.

16 Meuleau, N. and Dorigo, M. (2002). Ant colony optimization and stochastic gradient descent. *Artif. Life* 8 (2): 103–121.

17 Yang, X., Li, H., and Zhou, X. (2006). Nuclei segmentation using marker-controlled watershed, tracking using mean-shift, and Kalman filter in time-lapse microscopy. *IEEE Trans. Circuits Syst.* 53 (11): 2405–2414.

14

Automated Phenotypic Classification of Red Blood Cells

14.1 Introduction

Human blood contains different cell types. RBCs or erythrocytes are the most abundant cell type. They transport oxygen to the tissues and organs as well as carbon dioxide to be removed by the lungs. The biconcave shape of the erythrocyte is extremely important for RBC functionality as it increases the surface-area-to-volume (SAV) ratio and facilitates the large reversible elastic deformation of the RBC required to squeeze through tiny capillaries [1]. Pathological disorders can modify RBCs and lead to significant changes in their original shape [2]. The consequences of modified RBCs are often observed as clinical symptoms that range from the obstruction of capillaries and restriction of blood flow to necrosis and organ damage [2, 3]. Counting cell types in a blood sample during cytometry is an important task to investigate clinical status.

In the case of RBCs, a biconcave cell type accounts for a substantial portion of RBCs in a healthy person, although there are other RBC shapes with different percentages between healthy and non-healthy persons [4]. Accordingly, it is essential to determine the percentage of each RBC type in a blood sample that contains different RBC shapes to diagnose and determine the appropriate treatment of subjects. Typically, an image-based cell analysis for diagnosis is performed by experts. It has drawbacks, like being time-consuming and inaccurate. A sample is generally viewed through microscopy images based on the experts' subjective understanding of intensity, morphology, texture, and other characteristics. Usually, small-scale differences in features are overlooked by human eyes, especially in the case of a borderline diagnostic scenario.

Regarding conventional RBC classification problems, experts deal with 2D erythrocyte images obtained using a conventional microscopy. However, the

Artificial Intelligence in Digital Holographic Imaging: Technical Basis and Biomedical Applications, First Edition. Inkyu Moon.

intensity-based microscopy suffers from the loss of substantial quantitative information about cell structure and content. It also requires the use of dyes or stains, which could kill cells or destroy their structure. On the other hand, DHM is superior at the non-invasive imaging and analysis of biological cells. When studying RBCs, DHM enables the measurement of 3D features such as the mean corpuscular volume (MCV), surface area, SAV ratio, functionality factor, sphericity index, and sphericity coefficients. The chemical parameters of mean corpuscular hemoglobin (MCH) and MCH surface density (MCHSD) can also be measured using DHM. Accordingly, an automated RBC classification that can accurately distinguish between RBC types could benefit from DHM imaging.

In this chapter, we will introduce automated methods to classify four RBC shapes (biconcave, flat disks, stomatocyte, and echino-spherocyte) and quantify the percent of RBC types in human RBCs [5]. First, RBCs are obtained by DHM. Single RBCs are then extracted from phase images of multiple RBCs using the watershed segmentation algorithm (see Chapter 12). In the next step, 2D features of the projected surface area (PSA), perimeter, radius, elongation, and PSA-to-perimeter ratio are extracted from segmented RBC images. The volume, surface area, SAV ratio, average RBC thickness, sphericity index, sphericity coefficient, functionality factors, MCH, and MCHSD are also extracted from single RBCs. The latter feature set is related to the morphological and chemical properties of the RBC 3D profile. Along with these 3D features, two new features related to the ring section of RBC are introduced. These features can add significant information to the classification model and improve the discrimination power of the classifier. Each feature set is then separately fed into a neural network model. The classification results are compared using a 10-fold cross validation (CV) technique. To train our classifier, the Bayesian regulation back-propagation algorithm is used. It can adjust weights according to the Levenberg–Marquardt optimization technique. The hyperbolic tangent sigmoid activation function is used in mid-level layers.

To propose the best feature set, sequential forward feature selection (SFFS) is used to consider both 2D and 3D features. The best performance of a classification model is achieved by selecting the most informative features and removing noisy ones that are either redundant or irrelevant. Indeed, the reduced number of features can provide a shorter training time, decreased complexity of classifier, and a concise model for interpretation goals.

We extracted 108 biconcave RBCs from a healthy blood-bank sample, 106 samples of stomatocyte shaped RBCs from a sample with predominantly stomatocytes, 38 samples of flat-disk shaped RBCs, and 71 samples of echino-spherocyte shaped RBCs for training and testing our model. Flat-disk and echino-spherocyte cells were also extracted from RBC samples stored in the blood bank. The performance

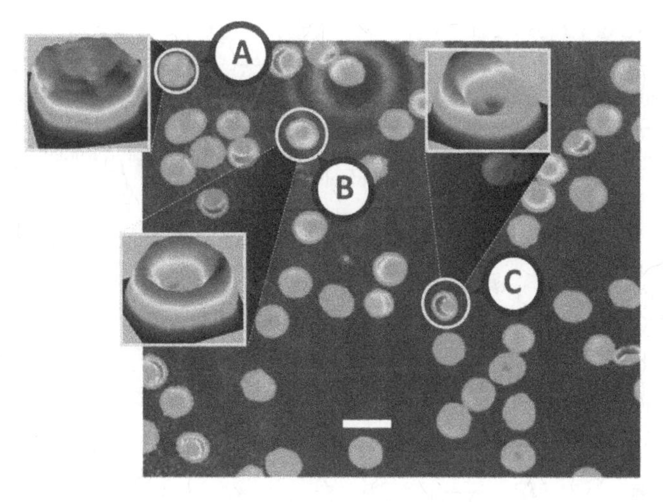

Figure 14.1 Reconstructed phase image with three RBC types; A is a flat disk RBC, B is a biconcave RBC, and C is a stomatocyte. The white scale bar is 10 μm [5] / SPIE / CC BY 3.0.

comparison was performed by calculating the misclassification rate of the 10-fold CV technique.

Our experimental results demonstrated that our model trained by 3D features performed well in automatically classifying and counting RBCs compared to 2D features. We also introduced the best set of features that blended 2D and 3D features to improve RBC classification accuracy. We believe that the final feature set, evaluated with the presented neural network classification, can offer better discrimination results. Figure 14.1 shows a reconstructed phase image of a sample consisting of biconcave, flat-disk, and stomatocyte shaped RBCs. It shows that within in a single sample, RBCs can have different morphologies.

14.2 Feature Extraction

14.2.1 Two-dimensional (2D) Features

After extracting many single RBCs, features can be extracted. We first started with 2D features (see Table 14.1). The elongation of the RBC is a measure of the width-to-length ratio for oblong RBCs. It can be calculated from the chain code by summing the number of each element type, 0–7, then combining 0 and 4, 1 and 5, 2 and 6, as well as 3 and 7 [6]. Average and standard deviation values of the above features are in the agreement with previously reported values (data not shown) [7, 8].

Table 14.1 Descriptions of 2D features.

	Feature name	Description
2D-F1	Projected surface area (*PSA*)	$PSA = N \times p^2$ (p is pixel size in the holographic image; N is the number of pixels within an RBC.)
2D-F2	Perimeter (*Pr*)	Length of the RBC boundary.
2D-F3	Circularity (*Ci*)	$Ci = \frac{Pr^2}{PSA}$
2D-F4	Elongation (*El*)	Orientation of chain code in the cell boundary.
2D-F5	Radios (*R*)	Radius is estimated by considering the radius of a circle with an area of the *PSA*. $R = \sqrt{\frac{PSA}{\pi}}$
2D-F6	*PSA*/Perimeter (*PSP*)	$PSP = \frac{PSA}{Pr}$

14.2.2 Three-dimensional (3D) Features

Since some 3D features require RBC thickness, we first need to convert the phase image into thickness image. Accordingly, the thickness value $h(i, j)$ for each pixel of (i, j) with the phase value $\varphi(i, j)$ in a phase image can be presented as

$$h(i,j) = \frac{\varphi(i,j) \times \lambda}{2\pi(n_{rbc} - n_m)},$$

where $\varphi(i, j)$ is the phase value in radians and the refractive index of RBCs, n_{rbc}, is measured with a dual-wavelength DHM. Here, n_{rbc} is 1.396, without a significant difference between groups of RBCs. The HEPA buffer index of refraction, n_m, is 1.3334. Surface area is another important property and is the main contribution to 3D features. Generally speaking, the RBC surface area is the surface area of the membrane mesh plus PSA. The method in this study splits and divides RBC surfaces into smaller regular areas (triangles) and adds these smaller areas to yield the entire surface area. The accuracy of such a calculation depends on the selected smaller area.

Table 14.2 is a list of 3D features. We will provide a brief description of the eight features related to the morphological properties of RBC. For more detailed descriptions of these features, refer to [9] and Chapter 15. Regarding the calculation of three features (F9–F11), we obtained many points over the ring section of RBC by applying two methods. First, we estimated the ring section (blue points in Figure 14.2) by computing the radius of a circle with the area of the RBC projection on the *x-y* plane (the ring is approximately three-fourths of the RBC radius). We

Table 14.2 Descriptions of 3D features.

	Feature name	Description
3D-F1	Average thickness (*AT*)	$AT = \frac{\sum_{i=1}^{k}\sum_{j=1}^{l} h(i,j)}{k \times l}$ ($h(i, j)$ is the thickness at $(i, j)^{th}$ pixel)
3D-F2	Volume (*MCV*)	$MCV \simeq p^2 \sum_{i=1}^{k} \sum_{j=1}^{l} h(i,j)$
3D-F3	Top-view surface area (*TVS*)	The surface area of the upper view of the RBC calculated by dividing the surface into small triangles and finding the sum of each triangle's surface area.
3D-F4	Top-view SAV ratio (*TVSV*)	$TVSV = \frac{TVS}{V}$
3D-F5	Total surface area (*SA*)	$SA = TVS + PSA$
3D-F6	Surface-area-to-volume ratio (*SAV*)	$SAV = \frac{SA}{V}$
3D-F7	Sphericity index (*SI*)	$SI = \frac{4\pi V^{\frac{2}{3}}}{\left(\frac{4\pi}{3}\right)^{\frac{2}{3}} SA}$
3D-F8	Functionality factor (*FF*)	$f = \frac{SA}{4\pi R^2}$
3D-F9	Sphericity coefficient (*SP*)	$SP = \frac{d_c}{d_r}$ (d_c and d_r are the thickness values in the RBC center and ring section, respectively.)
3D-F10	*STD* of thickness in ring section	The variation in *RP* thickness of the RBC ring.
3D-F11	Upper side of the ring/ lower side of ring	The division of four maximum *RP*s by four minimum *RP*s.
3D-F12	Mean corpuscular hemoglobin (*MCH*)	$MCH = \frac{10\lambda\overline{\varphi}(PSA)}{2\pi\alpha}$ ($\overline{\varphi}$ is the average phase value, λ is the wavelength of the light source of the setup, and $\alpha = 0.00196\ dl/g$ is the specific refraction increment constant, which is related to the protein concentration.)
3D-F13	MCH surface density (*MCHSD*)	$MCHSD = \frac{MCH}{PSA}$

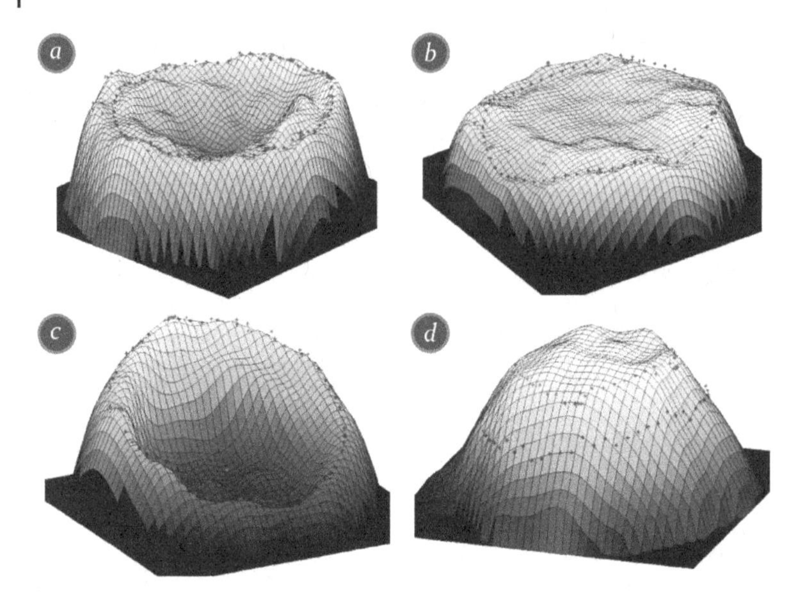

Figure 14.2 A 3D representation of four RBC types and points on the ring section. Typical (a) biconcave, (b) flat disk with a raised center, (c) stomatocyte, and (d) spherocyte RBCs [5] / SPIE / CC BY 3.0.

then updated the position of each point on the estimated ring (blue points) by finding the thickest point in a 3×3 neighbor (red points in Figure 14.2). A single green-point indicates the center of the RBC for the calculation of the sphericity coefficient.

We believe that the above 3D features can discriminate between different RBCs since they are related to the 3D profile of RBC. The statistic *t-test* using a two-sample Kolmogorov–Smirnov test revealed that some of these features were independent (data not shown). Table 14.3 shows the average and standard deviation values of each feature for each RBC type.

14.3 Pattern Recognition Neural Network

Artificial neural networks (ANN) are highly simplified mathematical models of biological neural networks that may learn and provide meaningful solutions to high-level complex and nonlinear problems. The ANN model is faster than its conventional techniques and more robust in noisy environments. It can solve a wide range of problems. A typical ANN is presented in Figure 14.3. An important application of neural networks is pattern recognition that can be implemented using a feed-forward neural network with a specific training function. During training, the

Table 14.3 Average and standard deviation values of 3D features (mean ± *STD*).

	Biconcave	Flat disk	Stomatocyte	Echino-spherocyte
Average thickness (μm)	2.18 ± 0.3	2.27 ± 0.25	2.75 ± 0.36	4.47 ± 0.36
Volume (μm^3)	93.23 ± 13	103.29 ± 14.72	95.85 ± 11.52	101.91 ± 16.4
Top-view SA (μm^2)	103.85 ± 15	94.50 ± 9.62	106.04 ± 12.39	95.65 ± 10.6
Top-view SAV ratio (μm^{-1})	1.12 ± 0.09	0.92 ± 0.05	1.11 ± 0.1	0.95 ± 0.06
Total SA (μm^2)	148.32 ± 16.38	143.21 ± 12.48	147.32 ± 16.11	120.76 ± 11.79
SAV ratio (μm^{-1})	1.61 ± 0.17	1.40 ± 0.13	1.55 ± 0.18	1.19 ± 0.09
Sphericity index (*SI*)	0.40 ± 0.18	0.74 ± 0.04	0.69 ± 0.06	0.87 ± 0.032
Functionality factor (*FF*)	0.86 ± 0.16	0.74 ± 0.07	0.93 ± 0.19	1.21 ± 0.14
Sphericity coefficient (*SP*)	0.67 ± 0.05	0.82 ± 0.06	0.5799 ± 0.21	1.16 ± 0.14
STD of ring thickness	0.0332 ± 0.021	0.054 ± 0.018	0.1321 ± 0.096	0.1254 ± 0.03
Upper side of the ring/ lower side of ring	1.31 ± 0.11	1.25 ± 0.09	1.93 ± 0.44	1.44 ± 1.14
MCH (pg)	31.29 ± 4.55	34.68 ± 5.01	32.17 ± 3.94	34.29 ± 5.51
MCHSD ($pg/\mu m^2$)	0.68 ± 0.1	0.71 ± 0.08	0.85 ± 11	1.39 ± 0.11

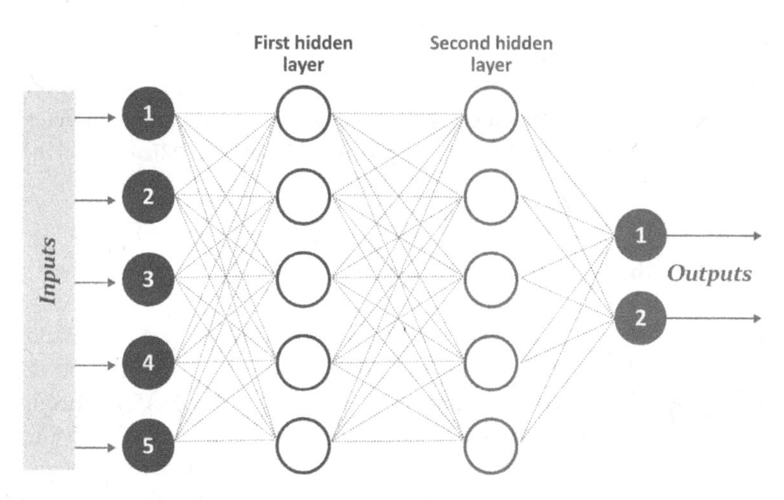

Figure 14.3 A feed-forward artificial neural network configuration with five input nodes, two output nodes, and two hidden layers.

ANN is trained to associate outputs with input patterns. After training, it can identify the input pattern and provide the associated output pattern.

The information processing in the ANN model starts from the input layer to the hidden layer (or from one hidden layer to another if there are multiple hidden layers) and from the hidden layer to the output layer. A synaptic weight is assigned to each link to indicate the relative connection strength of two nodes at both ends in predicting the input–output relationship. The output y_j of any node j is given as

$$y_j = f\left(\sum_{i=1}^{n} W_i X_i + b_i\right),$$

where W_i is the connection weight, X_i denotes the i_{th} input of node j, n is the number input(s) to node j, and b_j is the bias value. The activation function f determines the response of a node to the total input signal that is received. The hyperbolic tangent sigmoid function is used as the activation function for the hidden layer in this study.

For the output layer in the ANN model, we used the softmax function (normalized exponential) that could be interpreted as the posterior probability for a categorical target variable. It is highly desirable for those outputs to lie between zero and one with a sum of one. The purpose of the softmax activation function is to impose these constraints on outputs. Let the input to each output neuron be q_l, $l = 1,\ldots,k$, where k is the number of classes. The softmax output y_l is

$$y_l = \frac{e^{q_l}}{\sum_{m=1}^{k} e^{q_m}}. \tag{14.1}$$

According to Eq. (14.1), the sum of all outputs is equal to unity and can be interpreted as the posterior probability for the final decision. The training algorithm updates the weight and bias values according to the Levenberg–Marquardt optimization. It minimizes a combination of squared errors and weights and determines the correct combination to construct a network that generalizes well [10].

14.4 Experimental Results and Discussion

In our experiment, 108 RBCs labeled as biconcave, 106 RBCs labeled as stomatocytes, 38 RBCs labeled as flat-disk, and 71 RBCs labeled as echino-spherocytes were used. Figure 14.4 shows four samples of each group. The performance of the ANN model was assessed using a 10-fold CV check. The dataset was divided into 10 subsets, and the test was repeated 10 times. Each time, one of these 10 subsets was used as the test set and the other 9 subsets were put together to form a training set. The average misclassification error across all 10 trials could indicate the overall miss-classification error. Pattern recognition neural network (PRNN) has one input layer, one output layer, and three hidden layers. Each hidden layer

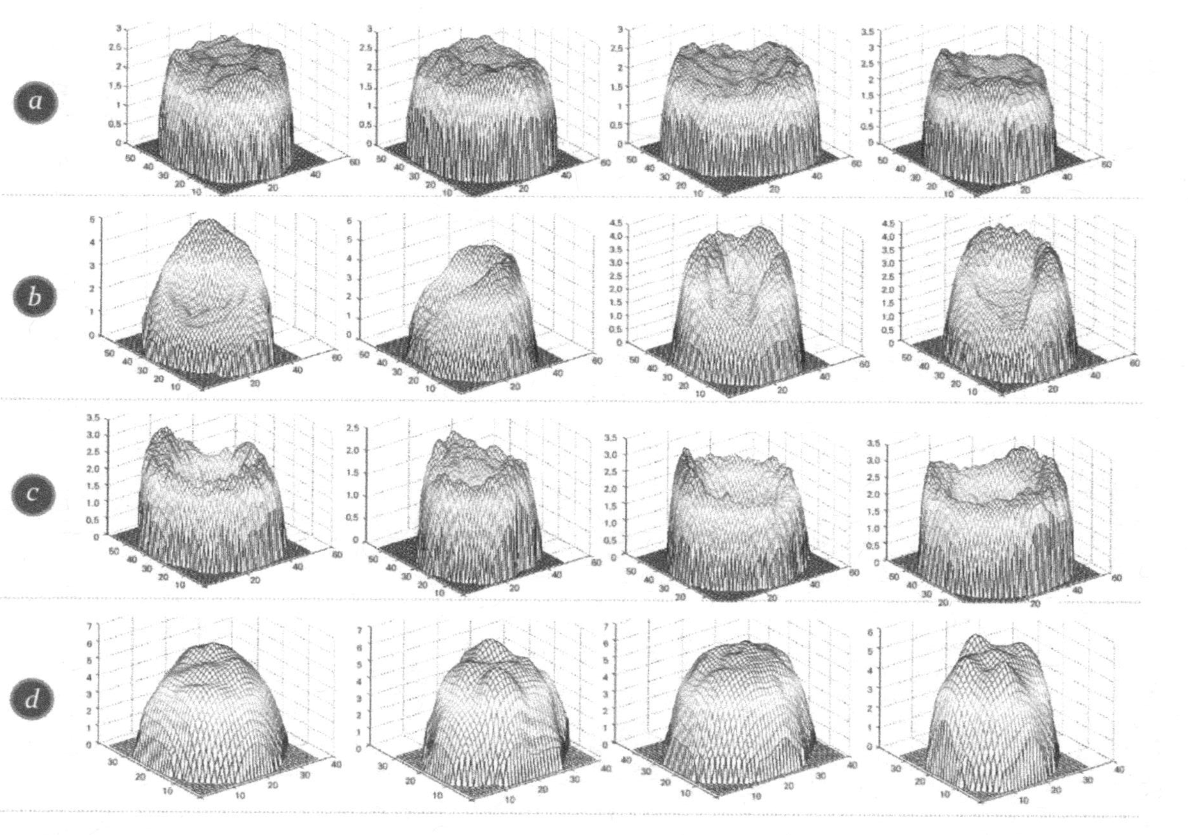

Figure 14.4 Samples of each RBC group used in the study. Four samples each of (a) flat disk, (b) stomatocyte, (c) biconcave, and (d) sphero-echinocyte RBC morphologies [5] / SPIE / CC BY 3.0.

has 5, 10, and 5 neurons, respectively. The number of neurons was obtained by the trial-and-error technique.

14.4.1 Comparison between 2D and 3D Features

In the case of 2D features (see Table 14.1), the 10-fold CV indicated that that the total misclassification rate was considerably high. The following are the percent of misclassification for each group: flat-disk, 64%; stomatocyte, 13.4%; biconcave, 32.3%; and echino-spherocyte, 4.2%. Only echino-spherocyte RBCs could be accurately classified using 2D features while other classes had significant errors. According to the confusion matrix, the ANN model became confused between biconcave and flat disks using 2D features (data not shown). In contrast, the 3D features explained in Table 14.2 offered more accurate and interesting results. According to the 10-fold CV results, the misclassification rates were 0, 1.6, 3.2, and 0% for flat-disk, stomatocyte, biconcave, and echino-spherocyte RBCs, respectively. Table 14.4 shows the classification errors for 2D and 3D features. Classification results obtained with the ANN model showed that 3D features were more effective than 2D features for RBC classification.

In another experiment, we evaluated the mutual information between each feature in 2D and 3D features. In the case of 2D features, Table 14.5 shows that there is considerable mutual information between some features. Mutual information is the quantity of information that two features share. If the mutual information between the two features is big (or small), the two features are closely (or not

Table 14.4 Misclassification results of 2D and 3D features.

	Flat disk	Stomatocyte	Biconcave	Echino-spherocyte
2D features	64%	13.4%	32.3%	4.2%
3D features	0%	1.6%	3.2%	0%

Table 14.5 Normalized mutual information between 2D features.

	2D-F1	2D-F2	2D-F3	2D-F4	2D-F5	2D-F6
2D-F1	1	0.61	0.13	0.37	0.31	0.33
2D-F2	0.61	1	0.057	0.25	0.071	0.37
2D-F3	0.13	0.057	1	0.12	0.7	0.014
2D-F4	0.37	0.25	0.12	1	0.15	0.14
2D-F5	0.31	0.071	0.7	0.15	1	0.04
2D-F6	0.33	0.37	0.014	0.14	0.04	1

closely) related. If the mutual information reaches zero, the two features are independent. For example, 2D-F1 had considerable mutual information with features 2D-F2, 2D-F4, 2D-F5, 2D-F6 (see first row of the Table 14.5). As a result, this feature was statistically redundant. It could not add significant information.

14.4.2 Combining 2D and 3D Features

The best classification model should consider both 3D and 2D features. The performance of any classification model can be improved using a feature selection (FS) technique. In FS, we try to find the best set of features with a strong ability to distinguish each class. In general, FS keeps original features intact. Features deemed unimportant, irrelevant, or redundant are removed from further consideration while only those features that significantly contribute to the classification are chosen. Therefore, FS can minimize the number of features in the classification problem, which makes the classification model simpler (or less complex) and reduces training time. We also implemented FS in this study using the SFFS technique. In SFFS, features are sequentially added to an empty candidate set until the addition of further features does not decrease the criterion. It has two components: an objective function (criterion) and a sequential search algorithm. Common criteria are the misclassification rate for classification and mean squared error for regression models. A sequential forward search algorithm adds features from a candidate subset while evaluating the criterion. Since an exhaustive comparison of the criterion value at all 2^n subsets of an n-feature dataset is typically infeasible (or time-consuming), sequential searches can increase the candidate set. The following features can provide better results in the classification of RBCs: average RBC thickness (3D-F1), top-view SAV ratio (3D-F4), sphericity coefficient (3D-F9), the upper side of the ring divided by lower side of the ring (3D-F11), and perimeter (2D-F2). Our experimental results also showed that adding more features did not provide significant discrimination ability to the final classification model. Figure 14.5 shows the data distribution of each RBC for each selected feature. Table 14.6 provides the misclassification rates of the final feature set and the ANN model.

The confusion matrix of the test set shows that the ANN model sometimes confuses stomatocytes and echino-spherocytes, since there are cases in which RBCs have a morphology similar to both stomatocytes and echino-spherocytes (see Figure 14.6). The posterior probabilities for belonging to stomatocyte and echino-spherocyte classes are 0.33 and 0.66, respectively. Even a human examiner may be challenged to put it in the correct category.

In another experiment, we tried to count different RBC types in five quantitative phase images with multiple RBCs. RBC images were also inspected visually by a human inspector. As shown in Figure 14.7a, stomatocyte RBCs were dominant (flat disks, 0/80; stomatocytes, 58/80; biconcave, 8/80; and echino-spherocyte,

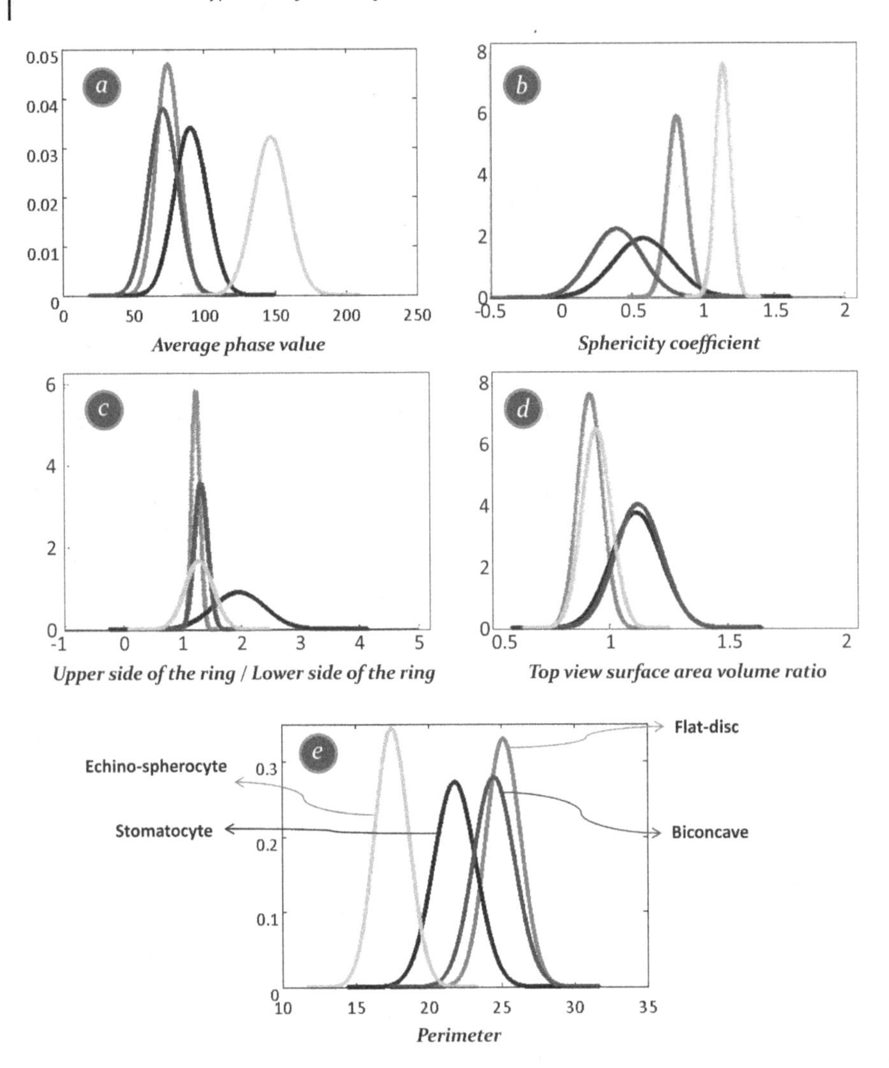

Figure 14.5 Data distribution of the best feature set [5] / SPIE / CC BY 3.0.

Table 14.6 Misclassification results for the best feature set obtained by sequential feature selection technique.

	Flat disk	Stomatocyte	Biconcave	Echino-spherocyte
Best feature set	0%	0.9%	3.1%	0%

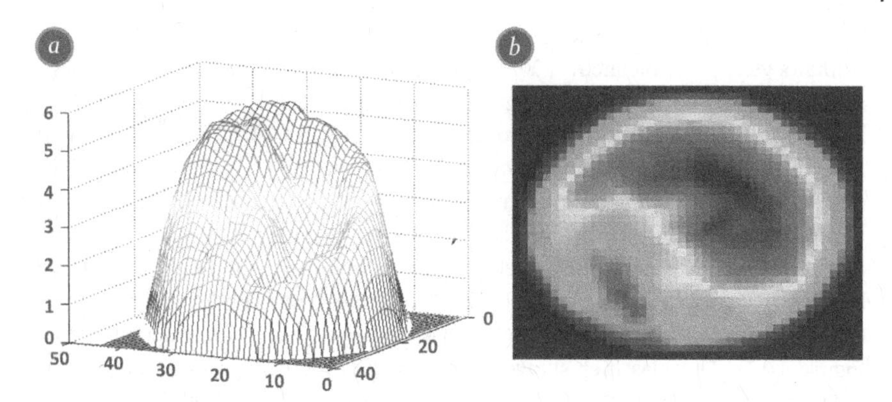

Figure 14.6 An RBC sample that confuses the neural network, which resembles both stomatocyte and spherocyte RBCs. (a) 3D representation, and (b) representation on the x–y plane [5] / SPIE / CC BY 3.0.

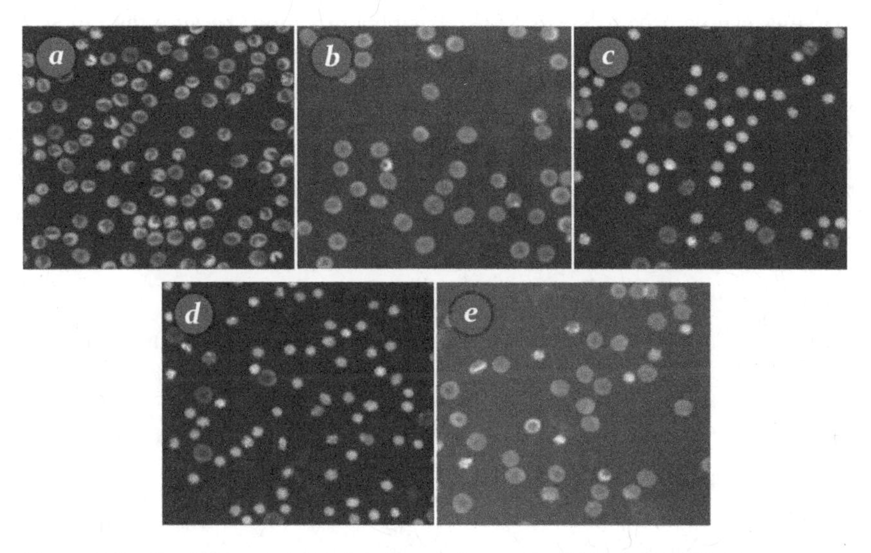

Figure 14.7 Five RBC samples and their counting results. (a) Flat-disk: 0%, Stomatocytes: 76.2% (61/80), Biconcave: 11.25% (9/80), and Echino-spherocyte: 10% (8/80). (b) Flat-disk: 7.40% (2/27), Stomatocytes: 18.51% (5/27), Biconcave: 74.07% (20/27) and Echino-spherocyte: 0%. (c) Flat-disk: 0%, Stomatocytes: 12.24% (6/49), Biconcave: 10.2% (5/49) and Echino-spherocyte: 77.55% (38/49). (d) Flat-disk: 0%, Stomatocytes: 7.94% (5/63), Biconcave: 1.59% (1/63) and Echino-spherocyte: 90.4% (57/63). (e) Flat-disk: 47.22% (17/36), Stomatocytes: 25% (9/36), Biconcave: 19.5.2% (7/36) and Echino-spherocyte: 8.3% (3/36) [5] / SPIE / CC BY 3.0.

14/80), whereas biconcave RBCs were dominant in Figure 14.7b (flat disks, 2/27; stomatocytes, 5/27; biconcave, 20/27; and echino-spherocyte, 0/27), and echino-spherocytes were dominant in Figure 14.7c (flat-disk, 0/49; stomatocytes, 7/49; biconcave, 6/49; and echino-spherocytes, 38/49). Figure 14.7d has the following RBC types: flat-disk, 0/63; stomatocytes, 3/63; biconcave, 1/63; and echino-spherocytes, 59/63. Figure 14.7e shows an RBC image with 40 days of storage time containing many flat-disk RBCs (flat-disk: 16/36; stomatocytes, 7/36; biconcave, 5/36; and echino-spherocytes; 8/36). Each image was first segmented into many RBCs because feature extraction should be applied at the single cell level. The percentages of each RBC type in RBC phase images were then calculated (see Figure 14.7). The classifier showed that stomatocyte RBCs were dominant in the first sample as expected. In contrast, biconcave and flat-disk RBCs were dominant types in the second and fifth figures. Echino-spherocytes were dominant RBCs in the third and fourth figures. Although there was a small error counting non-dominant RBCs, the most important outcome was counting and reporting the dominant RBC type.

According to these findings, our feature set and classifier could automatically count and categorize RBC types in humans using 2D and 3D RBC profiles. The classifier can be used to assess RBC-related abnormalities because the ratios of RBC types are associated with certain diseases. Figure 14.8a, for example, compares the density of the average echino-spherocyte cell thickness to those of other RBCs. Any binary classifier can be employed to separate these two groups. Figure 14.8b shows the scattering of two groups based on their average thickness value and surface area as well as the boundary region of an SVM classifier.

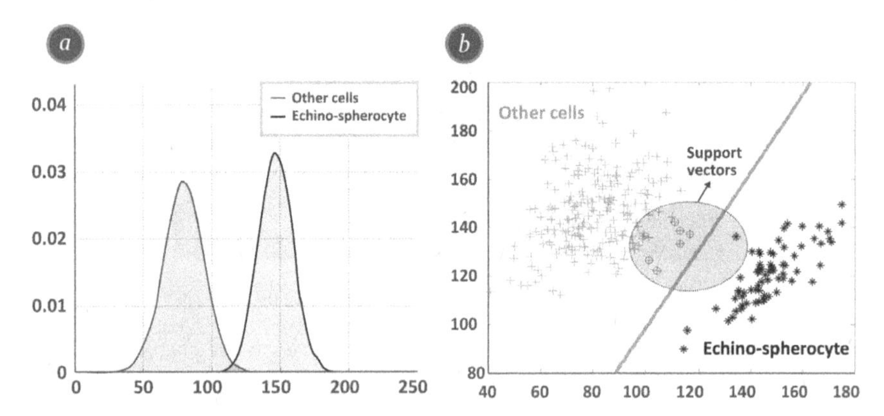

Figure 14.8 (a) The average thickness value distribution for echino-spherocytes and other RBCs. (b) An SVM classifier for separating echino-spherocytes and other cells using the average thickness and total surface area [5] / SPIE / CC BY 3.0.

14.5 Conclusions

We presented the PRNN applied to the 2D and 3D features of RBCs obtained by DHM to classify and count biconcave, stomatocyte, flat-disk, and echino-stomatocyte RBCs. Six 2D features and thirteen 3D features were extracted, and the classification results compared. Our findings showed that 3D features provided more useful information for RBC classification than 2D features. Furthermore, FS showed that the average RBC thickness, top-view SAV ratio, sphericity coefficient, the upper side of the ring divided by lower side of the ring, and RBC perimeter could be used to better classify RBCs into proper categories. The final feature set may help classify and count RBCs, which is important for analyzing RBC abnormalities and identifying shape-related diseases.

References

1. Uzoigwe, C. (2006). The human erythrocyte has developed the biconcave disc shape to optimise the flow properties of the blood in the large vessels. *Med. Hypotheses* 67: 1159–1163.
2. Bessis, M., Weed, R., and Leblond, P. (2012). *Red Cell Shape, Physiology, Pathology Ultrastructure*. Springer.
3. Kim, Y., Shim, H., Kim, K. et al. (2014). Profiling individual human red blood cells using common-path diffraction optical tomography. *Sci. Rep.* 4: 6659.
4. Buttarello, M. and Plebani, M. (2008). Automated blood cell counts. *Am. J. Clin. Pathol.* 130: 104–116.
5. Jaferzadeh, K. and Moon, I. (2016). Human red blood cell recognition enhancement with three-dimensional morphological features obtained by digital holographic imaging. *J. Biomed. Opt.* 21: 126015.
6. Bacus, J. and Weens, J. (1977). An automated method of differential red blood cell classification with application to the diagnostics of anemia. *J. Histochem. Cytochem.* 25: 614–632.
7. Yi, F., Moon, I., and Javidi, B. (2016). Cell morphology-based classification of red blood cells using holographic imaging informatics. *Biomed. Opt. Express* 7: 2385–2399.
8. Yi, F., Moon, I., and Lee, Y.H. (2015). Three-dimensional counting of morphologically normal human red blood cells via digital holographic microscopy. *J. Biomed. Opt.* 20: 016005.
9. Moon, I., Javidi, B., Yi, F. et al. (2012). Automated statistical quantification of three-dimensional morphology and mean corpuscular hemoglobin of multiple red blood cells. *Opt. Express* 20: 10295–10309.
10. David, J. and MacKay, C. (1992). Bayesian interpolation. *Neural Comput.* 4: 415–447.

15

Automated Analysis of Red Blood Cell Storage Lesions

15.1 Introduction

RBCs serve many purposes in the human body. One of the most important RBC functions is to transfer oxygen and carbon dioxide between the lungs and the rest of the body's tissues. Importantly, the shape of an RBC should be optimized for maximal deformation, a maximum surface at a given volume, rapid changes, and cell survival during repeated passages through narrow channels. Surface area is one of the most important properties of RBCs because the exchange of oxygen and carbon dioxide takes place at the cell surface. A larger surface area allows for greater oxygen and carbon dioxide exchange in the lungs and other bodily tissues. These conditions are satisfied by a biconcave disk shape, which is the medical norm for RBCs. The biconcave shape of RBCs has a flexible membrane with a high surface-area-to-volume ratio that facilitates extensive, reversible elastic deformation of the RBC as it repeatedly passes through narrow capillaries. The biconcave shape is thought to be the result of minimizing the free energy of membranes under area and volume constraints because there are no complicated inner structures in RBC. However, during the storage period, RBCs undergo metabolic, biochemical, biomechanical, and molecular changes, which are referred to as storage lesions [1]. Furthermore, many adverse effects related to RBC storage are reported in critically ill patients when the RBC storage period exceeds four weeks [2, 3]. Scientific evidence supports that the RBC structure undergoes essential changes in shape during the storage period, from biconcave, to flat, and finally an echino-spherocyte, which results in the inability of RBCs to carry oxygen [4–7].

In this chapter, we will introduce automated methods to analyze changes in 3D morphology and mean corpuscular hemoglobin (MCH) in RBCs caused by the length of storage for the 3D classification of RBCs with different storage periods using DHM [8, 9]. To analyze the morphological changes in RBCs induced by storage time, we use datasets from blood samples stored for 8, 13, 16, 23, 27, 30, 34, 37, 40, 47, and 57 days. These datasets were divided into 11 classes of RBCs stored for

Artificial Intelligence in Digital Holographic Imaging: Technical Basis and Biomedical Applications, First Edition. Inkyu Moon.

11 different periods. These 11 classes have more than 3300 blood cells with more than 300 blood cells per storage class.

Using RBCs donated by healthy persons, DHM reconstructed several RBC phase images from each class of blood sample. To automatically calculate characteristic features such as the averaged phase value, projected surface area (PSA), volume, surface-to-volume ratio (SVR), MCH, MCH surface density (MCHSD), sphericity index, morphological functionality factor, and sphericity coefficient of RBCs, the image-segmentation method explained in Chapter 12 was applied to remove the unnecessary background in the RBC phase image. All RBCs in the phase image were extracted to obtain characteristic RBC properties. More than 300 RBCs were extracted from segmented phase images for each class of blood sample. The sample size was large enough to derive statistical distributions of characteristic RBC features at a given storage period. Our main goal was to quantitatively analyze the relationship between RBC properties and their aging as well as to automatically analyze the morphological and chemical parameters of RBCs with different ages at the single-cell level. We also performed a correlation analysis between MCH or MCH concentration and the morphological parameters of RBCs at different storages. Our findings demonstrated that in storage lesions, the RBC structure underwent significant changes in terms of shape, like from a biconcave disc to an echino-spherocyte. Interestingly, our experimental results showed that the surface area values of normal RBCs obtained by DHM agreed with values reported by other methods [10–12].

15.2 Quantitative Analysis of RBC 3D Morphological Changes

To automatically analyze 3D morphological changes in RBCs induced by storage time, our off-axis DHM reconstructed several phase images from each class of blood sample. Individual RBCs were then extracted from the phase image using the segmentation method described in Chapter 12. For comparing RBC phase images with different storage times, we set the phase value of the phase image background to be 0°. After segmentation, the characteristic properties of RBCs were computed for all RBCs extracted from phase images. The following provides a short description of features related to the morphological properties of RBCs. For details of each feature, please refer to [8, 9, 13, 14]. The average phase value Φ induced by the whole RBC and the PSA are defined as

$$\Phi = \frac{1}{N}\sum_{i=1}^{N}\varphi_i, PSA = Np^2$$

where N is the total number of pixels within a RBC, p denotes the pixel size in the phase image, and φ_i is the phase value of each pixel within the RBC. The RBC volume or the size of RBC is a good indicator of RBC functionality. The volume of a single RBC or the corpuscular volume is denoted as

$$V \simeq \frac{p^2 \lambda \sum_i^N \varphi_i}{2\pi(n_{rbc} - n_m)}, \tag{15.1}$$

where n_{rbc} is the refractive index of RBCs, n_m is the index of refraction of medium, and λ is the wavelength of the light source [15]. When a population of RBCs is considered, Eq. (15.1) allows us to derive the mean corpuscular volume (MCV) [14]. Another important characteristic property is the dry mass that measures the weight of the cell after dehydration. The dry mass is a reliable biomass. It is widely used to compare cells since it is free from disturbance of water that exists in living beings [15]. The dry mass of a cell is related to the phase value and can be defined by

$$DryMass(DM) = \frac{10\lambda}{2\pi\alpha} \int \varphi ds = \frac{10\lambda}{2\pi\alpha} \Phi S, \tag{15.2}$$

where Φ is the average phase value induced by the whole cell and α is a constant known as the specific refraction increment (in m^3/kg or dl/g), which is related to the protein concentration [15, 16]. As far as RBCs are concerned, $\alpha = 0.00196$ dl/g is the hemoglobin refraction increment between 663 nm and 682 nm [15]. When an RBC population is considered, Eq. (15.2) provides the MCH. The MCH is an important parameter to investigate changes in RBC hemoglobin content. The MCHSD, which can show hemoglobin concentration, is defined as the ratio between the MCH and projected surface area S:

$$MCHSD = \frac{MCH}{S}.$$

Converting these phase images to thickness images allows an easier morphological computation. Accordingly, the thickness value $h(i, j)$ for each pixel of (i, j) with phase value $\varphi(i, j)$ in a phase image can be expressed as:

$$h(i,j) = \frac{\varphi(i,j) \times \lambda}{2\pi(n_{rbc} - n_m)}.$$

The sphericity coefficient, k, is the ratio of the RBC thickness at the center d_c to the thickness at half of its radius d_r (dimple area):

$$k = \frac{d_c}{d_r}. \tag{15.3}$$

The sphericity coefficient k can identify three types of RBCs. A value of k less than unity specifies a biconcave-shaped RBC. A value of k around unity denotes a flat disk-shaped RBC and a value of k greater than unity indicates an echinospherocyte.

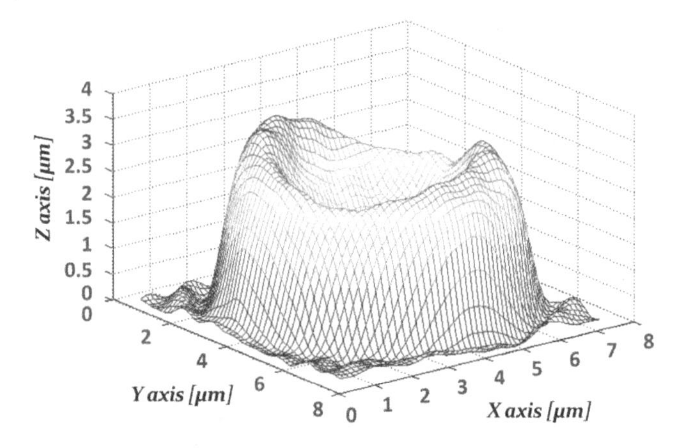

Figure 15.1 3D view of an RBC obtained by the DHM method [9].

The surface area of the top view of an RBC (see Figure 15.1) is the area of a surface or collection of surfaces that bind to a solid. There are several formulas to compute the surface areas of regular shapes. For irregular shapes, like RBCs, many methods are suggested [17]. In most cases, covering the entire surface area of an RBC with small triangles can provide high accuracy. In this study, the RBC surface area was calculated using Eq. (15.4), which was shown by our experimental results to be sufficiently accurate enough for calculating the RBC surface area:

$$Surface\ Area = \iint_S \sqrt{\left(\frac{\delta z}{\delta x}\right)^2 + \left(\frac{\delta z}{\delta y}\right)^2 + 1dA} \tag{15.4}$$

We assumed that the surface area had the form of a function $z = f(x, y)$ and took the integral of the entire surface. The total surface area (SA) of an RBC in DHM consists of the surface area of the top view added to the *PSA* of the top view:

$$SA = PSA + TVSA, \tag{15.5}$$

where *PSA* is the projected surface area and *TVSA* is the surface area of the top view. The RBC radius r can be estimated by considering the radius of a circle with the *PSA* of an RBC:

$$r \simeq \sqrt{\frac{PSA}{\pi}}.$$

We can also estimate the morphological functionality factor of RBCs, f, using the calculated surface area for RBC, which is the ratio of the RBC surface area to the surface area of an echino-spherocyte of the same volume [18]:

$$f = \frac{SA}{S_s} = \frac{SA}{4\pi r^2},$$

where SA can be determined by Eq. (15.5) and S_s is the surface area of the sphere with radius r. An f value around unity characterizes an echino-spherocyte. Moreover, we can compute the total surface area of all RBCs, S_t, which affects the oxygen capacity of blood. S_t is equal to the product of the number of RBCs in the blood, M, and the average surface area SA_m of RBCs [18]:

$$S_t = M \times SA_m. \tag{15.6}$$

Equation (15.5) can be applied to calculate the SVR, which is a critical parameter for an RBC since the greater the SVR, the more material a cell can exchange with its surroundings.

Another important parameter related to the shape of RBCs is the sphericity index, which indicates the extent to which the cell shape approaches a sphere. The sphericity index of each cell can determine the cell's degree of tolerance to deformation, for example, in passing through a small cylindrical channel. This parameter can also discriminate between healthy and pathological cells. For example, it can be used to identify pathological cells in some hereditary diseases such as spherocytosis. Thus, the sphericity index for determining RBC deformability can be defined as

$$SP = \frac{4\pi V^{\left(\frac{2}{3}\right)}}{\left(\frac{4\pi}{3}\right)^{\frac{2}{3}} SA}. \tag{15.7}$$

The mean value of the sphericity index for normal RBCs is reported as 0.79 ± 0.026 at room temperature. According to Eq. (15.7), the sphericity index has a maximal value of unity, which corresponds to a spherical cell.

15.3 Experimental Results and Discussion

To analyze morphological changes in RBCs caused by the length of storage time, we prepared 11 classes of blood samples stored for 8, 13, 16, 23, 27, 30, 34, 37, 40, 47, and 57 days. The off-axis DHM reconstructed several RBC phase images for each class of blood samples. After RBC phase images were obtained, they were segmented to remove unnecessary background in phase images. Figure 15.2a–f shows six original RBC phase images and Figure 15.2g–l their corresponding segmentation results. Once the image is segmented, each cell in the image is extracted individually and the corresponding properties are computed automatically.

After the segmentation and extraction of RBCs from phase images, characteristic properties, including the PSA, average phase value, MCV, MCH, SA, SVR, morphological functionality factor, cell diameter, sphericity coefficient, sphericity

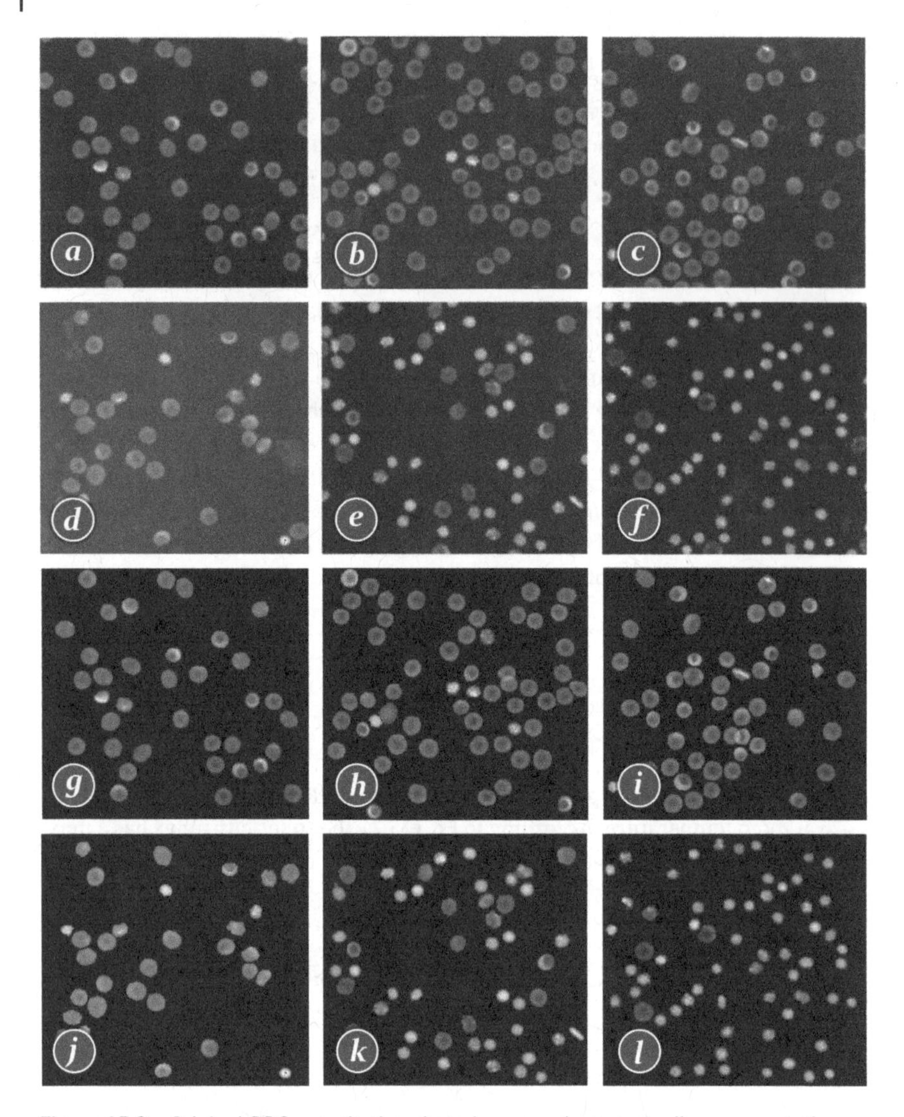

Figure 15.2 Original RBC quantitative phase image and corresponding segmentation results. (a), (b), (c), (d), (e), and (f) are RBC's stored for 8, 16, 30, 34, 47, and 57 days, respectively, while (g), (h), (i), (j), (k), and (l) are the corresponding segmented images of (a)–(f)[[8] / with permission of Optical Publishing Group.

index, and oxygen capacity for all samples were calculated as well as their standard deviations (STDs). To reduce noise and increase accuracy in Eq. (15.3), d_c and four adjacent values were averaged (d_r was the average of some values selected from different portions of the dimple). Table 15.1 shows the calculated mean and STD values for all characteristic properties of RBCs with different ages.

Table 15.1 Characteristic properties of RBCs across storage groups, estimated with the proposed method.

Property		Storage time (days)										
		8	13	16	23	27	30	34	37	40	47	57
Mean phase value (degree)	Mean	74	74	81	76	77	76	78	87	96	112	136
	STD	12	15	16	16	15	24	21	26	27	28	28
MCH (pg)	Mean	32.2	32.3	35.7	32.8	34.8	29.2	30.9	32.7	33.6	34.6	32.5
	STD	5.0	5.5	7.7	7.3	8.6	7.6	6.1	5.7	5.0	6.8	5.1
MCH surface density (pg/μm^2)	Mean	0.70	0.70	0.76	0.71	0.72	0.71	0.73	0.82	0.90	1.05	1.28
	STD	0.11	0.13	0.14	0.15	0.14	0.23	0.20	0.24	0.25	0.26	0.26
Projected surface area (PSA) (μm^2)	Mean	45	46	47	45	47	42	43	41	39	34	26
	STD	5	6	7	7	10	8	9	8	9	9	6
Top-view surface area (TVSA) (μm^2)	Mean	81	81	89	83	84	85	84	81	81	79	81
	STD	8	8	9	8	8	9	7	6	8	8	10
Surface area (μm^2)	Mean	126	127	136	128	131	127	127	122	120	113	107
	STD	11	14	16	16	17	16	11	12	13	14	15
Volume (μm^3)	Mean	91	92	102	94	98	88	86	93	98	98	94
	STD	9	12	14	14	15	20	12	11	12	13	11
SVR (per μm)	Mean	1.38	1.38	1.33	1.36	1.34	1.44	1.48	1.31	1.22	1.15	1.13
	STD	0.19	0.22	0.19	0.20	0.21	0.26	0.21	0.22	0.20	0.20	0.16
Cell diameter (μm)	Mean	7.76	7.77	7.76	7.63	7.70	7.45	7.65	7.52	7.25	6.73	6.04
	STD	0.49	0.55	0.53	0.58	0.58	0.69	0.75	0.75	0.86	0.89	0.68
k factor	Mean	0.88	0.78	0.66	0.84	0.70	0.70	0.93	0.93	1.03	1.20	1.35
	STD	0.18	0.19	0.21	0.22	0.23	0.37	0.30	0.56	0.34	0.36	0.41
f factor	Mean	0.66	0.69	0.72	0.70	0.70	0.73	0.69	0.69	0.73	0.79	0.93
	STD	0.07	0.09	0.10	0.10	0.10	0.13	0.17	0.14	0.15	0.16	0.16
Sphericity index	Mean	0.78	0.78	0.78	0.78	0.79	0.75	0.74	0.81	0.86	0.91	0.94
	STD	0.07	0.08	0.07	0.08	0.07	0.08	0.08	0.09	0.09	0.10	0.09

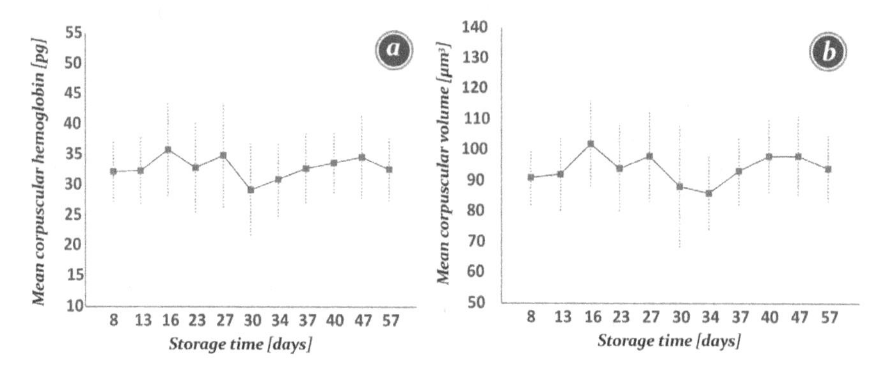

Figure 15.3 Relationship between the MCH or MCV of RBCs and storage durations. (a) Relationship between the MCH and storage time. (b) Relationship between the MCV of RBCs and storage time. square = mean; bar = standard deviation.

Figure 15.3 illustrates the relationship between RBC MCH or MCV and the storage time of RBC samples. Results revealed that the trends for RBC MCH and MCV values were almost identical with increasing days of storage. Both MCH and MCV values seemed to swing around their respective mean value. Figure 15.3a shows the relationship between MCH and storage time; although the MCH fluctuated around a value of 32 pg, the MCH was nearly stable, even with increasing storage time. Figure 15.3b shows the relationship between MCV and storage time. It was noted that the MCV hovered around of 94 μm^3, even with increasing storage day. The MCV was not significantly affected by the storage time. As a result, we can conclude that the hemoglobin content within RBCs does not change as a function of storage time. This is an expected outcome because MCV and n_{rbc} do not vary over storage time.

Figure 15.4 shows the relationship between MCHSD and storage time. Although the MCH in Figure 15.3a showed little fluctuation over the storage time, the MCHSD tended to increase as shown in Figure 15.4. This could be explained by the fact that the MCH remained constant while the average projected surface area of RBCs tended to decrease with increasing storage periods. We also noted that the MCHSD was nearly constant when RBCs were stored for less than 34 days. When the storage period was longer than 34 days, there was a noticeable increase in MCHSD.

Figure 15.5 shows is a graphical representation of the central tendency of RBC 3D geometric characteristics as a function of age. Figures 15.6 and 15.7 show RBC samples at different storage days and their respective histograms of the surface area, which shows the variation in size and shape of stored RBCs. As shown in Figures 15.6 and 15.7, the leftward shift of the normal distribution for the surface area indicates that the RBC surface area decreases as storage increases. Considering

Figure 15.4 Relationship between the MCHSD of RBCs and storage duration. square = mean of the dry mass surface density; bar = standard deviation.

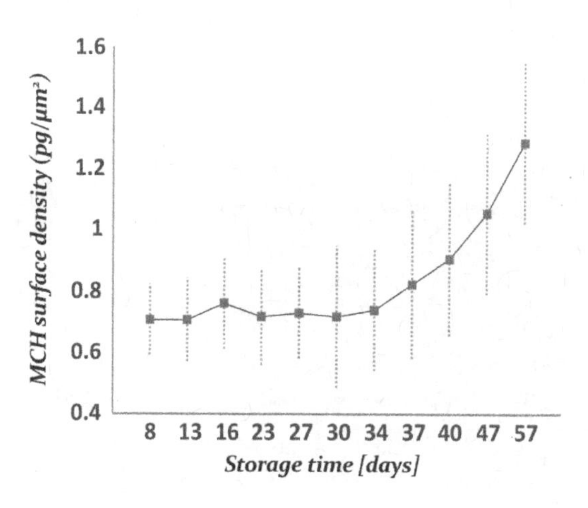

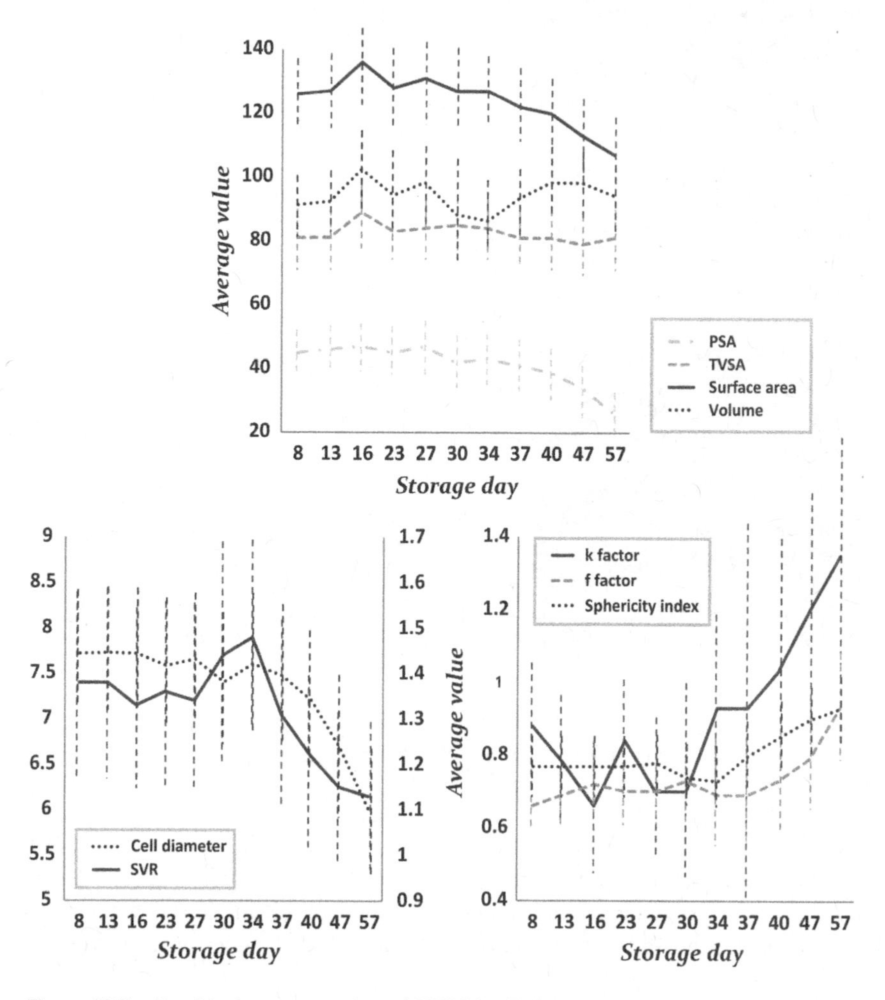

Figure 15.5 Graphical representation of RBC' PSA, TVSA, surface area, volume, cell diameter, SVR, *k* factor, *f* factor, and sphericity index trends across storage duration. bar = standard deviation [9] / with permission of SPIE.

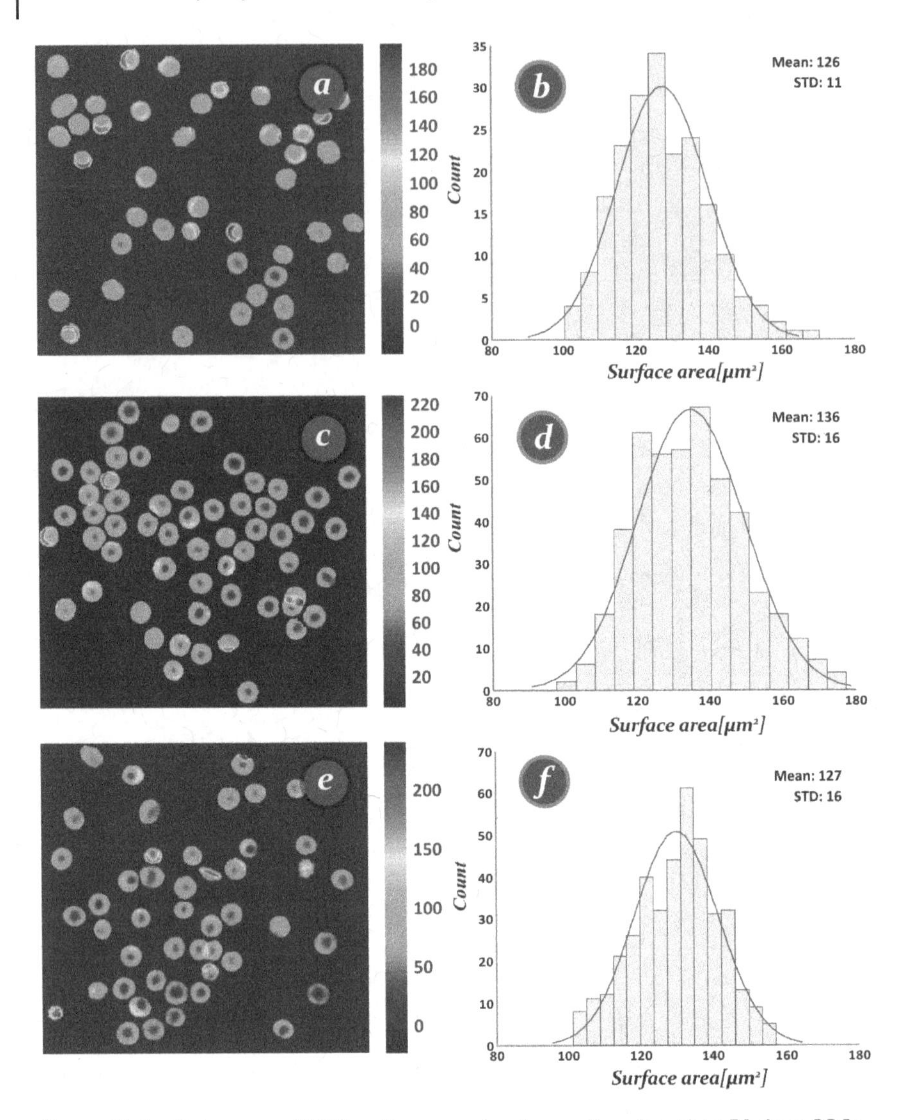

Figure 15.6 Histograms of RBC' surface areas by storage time, less than 30 days; RBCs stored for (a,b) 8, (c,d) 16, and (e,f) 23 days [9] / with permission of SPIE.

Table 15.1 and Figures 15.5, 15.6, and 15.7, the SA fluctuated during the first four weeks of the storage period, then began to decline. The same trend was also observed for changes in cell diameter while there were only slight fluctuations in volume. As the diameter decreased with an almost fixed volume, the thickness value increased.

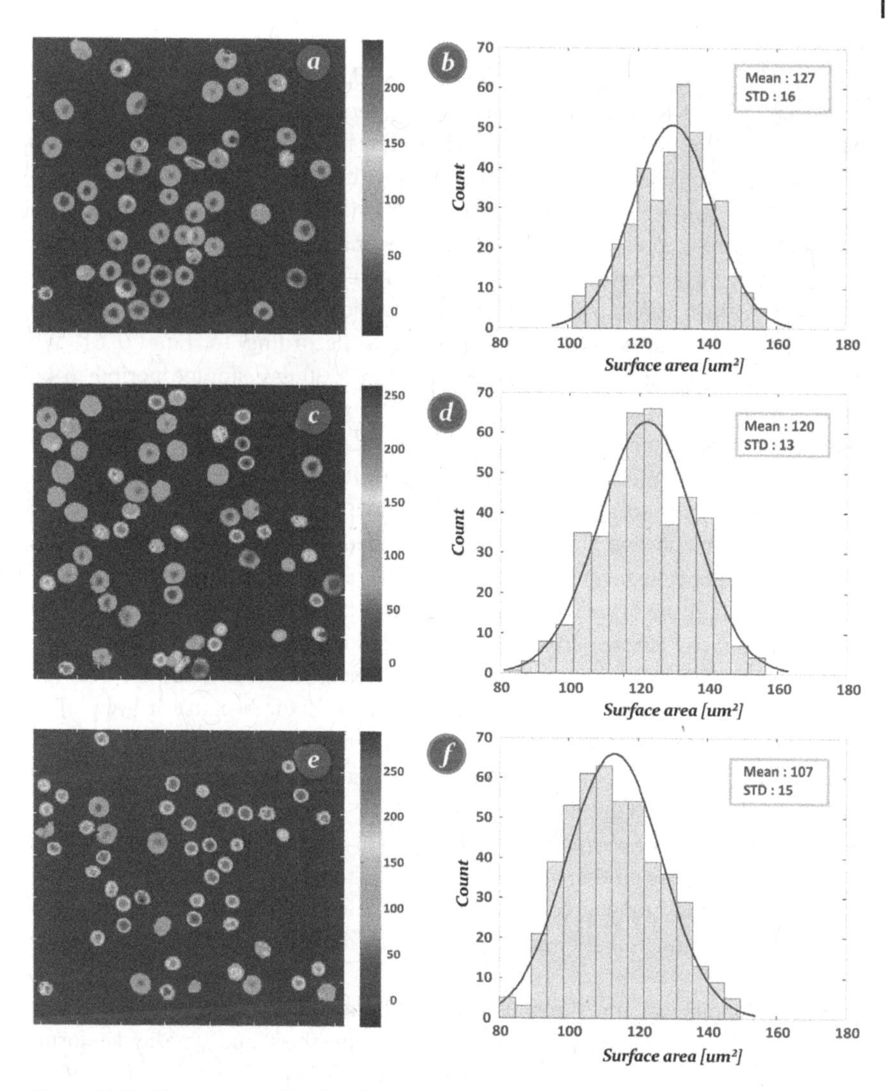

Figure 15.7 Histograms of RBC surface areas by storage time, at least 30 days. RBCs stored for (a,b) 30, (c,d) 40, and (e,f) 57 days [9] / with permission of SPIE.

The cell diameter for healthy RBCs was reported to be 7.7 ± 0.5 μm [18]. Although cell diameters for the first five weeks were within this reported range, they began decreasing when storage exceeded five weeks. It was interesting to note that the RBC SA values within the first four weeks of storage time (see Table 15.1) were close to those obtained by other methods [10–12]. We also noted that the top-view surface area neither decreased nor increased significantly and remained within its STD during the entire storage period. The PSA dropped as storage increased. The SVR decreased as the volume remained relatively constant and surface area dropped. Most biological cells can maximize the SRV to preserve their biological processes. The SVR in RBCs within the first four weeks of storage was approximately 1.36. As mentioned previously, aside from the presence of hemoglobin, the cell surface area or SVR is the most important factor in the oxygen-carrying processes of erythrocytes. The effect of a smaller surface area is clearly seen in terms of oxygen capacity, which can be estimated using the averaged surface area values across storage, according to Eq. (15.6). We observed the oxygen capacity for RBCs within a 30-day storage period to be approximately 15% larger than that for RBCs with a storage period greater than six weeks, assuming an equal number of RBCs for each group.

The k factor (or sphericity coefficient) began with a value less than one. It then increased gradually to a value greater than one. This indicated that the RBCs were initially biconcave and as storage exceeded five weeks, they became flat disks and finally transformed into echino-spherocytes. In contrast, the functionality factor f had the same trend as the k factor in that f approached unity. This indicated that the surface area of RBCs was almost equal to a sphere with the same radius. The sphericity index, which determines deformability, had the same trend as the functionality factor and the k factor because it grew after the first five weeks of storage. Our finding regarding the sphericity index was that over storage, the sphericity index grew, implying less tolerance during the passage through small channels.

Modifications in the morphological functionality factor, along with the sphericity coefficient, consistently showed that RBCs changed from biconcave into an echino-spherocyte under a constant volume. These substantial changes caused a decrease in the functionality with respect to materials exchanged between tissues and lungs.

The morphological functionality factor, sphericity coefficient, and sphericity index are very important because they can determine the shape, type, and deformability of RBCs as they age. An increase in the sphericity coefficient corresponds to a decrease in the RBC surface area under its given volume. Geometric modifications related to decreased RBC surface area can also decrease their functionality with respect to the supply of oxygen to the tissues and organs. Variations in k factor during the first five weeks of storage might have been due to inaccuracies in determining correct values of r_c and r_d, despite averaging some pixels or the presence of stomatocyte RBCs. Many studies have established that normal or biconcave RBCs will undergo transformation in their shapes upon variation in some of their chemical components. A morphological functionality factor close to unity confirms that

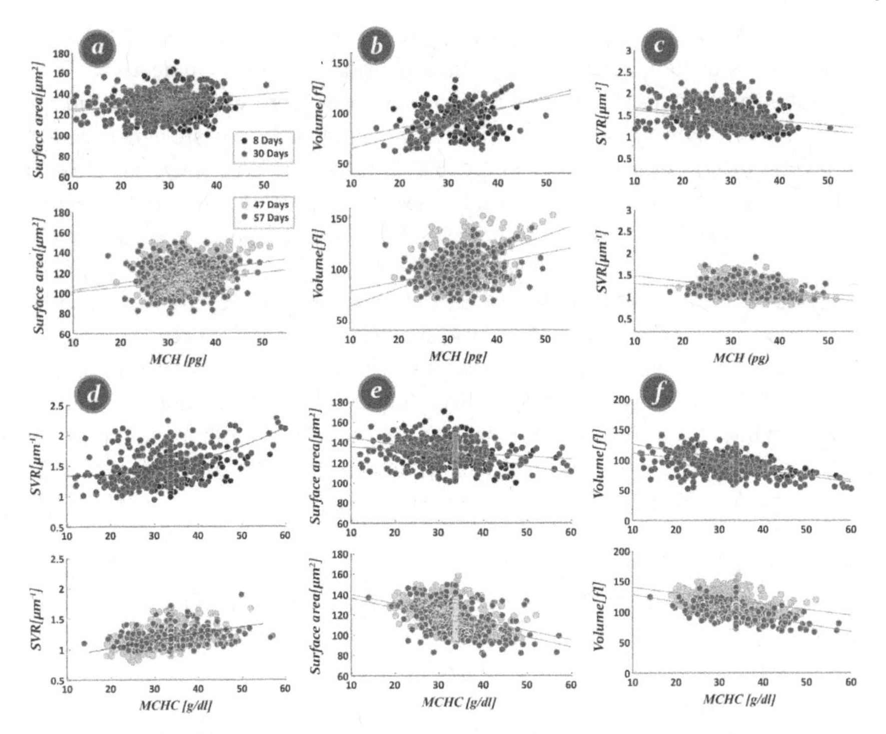

Figure 15.8 Correlation between chemical-morphological and morphological-morphological properties. Gray lines show a linear relationship. (a) Surface area and MCH, (b) volume and MCH, (c) SVR and MCH, (d) SVR and MCHC, (e) surface area and MCHC, (f) volume and MCHC.

the RBC surface area is nearly equal to the surface area of a sphere with the same radius. Our findings about k and f are consistent, showing that RBCs become flat disks and finally transform into echino-spherocytes. The sphericity index that indicates the degree of cell tolerance for deformation increased, implying that the deformability required for passing through a narrow cylindrical channel was decreasing.

Since chemical and morphological parameters of RBCs are simultaneously measured at the single-cell level using DHM, we can perform correlation analyses for these parameters. We divided our samples into two groups according to their storage duration for clear visualization. The first group had DHM data from 8 and 30 days of storage, and the second group had DHM data from 47 and 57 days of storage. Figure 15.8 shows the correlation analysis results between the chemical and morphological parameters of the two groups. Correlations between RBC parameters allow us to understand the cellular physiology of RBCs in detail.

As shown in Figure 15.8, there was a positive correlation between MCH and the surface area (Figure 15.8a) as well as a significantly strong positive correlation between MCH and volume (Figure 15.8b), consistent with previous reports [19, 20]. The correlation coefficient fluctuated slightly, and the slope of the regression line changed in the case of storage lesions. It was interesting to observe that RBCs with higher MCHC had lower surface area and volume values (Figures 15.8e,f) in all cases. We noted that the correlation coefficient between the surface area and MCHC value was smaller than that between the volume and MCHC in the same group. Our analysis also showed that there was no significant correlation between functionality factor-MCH and functionality factor-MCHC while RBCs were aging. Furthermore, the sphericity index followed the same trend as the functionality factor. Interestingly, RBCs with higher MCH had lower SVR values in all storage groups (Figure 15.8c). However, higher MCHC RBCs had higher SVR values (Figure 15.8d). We noted that in all cases, there were linear relationships between two variables according to the computed root mean squared error of residuals (RMSE) values. However, the RMSE of polynomial line of degree 2 for SVR-MCHC was significantly lower than that of the linear relationship.

We also computed correlation coefficients between geometrical parameters of RBCs to help us understand storage lesions better. Table 15.2 shows correlation results between geometrical parameters of RBCs. The correlation between surface area and diameter was more significant than that between volume and diameter. In addition, we found that the surface area and volume had a significant positive correlation, indicating that RBCs with larger volumes had larger surface areas. Interestingly, when the storage duration was 57 days, the correlation coefficient between volume and surface area was larger than those on other days.

The biconcave shape provides RBCs with sufficient surface area to exchange metabolic products across the membrane and cytoskeleton, resulting in their slight deformation. RBCs must be deformable to moderately stretch as they undergo distortions under mechanical stress during their circulation through small channels. However, echino-spherocytes with smaller surface area have less deformability. The viscosity of spherocytic RBCs is also higher than that of RBCs with a biconcave shape, which leads to a high resistance during transit through narrow channels. Findings of this study showed that when the storage time of RBCs exceeded five weeks, the dominant morphology was echino-spherocyte, they lost surface area, and became roughly 20% less deformable.

Table 15.2 Correlation results between RBC geometrical parameters.

	Storage time (days)										
Properties	**8**	**13**	**16**	**23**	**27**	**30**	**34**	**37**	**40**	**47**	**57**
Surface area vs. diameter	0.7109	0.6722	0.5783	0.6145	0.6228	0.5105	0.6974	0.6513	0.7086	0.7162	0.5780
Volume vs. diameter	0.2582	0.3331	0.3518	0.283	0.4336	0.1413	0.272	0.1476	0.1679	0.2726	0.2331
Volume vs. surface area	0.4127	0.4617	0.5589	0.5433	0.5882	0.4701	0.6064	0.4071	0.4874	0.5449	0.6359

15.4 Conclusions

We quantitatively analyzed the 3D morphology and MCH of RBCs with different storage durations. Statistical analyses showed that 3D morphological changes in RBCs were induced by the length of storage, while the hemoglobin content within RBCs did not change substantially. In addition, we found that 34 days of storage might be a threshold beyond which the RBC morphology begins to change significantly and possibly alters their functionality.

Furthermore, 3D geometric changes of RBCs during storage were analyzed using the surface area values of RBCs. Investigations of the sphericity coefficient and functionality factor along with the sphericity index demonstrated that RBCs could transform from biconcave to spherical when the storage time exceeded five weeks. Our experimental results showed that the transition from biconcave to echino-spherocytes was accompanied by a significant loss of surface area and an increase in the sphericity index. We performed correlation analyses between morphological and chemical properties. Results showed that surface area had a significant negative correlation with MCHC values. Interestingly, sphericity index did not correlate with either MCH or MCHC values, similar to the functionality factor. Furthermore, the correlation between the surface area and diameter was stronger than the relationship between volume and diameter. Automated analysis of the relationship between characteristic RBC properties and storage time might aid the quality assessment of stored RBCs.

References

1 Card, R. (1988). Red cell membrane changes during storage. *Transfus. Med. Rev.* 2 (1): 40–47.

2 Robinson, S., Janssen, C., Fretz, E. et al. (2010). Red blood cell storage duration and mortality in patients undergoing percutaneous coronary intervention. *Am. Heart J.* 159 (5): 876–881.

3 Alfano, K. and Tarasev, M. (2008). Investigating direct non-age metrics of stored blood quality loss. *Internet J. Med. Technol.* 5 (1).

4 Laurie, J., Wyncoll, D., and Harrison, C. (2010). New versus old blood - the debate continues. *Crit. Care* 14 (2): 130.

5 Alessandro, A., Liumbruno, G., Grazzini, G., and Zolla, L. (2010). Red blood cell storage: the story so far. *Blood Transfus.* 8 (2): 82–88.

6 Aubron, C., Nichol, A., Cooper, D., and Bellomo, R. (2013). Age of red blood cells and transfusion in critically ill patients. *Ann. Intensive Care* 3 (2): 1–11.

7 Bosman, G., Werre, J., Willekens, F., and Novotny, V. (2008). Erythrocyte ageing in vivo and in vitro: structural aspects and implications for transfusion. *Transfus. Med.* 18 (6): 335–347.
8 Moon, I., Yi, F., Lee, Y. et al. (2013). Automated quantitative analysis of 3D morphology and mean corpuscular hemoglobin in human red blood cells stored in different periods. *Opt. Express* 21: 30947–30957.
9 Jaferzadeh, K. and Moon, I. (2015). Quantitative investigation of red blood cell three-dimensional geometric and chemical changes in the storage lesion using digital holographic microscopy. *J. Biomed. Opt.* 20.
10 Tomaiuolo, G., Rossi, D., Caserta, S. et al. (2012). Comparison of two flow-based imaging methods to measure individual red blood cell area and volume. *Cytometry A* 81 (12): 1040–1047.
11 Evans, E. and Fung, Y. (1972). Improved measurements of the erythrocyte geometry. *Microvasc. Res.* 4 (4): 335–347.
12 Waugh, R., Narla, M., Jackson, C. et al. (1992). Rheologic properties of senescent erythrocytes: loss of surface area and volume with red blood cell age. *Blood* 79 (5): 1351–1358.
13 Moon, I., Javidi, B., Yi, F. et al. (2012). Automated statistical quantification of three-dimensional morphology and mean corpuscular hemoglobin of multiple red blood cells. *Opt. Express* 20: 10295–10309.
14 Rappaz, B., Barbul, A., Emery, Y. et al. (2008). Comparative study of human erythrocytes by digital holographic microscopy, confocal microscopy and impedance volume analyzer. *Cytometry A* 73: 895–903.
15 Rappaz, B., Cano, E., Colomb, T. et al. (2009). Noninvasive characterization of the fission yeast cell cycle by monitoring dry mass with digital holographic microscopy. *J. Biomed. Opt.* 14: 034049.
16 Barer, R. (1952). Interference microscopy and mass determination. *Nature* 169: 366–367.
17 Cesari, L. (1956). *Surface Area*. New Jersey: Princeton University Press.
18 Tishko, T., Dmitry, T., and Vladimir, T. (2011). *Holographic Microscopy of Phase Microscopic Objects*. World Scientific.
19 Long, X., Fang, J., Yao, L. et al. (2014). Correlation analysis between mean corpuscular hemoglobin and mean corpuscular volume for thalassemia screening in large population. *Am. J. Anal. Chem.* 5: 901–907.
20 Rao, G., Morghom, L., and Mansori, S. (1985). Negative correlation between erythrocyte count and mean corpuscular volume or mean corpuscular haemoglobin in diabetic and non-diabetic subjects. *Horm. Metab. Res.* 17: 540–541.

16

Automated Red Blood Cell Classification with Deep Learning

16.1 Introduction

RBC transfusions are a life-saving clinical procedure for patients with severe bleeding before or during surgery. For people who need RBC transfusions, the blood bank collects blood, isolates RBCs, and stores them at 4 °C for 42 to 49 days, depending on the additive used to improve oxygen saturation and product quality. However, stored RBCs are continuously degraded over time, resulting in structural and biochemical changes known as RBC storage lesion, which may cause problems in transfused patients due to altered RBC's functionality [1–3]. Multiple studies have shown that RBCs undergo crucial changes in shape from biconcave disks into a spherocyte during storage to maintain optimal functionality, which may lead to an inability to transfer oxygen [4–6].

Therefore, rapid automated screening of RBC storage lesions could be beneficial for safe transfusions. However, the traditional intensity-based microscopy does not provide robust quantitative information about the morphological properties of RBCs, which may affect cell investigation. Furthermore, staining contrast agents are needed to visualize cells, which can destroy the original shape or structure of RBCs. In Chapter 15, we introduced automated methods to investigate 3D morphological alterations in RBC storage lesions using phase images obtained with label-free DHM. However, the presented methods require multiple image processing algorithms to extract a single RBC, compute important morphological parameters at the single cell level, and so on, which creates a high computation burden.

In this chapter, we will introduce new deep learning–based methods for efficient RBC segmentation and classification to reduce the computational burden while achieving a high classification accuracy [7]. We will also show that the presented deep-learning models can classify RBCs stored for different durations, identify dominant shapes in each storage group, and evaluate storage lesions in RBCs for safe transfusions.

Artificial Intelligence in Digital Holographic Imaging: Technical Basis and Biomedical Applications, First Edition. Inkyu Moon.

Recently, deep-learning technology was successfully used to analyze natural and medical images [8, 9]. CNNs are widely used in computer vision and visual recognition problems with remarkable results [10–21]. Long et al. proposed FCNs for semantic segmentation [10] with good performance in benchmark tests. These methods have been applied to RBC segmentation with good performance outcomes [11]. U-Net was developed based on FCNs and applied to medical images using skip connections between an encoder and a decoder [12].

Some challenges in RBC segmentation include independent boundary detection and the separation of overlapping RBCs. Conventional CNN-based segmentation generally produces blurry outputs since the optimization function aims to minimize the Euclidean distance between the actual value and predicted results during training. Therefore, standard models with traditional pixel-based loss function (e.g. mean square error (MSE)) struggle to segment overlapped RBCs precisely. For example, several overlapped RBCs can be detected as one cell. To overcome this, we will use generative adversarial networks (GANs) with a generator and a discriminator. In particular, conditional GANs and Pix2Pix have demonstrated that adversarial loss could help us obtain sharp outcomes [13–15]. The structure of adversarial loss is essential for detecting boundaries of individual RBCs among overlapping RBCs. Many recent studies have shown promising results using GAN models in medical-image segmentation [16–21].

We found that the GAN model could be used to simultaneously segment and classify RBC types, including overlapping RBCs in phase images obtained with DHM. It can also identify morphological changes that occur in RBCs during storage. Accordingly, we could extract and classify RBCs stored for different durations using GAN-based deep-learning models to accurately assess RBC storage lesions. We will also present a method to generate additional markers to separate overlapping RBCs. Separation of connected RBCs is very important for automated RBC analysis.

We will provide a new approach that combines GAN [22] and a watershed-segmentation algorithm to effectively segment RBCs according to their shape as well as separate overlapping RBCs (see Figure 16.1). Holograms of RBCs are recorded using DHM, and RBC phase images are reconstructed using a numerical reconstruction algorithm (see Chapter 5). We manually segmented RBCs from phase images to obtain RBC masks with multiple labels. We then applied a distance-transform algorithm to RBC masks to determine RBC markers. RBC mask and marker images were used as true labels to train our deep-learning models (see Figure 16.1). Our model received RBC phase images and generated multi-class segmentation and binary marker maps. Finally, a marker-controlled watershed

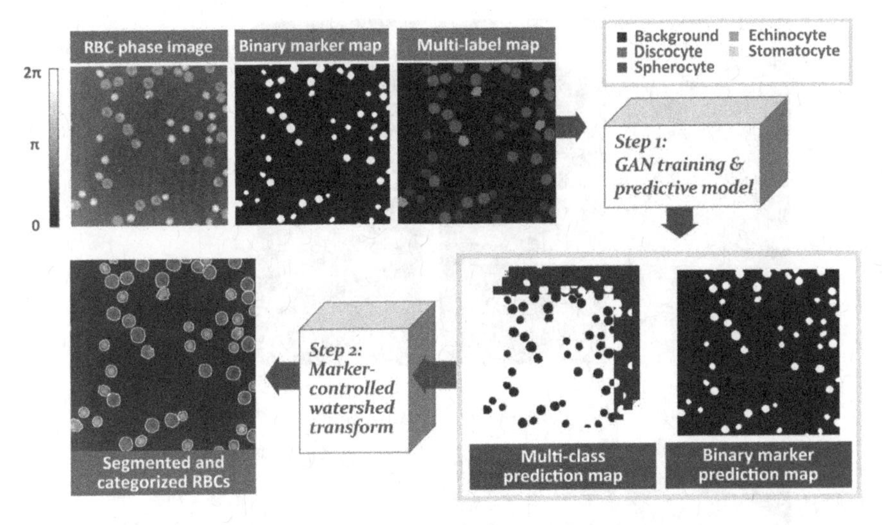

Figure 16.1 Presented method for segmentation and classification of RBCs, including overlapped RBCs. RBC phase images are used as input images. Multi-label maps (background, discocytes, spherocytes, echinocytes, and stomatocytes) and binary marker maps are used as ground truth to train the proposed deep-learning method. GAN provides two outputs of multi-class labels for semantic segmentation and binary markers. These two outputs are used in marker-controlled watershed transformation to find individual RBCs. These multi-labeled individualized RBCs are then visualized.

algorithm was employed to efficiently separate overlapping RBCs using the predicted multi-class map and binary marker map as inputs.

Furthermore, we compared our method with recent deep learning–based segmentation algorithms. Performance comparisons for the segmentation of different RBC shapes and overlapping RBCs were performed with the aggregated Jaccard index (AJI) [23] and Dice coefficient [24]. Experimental results showed that our method outperformed other deep-learning models in terms of semantic segmenting and separation of overlapping RBCs.

Our models were tested to classify different RBC shapes stored for different lengths of time (days). Ground-truth images were manually annotated and classified by an expert biologist. Four typical RBC shapes were annotated with different values (0: background in black, 1: discocytes in red, 2: spherocytes in blue, 3: echinocytes in green, and 4: stomatocytes in yellow, see Figure 16.2). According to Bessis' nomenclature [25] and previous studies [26, 27], there are multiple RBC types. Stomatocytes are divided into four subclasses: I, II, III, and IV. Echinocytes includes subclasses of I, II, III, and IV. Echinocyte IV is also defined as an echino-spherocyte. In this study, echino-spherocytes were grouped and annotated as echinocytes [25–27].

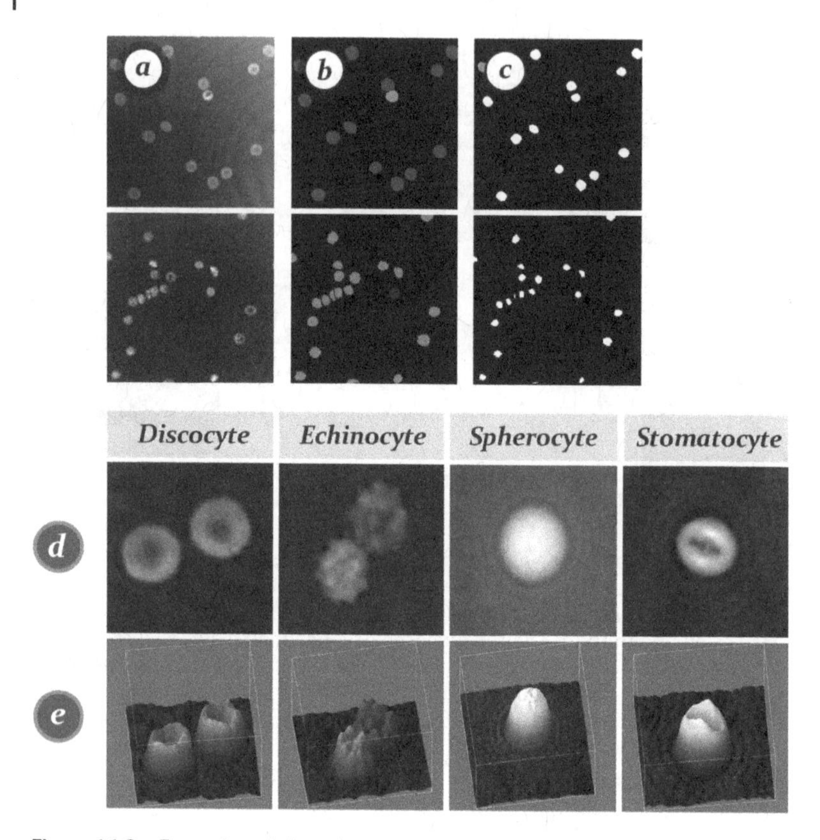

Figure 16.2 Examples of phase images and training maps. (a) RBC phase images obtained with DHM. (b) Annotated images and color-coded cell types. (c) Binary marker maps obtained from annotated images. (d) Phase images of four RBC types and (e) corresponding 3D visualizations. Note that the two examples for echinocyte cells are defined as echino-spherocytes. We assume that these echino-spherocytes are grouped and annotated as echinocytes [7] / with permission of IEEE.

The total number of phase images used to train our model was 219. Of the total dataset, 13% had overlapping RBCs. We used about 20, 70, and 10% of the phase images for testing, training, and validation, respectively. Annotated images were generated with five channels including the background through one-hot encoding to allow our models to separate each type of RBC from the background. To generate marker maps to be learned as seeds for separating overlapping RBCs, annotation images were converted to binary maps. The distance transform function was then applied to each cell. Finally, the image was subjected to thresholding (the threshold value was 0.3). Figure 16.2c shows results of the binary marker map method.

16.2 Proposed Deep-learning Model

16.2.1 Model Architecture

Multi-label segmentation requires incorporating both local low-resolution and high-resolution information in images. Therefore, we employed a deep CNN with nonlinear processing units that could extract low-level to high-level features of each RBC in a hierarchical manner. Adding skip connections in each layer could help directly transfer data from one layer to another one. Furthermore, since detecting and predicting the inner part of overlapping RBCs are difficult tasks, we developed a new two-stage procedure that combined a GAN-based model with a marker-controlled watershed method to separate the overlapped RBCs as shown in Figure 16.1.

In the training step, we generated eight patch images from the original phase image with a size of 768 × 768 pixels by randomly selecting an arbitrary location in the image. The size of the patch images were 256 × 256 pixels. Patch images were also rotated randomly by 90, 180, or 270° for each epoch. All patch images were fed into our model. We obtained two output images from the model.

Our deep-learning model was based on GAN with a generator and discriminator (see Figure 16.3) [15, 22]. The generator was trained to take an input image, extract the RBCs from the background, segment and classify them by RBC type, and generate markers representing internal RBCs (Figure 16.3a). Two generator's output images (i.e. multi-class probability map and binary marker probability map) were combined and fed to the input of the discriminator.

The training goal of the discriminator was to distinguish whether the input image was fake or real (see Figure 16.3b). In this way, the discriminator forces the generator to produce a desired image. Additionally, our generator needs to produce separate markers for tightly overlapped RBCs. Therefore, we built the generator model based on a deep residual U-Net that could take advantage of both residual network and U-Net architecture. The main feature is the use of residual units as basic blocks instead of standard convolutional layer units. Figure 16.3c shows the residual block construction. The advantage of this structure is that it allows us to design a neural network with fewer parameters. Even if the network layer is deep, the residual unit can facilitate training [28].

In our network, the encoder has one block for input image and three residual blocks. The output of the last residual block goes into the residual block in the bridge. The outcome of the bridge block then enters the decoder as shown in Figure 16.3a. The decoder has upsampling, concatenate, and residual block layers. This structure is repeated four times to restore the desired output size with features

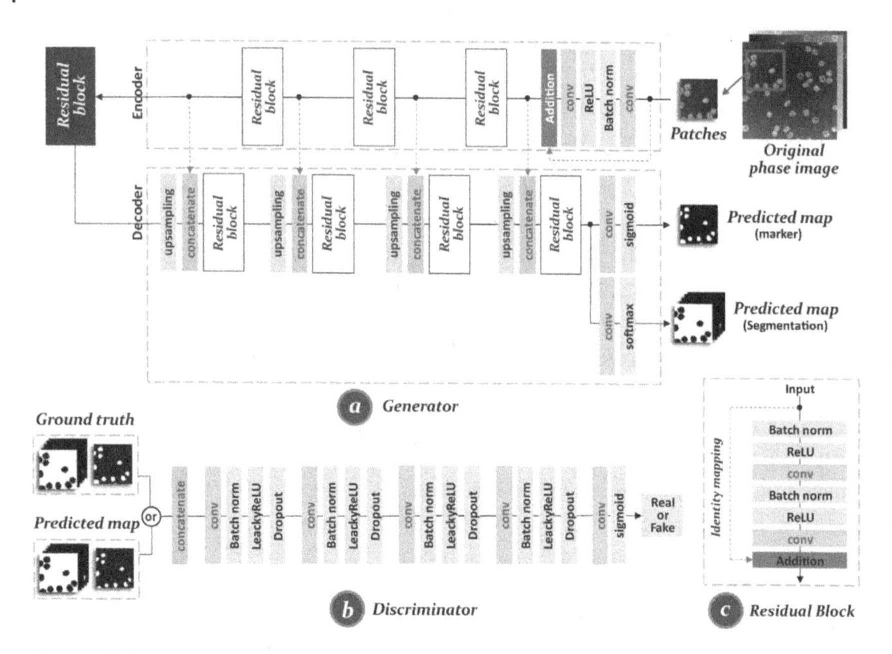

Figure 16.3 An overview of the proposed network for multi-class RBC segmentation and separation of overlapping RBCs. (a) A generator based on a deep ResUnet [23], which feeds a random patch from an RBC phase image and generates a marker prediction map and semantic-segmentation prediction map. (b) A discriminator that takes the predicted map from the generator or ground truth and estimates whether it is real or fake in units of patches. (c) A residual unit with identity mapping.

extracted from the encoder. Finally, our decoder generates two types of predictive maps through different 1×1 convolutional layers. The first type is a multi-class probability map to segment and classify RBCs according to type. The second type is a binary marker probability map to separate overlapping RBC. Therefore, the activation function of the first output is softmax, whereas the function of the second output is sigmoid.

Our discriminator model is based on a general Markovian discriminator that consists of the following sequence: a convolution layer with 3×3 filter size, a batch normalization with a momentum of 0.8, a leackyReLU with an incline of 0.2, and a dropout. The discriminator concatenates the marker prediction map with $256 \times 256 \times 1$ and the multi-class prediction map with $256 \times 256 \times 5$ and generates the patch image with a size of 16×16 pixels (see Figure 16.3b). Our GAN model learns how to distinguish real (ground truth) and fake patch images by overlapping patch images. It has fewer parameters that facilitate learning more than a full image discriminator.

16.2.2 Objective Function

The goal of our model is to find weight values to minimize the loss function between the two types of output and the corresponding ground truth. In addition, an adversarial loss function ensures that high-frequency features are completely preserved while the output is being generated [13, 15]. Accordingly, the overall objective function can be defined as

$$G^* = arg\,min_G\, max_D L_{CGAN}(G, D) + \lambda_S L_S(G) + \lambda_M L_M(G),$$

where G denotes the generator and D is the discriminator network. λ_S and λ_M are two empirical weight parameters. G tries to minimize this objective, while adversarial D tries to maximize it. L_S is the cost associated with the supervised RBC semantic segmentation, and L_M is the loss associated with the supervised task of RBC binary marker segmentation. L_{CGAN} is an adversarial loss between the generator and the discriminator [22]. It can be expressed as

$$L_{CGAN}(G, D) = E_y[logD(y)] + E_x[\,log\,(1 - D(G(x)))],$$

where x is the RBC phase image and y is the concatenation of the multi-label ground-truth segmentation map and binary marker map ground truth. $G(x)$ is the concatenation of the multi-label predicted segmentation map and the predicted binary marker map.

Accordingly, the size of $G(x)$ is $H \times W \times (C + 1)$ and the size of y is $H \times W \times (C + 1)$. Given our class-imbalance segmentation task, we added an additional generalized multi-class Dice loss function L_{Dice} [29] and a cross-entropy (CE) cost function L_{CE} for pixel classification. The objective function (L_S) is defined as

$$L_S = L_{Dice}(y_1, G_1(x)) + 10L_{CE}(y_1, G_1(x)),$$

where y_1 and $G_1(x)$ are the multi-class ground truth segmentation map and the multi-class predicted segmentation map, respectively. The loss function of L_M is expressed by

$$L_M = L_{Dice}(y_2, G_2(x)),$$

where y_2 and $G_2(x)$ are the ground truth binary marker map and the predicted binary marker map, respectively.

16.2.3 Deep Learning-based RBC Image Segmentation

To test the trained model for RBC image segmentation, test images with a size of 768 × 768 pixels were zero-padded and turned into images of 1024 × 1024 pixels. We then generated test patch images with a size of 256 × 256 pixels by moving the first pixel point of the patch image over the zero-padded test image of size 1024 × 1024. For example, if the first pixel point at the first test patch image was (1, 1), the first pixel point at the next patch image would be (129, 129). These patch images were fed into the trained model. Finally, we obtained the two types of prediction images (marker and segmentation) with the original size of 768 × 768 pixels by extracting the center area of size 128 × 128 from each output patch image and stitching all extracted patch images. Note that any image with a large size can be smoothly reconstructed from the trained model. The reconstructed output maps can be used as inputs to the watershed-segmentation algorithm to further separate overlapping RBCs. We can achieve complete separation of overlapping RBCs by combining marker-controlled watershed transformation (at the post-processing level) with the binary marker probability maps and the multi-class probability maps obtained from our model (see Figure 16.4). As shown in Figures 16.4b,c, our deep learning model was unable to completely separate overlapping RBCs. However, our model provided a more accurate marker map by learning valuable features from RBC phase images than did the marker-controlled watershed transform. Therefore, the marker map can be used to obtain much better performance in the marker-controlled watershed segmentation. Figure 16.4d shows that the overlapped RBCs can be segmented and classified by combining the deep-learning model and the watershed-segmentation method.

16.2.4 Evaluation Metrics

In general, cell-segmentation methods are evaluated at the pixel level (the shape and size of RBCs) or at the object level (RBC detection). Due to class imbalance, we computed the Dice coefficient to measure segmentation and classification performance rather than the pixel accuracy at the pixel level. The Dice coefficient is widely used in segmentation applications. It is defined as

$$DiceCoef(G, S) = \frac{2|G \cap S|}{|S| + |G|},$$

where $|G|$ and $|S|$ are pixels of the ground-truth image and the corresponding segmented image, respectively. $|G \cap S|$ represents the intersection between the pixels from the two images. We also calculated the AJI to evaluate the performance at the object level. The AJI is defined as an extension of the global Jaccard index:

$$AJI = \frac{\sum_{i=1}^{N} |G_i \cap S_i|}{\sum_{i=1}^{N} |G_i \cup S_i| + \sum_{i \in R} |S_i|},$$

where S_i is the predicted object that maximizes the Jaccard index with the ground truth object G_i, and R is the set of segmented objects that do not match with the ground truth. The AJI reflects the ratio between the intersection area of the matching element and segmentation results. Therefore, any inaccurate segmentation (under- or over-segmentation) will result in a decreased AJI.

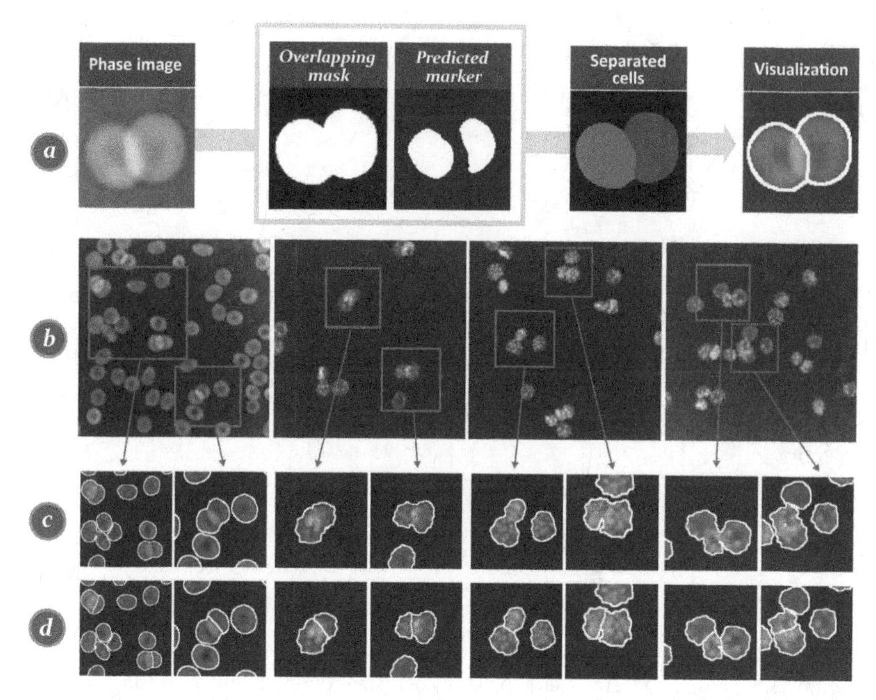

Figure 16.4 Gallery of segmentation examples with or without markers. (a) The RBC separation process using predicted multi-class maps with overlapping cells and predicted markers. The marker-controlled watershed algorithm floods basins from markers until basins from another marker meet the watershed line. The flooding of basins from predicted markers separates the overlapping mask along a watershed line. (b) Samples of RBC phase image. Enlargements of red boxes: (c) Segmentation results when there are no markers and (d) results of separating overlapping RBCs by our method [7] / with permission of IEEE.

16.3 Experimental Results

To evaluate our model performance, we prepared an unobserved test dataset, which consisted of 40 phase images. Of them, nine images included overlapping RBCs. Due to an imbalance in the number of cells in each image, the calculation for the evaluation was performed for the test dataset, not per image. For performance comparison, our method was compared with the following: transfer learning-based FCN-VGG16 [11], standard U-Net [13], Pix2Pix [15], Resnet [28], GANs with Resnet generator [30], DeepLab v3+ [31], and MAnet [32]. All methods were trained with enough epochs to allow convergence.

16.3.1 Implementation Details

Hyper-parameter values used in the experiment were determined through trial and error. Objective functions of our model were set at $\lambda_S = 50$ and $\lambda_M = 25$ to reduce visual artifacts and balance learning. An Adam optimizer with an adaptive

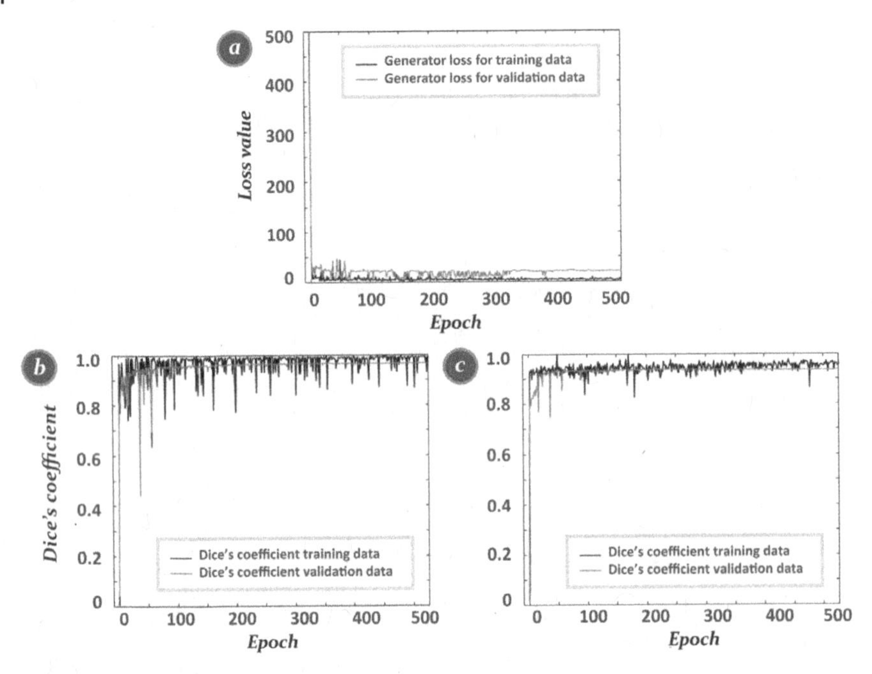

Figure 16.5 Loss of generator and the Dice coefficient during training. (a) Our generator loss values for training and validation dataset, (b) Dice coefficients for training and validation dataset of multi-class segmentation, (c) Dice coefficients for training and validation dataset of markers [7] / with permission of IEEE.

momentum was used. Epochs of the model stopped learning when the loss function of the generator was stabilized (no longer decreasing). The batch size was 16. The initial learning rate was 0.0002. It was reduced by one tenth for every 30 epochs.

Figure 16.5 shows generator loss and the Dice coefficient in each epoch. The average and variance of the generator loss were decreased with the increase of training epochs. The best prediction model was obtained when the Dice coefficient for the validation task reached its maximum after 197 epochs. The prediction time for a 256 × 256 patch image was 0.026 seconds. The merging time from patch to full-size image was 0.060 seconds.

16.3.2 Evaluation of RBC Segmentation and Classification Performance

Figure 16.6 shows the final segmentation, classification, and separation results. Different colors of results represent RBC types. The segmentation boundary is shown in white (Figure 16.6 [second to last row]). Seven other evaluated methods

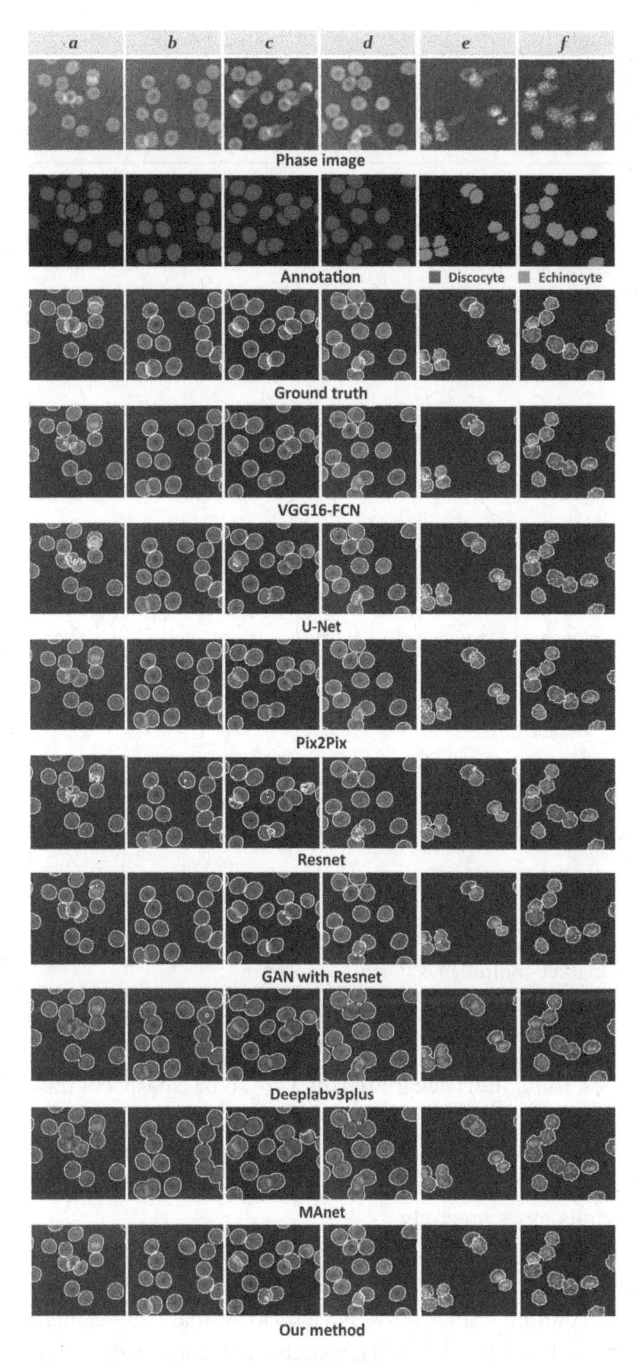

Figure 16.6 Visualization of RBC semantic segmentation and separation results for phase images. (a)–(f) Examples of tested phase images. Grey arrows in phase images indicate overlapping cells [7] / with permission of IEEE.

Table 16.1 Results of quantitative performance analysis of multi-class RBC semantic segmentation.

	Type	Discocytes	Spherocytes	Echinocytes	Stomatocytes	Overall
Dice (Pixel based)	FCN	0.9519	0.9442	0.9401	0.8904	0.9317
	U-Net	0.9520	0.9535	0.9397	0.8895	0.9337
	Pix2Pix	0.9558	0.9495	0.9442	0.9009	0.9376
	Resnet	0.9485	0.9452	0.9377	0.8238	0.9138
	GAN with Resnet	0.9533	0.8551	0.9435	0.8727	0.9062
	Deeplab v3+	0.9284	0.8674	0.8457	0.8051	0.8617
	MAnet	0.8941	0.8911	0.7993	0.8220	0.8516
	Proposed	0.9548	0.9575	0.9466	0.9064	0.9413
AJI (Object based)	FCN	0.8071	0.8941	0.7545	0.7977	0.8133
	U-Net	0.8698	0.9112	0.7330	0.8008	0.8287
	Pix2Pix	0.8582	0.9039	0.7713	0.8197	0.8383
	Resnet	0.8922	0.8962	0.8827	0.7003	0.8429
	GAN with Resnet	0.8876	0.7469	0.8796	0.7742	0.8221
	Deeplab v3+	0.8651	0.6822	0.7913	0.7348	0.7684
	MAnet	0.8504	0.6876	0.8214	0.6176	0.7442
	Proposed	0.8932	0.9169	0.8915	0.8288	0.8826
Throughput rate (cells/second)	183: Before marker-controlled watershed transform 153: After marker-controlled watershed transform					

could segment RBC types from the background due to some sharp contrast between the background and RBCs in the phase image, as shown in Figure 16.6. Table 16.1 shows the Dice coefficients for different types of RBCs in the seven models. Our method showed a slightly higher Dice coefficient because it could generate RBC details more precisely.

In the case of U-Net, we could see that it wrongly predicted discocytes (red) as stomatocytes (yellow), which might be due to the loss of the inner part of discocytes (see Figure 16.6 [first column]). Since it was not easy to evaluate the segmentation performance of overlapped RBCs by the Dice coefficient, we calculated the

AJI metric. Our method showed better performance in dividing overlapping RBCs than other models (see Figure 16.6a–f). As shown in Table 16.1, with our method, AJIs were high for all RBC types used in the experiment and RBCs were well-divided. Since there were a few overlapping RBCs among spherocytes and stomatocytes, there was a small difference in the AJI evaluation compared to that of discocytes and echinocytes, as shown in Table 16.1. Comparison results with two recent networks (DeepLab v3+ and MAnet, which are widely used for semantic segmentation) also showed that our method provided the best results for segmentation and classification of RBCs.

Moreover, our method is suitable for high-throughput screening applications. It could classify around 183 RBCs per second. Its throughput for counting RBCs was slightly lower (around 153 RBCs per second) due to computation of the marker-controlled watershed transformation.

16.3.3 Automated Assessment of RBC Aging Markers

In this section, we will demonstrate how our image-based, deep-learning models can perform an instant phenotypic assessment of RBC storage lesions, which has the potential to lead to new efficient tools for safe transfusions and measurement of stored RBC quality. To automatically identify any changes in RBC shape due to storage, without computing any feature vectors such as PSA, MCH, or MCH surface density in RBC phase images, we tested our trained models with datasets from RBC samples stored for different periods (8, 13, 16, 23, 27, 30, 34, 37, 40, 47, and 57 days). These datasets contained more than 2500 RBCs across 11 storage durations, with more than 200 blood cells for each storage duration. The total number of phase images under all storage periods was 66 (6 for each of 11 classes). Figure 16.7a–k show examples of the semantic segmentation and classification results from RBC phase images with differing storage durations (8, 13, 16, 23, 27, 30, 34, 37, 40, 47, and 57 days) achieved with our model. Figure 16.8 shows the accuracies of our trained model for classifying four RBC types (discocytes, spherocytes, echinocytes, and stomatocytes) using all 66 RBC phase images. Our model showed an accuracy of over 90% for all RBC classification types. In particular, the accuracy for discocyte and spherocyte classification was over 97%.

To see the trend of changes in the 3D morphology of RBCs as a function of storage duration, we also applied our trained models to phase images of RBCs from each storage duration for classification. Figure 16.9 shows the relationship between the ratio of RBC types and storage duration of RBC samples. It demonstrated that the dominant shape of RBCs began to change when stored longer than 40 days. Increased storage duration was strongly related to the transformation of RBCs from discocytes to transitory echinocytes and finally to spherocytes. To confirm these classification results, we manually counted each RBC type for each

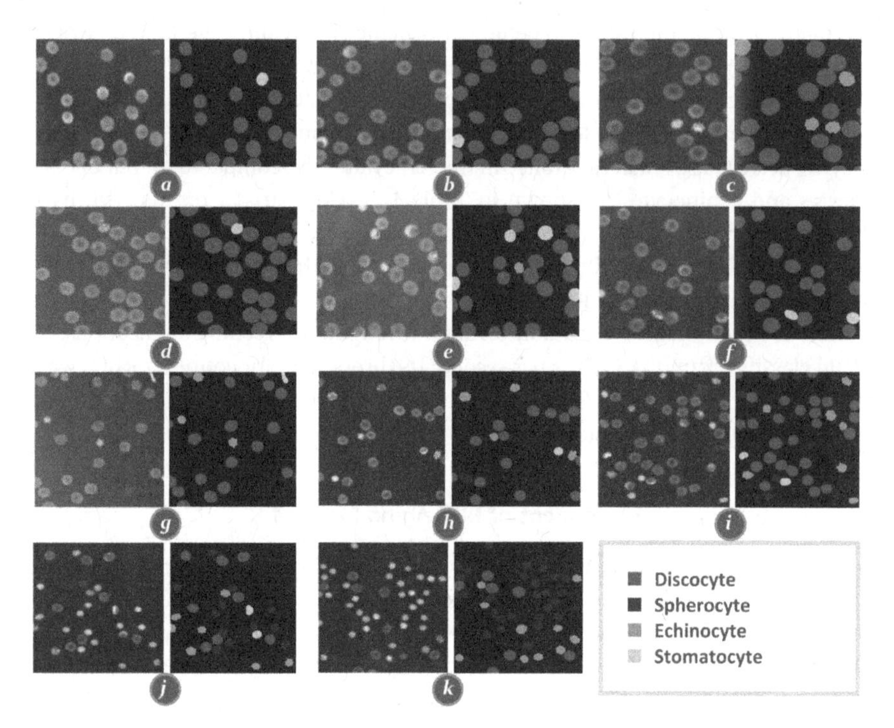

Figure 16.7 Visualization of RBC segmentation and classification results achieved with our deep-learning models. (a)–(k) Original RBC phase images and corresponding segmentation and classification results with 8, 13, 16, 23, 27, 30, 34, 37, 40, 47, and 57 days of storage, respectively.

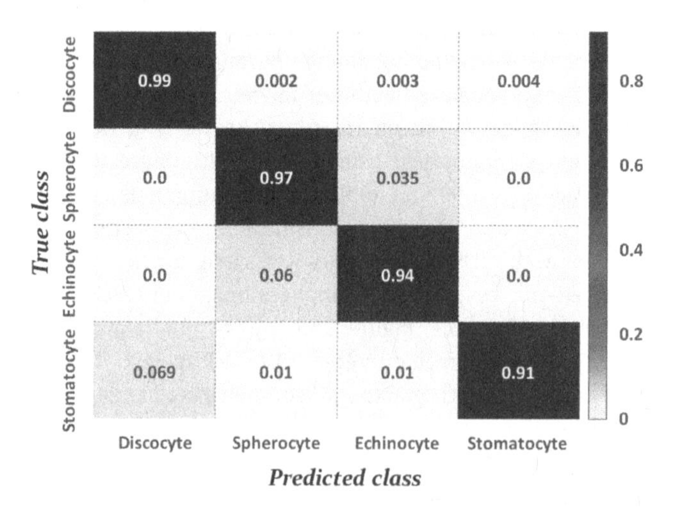

Figure 16.8 Confusion matrix of RBC type classifications for all 66 phase images of RBCs stored across 11 storage periods.

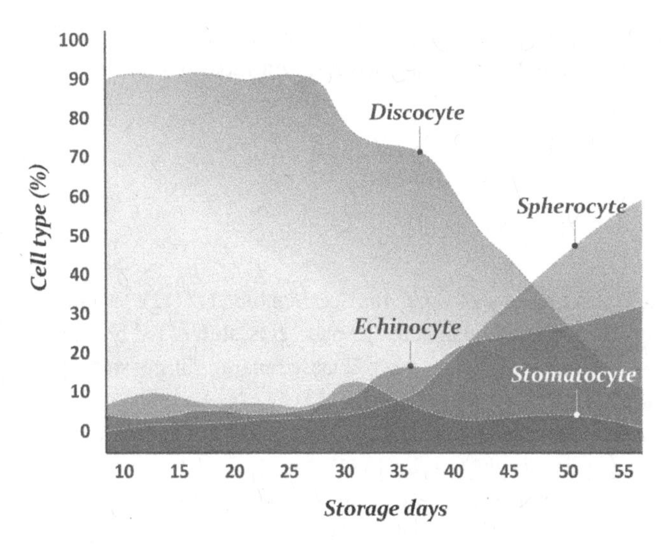

Figure 16.9 Relationships between the ratios of RBC types and varied storage time of RBC samples. Note that the dominant shape of RBCs started to change when the storage time exceeded 40 days.

storage period. These two methods provided similar classification results with a high concordance rate of over 95%.

Furthermore, our classification results were consistent with previous findings showing that the dominant shape in RBCs changed as storage time increased [2, 3, 33–35]. In this experiment, we developed a learning model to calculate the percent of spherocytes in stored RBCs, which provided information about morphological perturbation. Therefore, this method could be used to monitor cell distribution and assess stored blood quality.

16.4 Conclusions

We presented an approach for classification and segmentation of RBCs and demonstrated the potential of this approach for assessing RBC storage lesions for safe transfusion. Our model combined a GAN with marker-controlled watershed-segmentation method. Our approach obtained good segmentation and classification accuracy with a Dice coefficient of 0.94 and a high-throughput rate of more than 152 RBCs per second. Our model, with fewer parameters, outperformed standard U-Net, FCN-VGG16, Pix2Pix, Resnet, GANs with Resnet generator, DeepLab v3+, and MAnet models. Moreover, our deep-learning model identified morphological changes that occur in RBCs during storage. We believe that our deep-learning models can be applied to the automated assessment of RBC quality, identification of storage lesions for safe transfusions, and identification of RBC-related diseases.

References

1 Hess, J. (2016). RBC storage lesions. *Blood, J. Am. Soc. Hematol.* 128 (12): 1544–1545.

2 Roussel, C., Dussiot, M., Marin, M. et al. (2017). Spherocytic shift of red blood cells during storage provides a quantitative whole cell-based marker of the storage lesion. *Transfusion* 57 (4): 1007–1018.

3 Bardyn, M., Rappaz, B., Jaferzadeh, K. et al. (2017). Red blood cells ageing markers: a multi-parametric analysis. *Blood Transfus.* 15 (3): 239.

4 Koch, C., Li, L., Sessler, D. et al. (2008). Duration of red-cell storage and complications after cardiac surgery. *N. Engl. J. Med.* 358: 1229–1239.

5 Doan, M., Sebastian, J., Caicedo, J. et al. (2020). Objective assessment of stored blood quality by deep learning. *Proc. Natl. Acad. Sci.* 117 (35): 21381–21390.

6 Yoshida, T., Prudent, M., and D'Alessandro, A. (2019). Red blood cell storage lesion: causes and potential clinical consequences. *Blood Transfus.* 17 (1): 27.

7 Kim, E., Park, S., Hwang, S. et al. (2022). Deep learning-based phenotypic assessment of red cell storage lesions for safe transfusions. *IEEE J. Biomed. Health Inform.* 26 (3): 1318–1328.

8 Shen, D., Wu, G., and Suk, H. (2017). Deep learning in medical image analysis. *Annu. Rev. Biomed. Eng.* 19 (1): 221–248.

9 Ahmadzadeh, E., Jaferzadeh, K., Shin, S., and Moon, I. (2020). Automated single cardiomyocyte characterization by nucleus extraction from dynamic holographic images using a fully convolutional neural network. *Biomed. Opt. Express* 11 (3): 1501–1516.

10 Shelhamer, E., Long, J., and Darrell, T. (2017). Fully convolutional networks for semantic segmentation. *IEEE Trans. Pattern Anal. Mach. Intell.* 39 (4): 640–651.

11 Yi, F., Moon, I., and Javidi, B. (2017). Automated red blood cells extraction from holographic images using fully convolutional neural networks. *Biomed. Opt. Express* 8 (10): 4466–4479.

12 Ronneberger, O., Fischer, P., and Brox, T. (2015). U-net: Convolutional networks for biomedical image segmentation. In: *Medical Image Computing and Computer-Assisted Intervention – MICCAI 2015 (Lecture Notes in Computer Science)* (ed. N. Navab, J. Hornegger, W. Wells and A. Frangi), Munich, Germany (5–9 October 2015), 234–241. New York (NY): Springer, Cham.

13 Pathak, D., Krähenbühl, P., Donahue, J. et al. (2016). Context encoders: Feature learning by inpainting. In: *2016 IEEE Conference on Computer Vision and Pattern*

Recognition (CVPR), Las Vegas, NV, USA (27–30 June 2016), 2536–2544. New York (NY): IEEE.

14 Moon, I., Jaferzadeh, K., Kim, Y., and Javidi, B. (2020). Noise-free quantitative phase imaging in Gabor holography with conditional generative adversarial network. *Opt. Express* 28 (18): 26284–26301.

15 Isola, P., Zhu, J., Zhou, T., and Efros, A. (2017). Image-to-image translation with conditional adversarial networks. In: *2017 IEEE Conference on Computer Vision and Pattern Recognition (CVPR)*, Honolulu, HI, USA (21–27 July 2017), 5967–5976. New York (NY): IEEE.

16 Moeskops, P., Veta, M., Lafarge, M. et al. (2017). Adversarial training and dilated convolutions for brain MRI segmentation. In: *2017 Deep Learning in Medical Image Analysis and Multimodal Learning for Clinical Decision Support (Lecture Notes in Computer Science)*, Québec City, QC, Canada (14 September 2017), 56–64. New York (NY): Springer, Cham.

17 Xue, Y., Xu, T., Zhang, H. et al. (2018). Segan: adversarial network with multi-scale l 1 loss for medical image segmentation. *Neuroinformatics* 16 (3): 383–392.

18 Yang, D., Xu, D., Zhou, S. et al. (2017). Automatic liver segmentation using an adversarial image-to-image network. In: *Medical Image Computing and Computer Assisted Intervention – MICCAI 2017 (Lecture Notes in Computer Science)* (ed. M. Descoteaux, L. Maier-Hein, A. Franz, et al.), Québec City, QC, Canada (14 September 2017), 507–515. New York (NY): Springer, Cham.

19 Zhang, Y., Yang, L., Chen, J. et al. (2017). Deep adversarial networks for biomedical image segmentation utilizing unannotated images. In: *Medical Image Computing and Computer Assisted Intervention – MICCAI 2017 (Lecture Notes in Computer Science)* (ed. M. Descoteaux, L. Maier-Hein, A. Franz, et al.), Québec City, QC, Canada (14 September 2017), 408–416. New York (NY): Springer, Cham.

20 Zhang, D., Song, Y., Liu, S. et al. (2018). Nuclei instance segmentation with dual contour-enhanced adversarial network. In: *2018 IEEE 15th International Symposium on Biomedical Imaging (ISBI 2018)*, Washington, D.C., USA (4–7 April 2018), 409–412. New York (NY): IEEE.

21 Mahmood, F., Borders, D., Chen, R. et al. (2020). Deep adversarial training for multi-organ nuclei segmentation in histopathology images. *IEEE Trans. Med. Imaging* 39 (11): 3257–3267.

22 Mirza, M. and Osindero, S. (2014). Conditional generative adversarial nets. arXiv. https://doi.org/10.48550/arXiv.1411.1784.

23 Kumar, N., Verma, R., Sharma, S. et al. (2017). A dataset and a technique for generalized nuclear segmentation for computational pathology. *IEEE Trans. Med. Imaging* 36 (7): 1550–1560.

24 Dice, L. (1945). Measures of the amount of ecologic association between species. *Ecology* 26 (3): 297–302.

25 Bessis, M. (1973). *Red Cell Shapes. An Illustrated Classification and Its Rationale.* Berlin, Heidelberg: Springer.

26 Lim, H., Wortis, M., and Mukhopadhyay, R. (2008). Red blood cell shapes and shape transformations: newtonian mechanics of a composite membrane: sections 2.1-2.4. *Soft Matter: Lipid Bilayers Red Blood Cells* 4: 83–139.

27 Geekiyanage, N., Balanant, M., Sauret, E. et al. (2019). A coarse-grained red blood cell membrane model to study stomatocyte-discocyte-echinocyte morphologies. *PLoS One* 14 (4): e0215447.

28 He, K., Zhang, X., Ren, S., and Sun, J. (2016). Deep residual learning for image recognition. In: *2016 IEEE Conference on Computer Vision and Pattern Recognition (CVPR)*, Las Vegas, NV, USA (27–30 June 2016), 770–778. New York (NY): IEEE.

29 Sudre, C., Li, W., Vercauteren, T. et al. (2017). Generalized Dice overlap as a deep learning loss function for highly unbalanced segmentations. In: *2017 Deep Learning in Medical Image Analysis and Multimodal Learning for Clinical Decision Support (Lecture Notes in Computer Science)* (ed. M.J. Cardoso, T. Arbel, G. Carneiro, et al.), Québec City, QC, Canada (14 September 2017), 240–248. New York (NY): Springer, Cham.

30 Zhu, J., Park, T., Isola, P., and Efros, A. (2017). Unpaired image-to-image translation using cycle-consistent adversarial networks. In: *2017 IEEE International Conference on Computer Vision (ICCV)*, Honolulu, HI, USA (21–27 July 2017), 2223–2232. New York (NY): IEEE.

31 Chen, L., Zhu, Y., Papandreou, G. et al. (2018). Encoder-decoder with atrous separable convolution for semantic image segmentation. In: *Computer Vision – ECCV 2018 (Lecture Notes in Computer Science)* (ed. V. Ferrari, M. Hebert, C. Sminchisescu and Y. Weiss), Munich, Germany (8–14 September 2018), 801–818. New York (NY): Springer, Cham.

32 Fan, T., Wang, G., Li, Y., and Wang, H. (2020). MA-Net: a multi-scale attention network for liver and tumor segmentation. *IEEE Access* 8: 179656–179665.

33 Moon, I., Yi, F., Lee, Y. et al. (2013). Automated quantitative analysis of 3D morphology and mean corpuscular hemoglobin in human red blood cells stored in different periods. *Opt. Express* 21 (25): 30947–30957.

34 Hess, J., Sparrow, R., Meer, P. et al. (2009). Blood components: red blood cell hemolysis during blood bank storage: using national quality management data to answer basic scientific questions. *Transfusion* 49 (12): 2599–2603.

35 Blasi, B., D'alessandro, A., Ramundo, N., and Zolla, L. (2012). Red blood cell storage and cell morphology. *Transfus. Med.* 22 (2): 90–96.

17

High-throughput Label-free Cell Counting with Deep Neural Networks

17.1 Introduction

Previous chapters introduced a digital holographic microscopy (DHM) system to reconstruct phase images of red blood cells (RBCs) using a numerical reconstruction method. This numerical reconstruction algorithm includes processes such as spatial filtering, phase unwrapping, and numerical propagations of complex diffraction waves. Since cell studies based on phase images need a numerical reconstruction step, cell analysis may benefit from this new scheme that can completely eliminate the numerical reconstruction step.

Reconstructed phase image can clearly reveal targets while those in the raw hologram are much more vague. The cell edge in the diffraction pattern recorded by DHM is usually not clearly defined. Therefore, it may be difficult to perform cell analyses based on a raw hologram using traditional image processing due to ambiguity of biological cells in the diffraction pattern. Fortunately, the diffraction pattern recorded by DHM can be analyzed using deep neural networks, which do not require defined specific features within the raw hologram. They can automatically extract useful features according to the goal of the networks.

In this chapter, we will introduce the U-Net algorithm for cell detection and counting using the DHM diffraction pattern [1]. The U-Net algorithm is an end-to-end segmentation method that uses images as input and output. It can overcome disadvantages of the CNN algorithm such as time consuming and difficulty in patch size selection for pixel-based image segmentation. In addition, the U-Net algorithm is invariant to the size of the input image while the CNN algorithm requires a fixed-size input image. After the U-Net algorithm is trained, cell targets inside the diffraction pattern are directly investigated without any numerical reconstruction processing. Similar to U-Net, other fully convolutional neural networks (FCNs) such as the pyramid scene parsing network (PSPNet) [2] and

Artificial Intelligence in Digital Holographic Imaging: Technical Basis and Biomedical Applications, First Edition. Inkyu Moon.

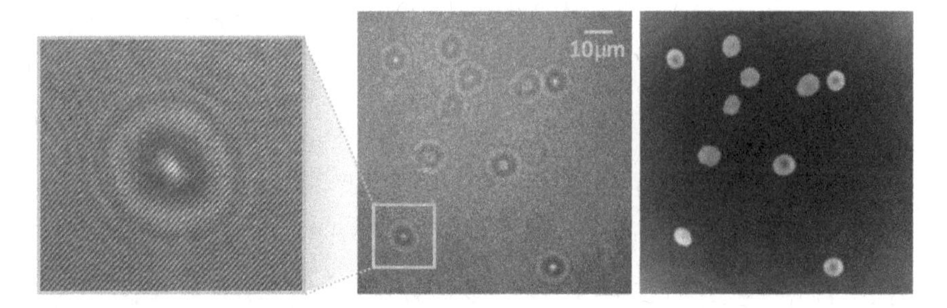

Figure 17.1 Hologram and phase image. Left: digital hologram of RBCs recorded by DHM. Right: RBC phase image reconstructed from the digital hologram.

Deeplab V3+ may also work for cell detection and counting in hologram images. Figure 17.1 shows one hologram of RBCs and its corresponding reconstructed phase images of RBCs.

17.2 Materials and Methods

17.2.1 U-Net Algorithm

The U-Net is a kind of deep FCN. It does not have any fully connected layers. Thus, it can handle images of different sizes. Figure 17.2 shows U-Net's structure for cell detection and counting. Basically, there are convolution, rectified linear unit (Relu) activation, up-convolution, max pooling, and concatenate layers in our U-Net's structure. The convolution layer extracts feature from the input image.

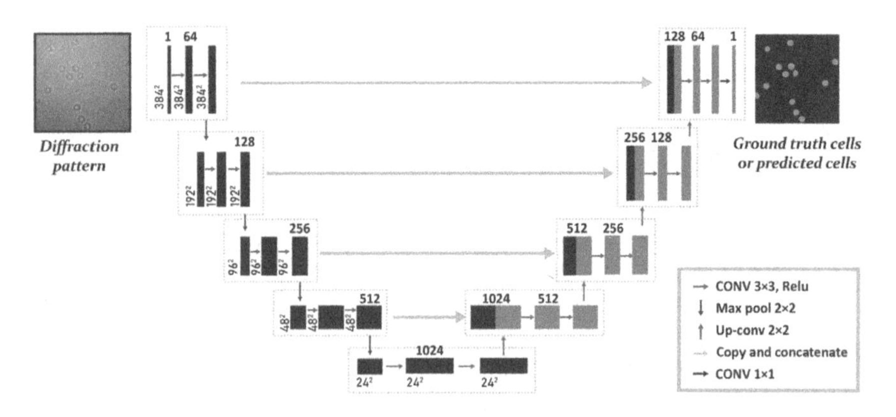

Figure 17.2 U-Net structure for cell detection and counting in diffraction pattern.

It is a mathematical operation between two inputs such as image or feature map and a kernel with learning parameters. The convolution operation can preserve the spatial relationship among pixels and learn image features based on small squares of input image or feature map [3]. The Relu activation is defined as $y = \max(0, x)$, where x is the input image or feature, y is the output, and max() is the maximum operation function. The Relu activation function is used to introduce nonlinearity into the networks. Compared to the sigmoid activation function in deep neural networks, the Relu activation function can alleviate the gradient vanish problem during training [3]. The up-convolution layer is commonly used in encoder-decoder architecture networks such as FCN and U-Net algorithms. The up-convolution operation reconstructs the previous spatial resolution before performing a convolution. It can be implemented using a combination of image upscaling and convolution layer. The up-convolution layer also refers to as a deconvolution layer or a transposed convolution layer [3, 4]. Max pooling layer is one kind of pooling operation. It extracts the largest element from the rectified feature map [3, 4]. Pooling layer in deep FCNs can reduce the dimensionality of each feature map, which improves the network's computation. Although the size of the feature map is reduced in the max pooling layer, important feature information can still be retained. To some extent, the max pooling layer can also make the algorithm robust to perform a translation of the target in the input image. The concatenate layer is a utility layer that can merge multiple feature maps into one group of feature map.

In Figure 17.2, single numbers such as 64 and 128 denote the number of feature map in that layer, whereas a squared number like 384^2 indicates the spatial size of the feature map. In our U-Net algorithm, the diffraction pattern from DHM is used as the input image and the size used to feed the U-Net is 384^2. The ground-truth image is a binary image where the foreground represents target cells. The sigmoid function is applied to the last layer of U-Net to ensure that each pixel in the output image has a value between 0 and 1. The loss function is defined between the ground-truth image and the U-Net output. Equation (17.1), which combines a soft Dice coefficient and cross-entropy loss (logarithmic loss for the two classes case) L is used to learn parameters in U-Net's architecture using the back-propagation algorithm [3, 4]:

$$L = \left(1 - \frac{\sum_{i=1}^{N} 2(y_i \overline{y_i})}{\sum_{i=1}^{N} (y_i + \overline{y_i})}\right) - \frac{1}{N}\sum_{i=1}^{N} [y_i \, log\,(\overline{y_i}) + (1 - y_i)\, log(1 - \overline{y_i})], \tag{17.1}$$

where N is the total number of pixels in the output map, y_i is the true category of ith pixel in the ground truth image, and $\overline{y_i}$ is the probability of belonging to the foreground category for the ith pixel in the output map from U-Net algorithm.

17.2.2 Multiple Cell Detection and Counting Procedure

There are two main steps for cell detection and counting in diffraction patterns: a learning step and a prediction step. In the learning stage, the input is the diffraction pattern or raw hologram, and the output is a probability map. The output is then compared to the ground truth image. Parameters in our U-Net are then learned using a back-propagation algorithm based on the cross-entropy loss function. In the prediction phase, the output of our trained U-Net is considered the prediction result for a new input diffraction pattern. We set the threshold value at 0.5, indicating that the pixel is categorized into cells if the probability value is larger than 0.5. Otherwise, the corresponding pixel is labeled as the background. Before the learning starts, parameters of the U-Net in the encoder layer are initialized with pre-trained VGG networks [3, 5] while the first layer and other layers in the decoder part are initialized using the He initialization method [6]. During training, input images are augmented to increase the training dataset and reduce the over-fitting problem. Some augmentation methods are described in the data preparation section.

Multiple cells were segmented from our U-Net model. We obtained a binary image where the foreground represented cells. Before counting cells, the binary mask image was processed using morphological opening operation where the structuring element was a disk with a radius of 3 [7]. This morphological operation could remove some small, isolated targets that might be noise while smoothing the target edge. Moreover, we excluded these cells on the frame boundary whose area was less than 40 pixels from both the ground truth and predicted image for cell counting. The object in the processed binary image was then labeled. A unique label was assigned to every pixel of each cell in the binary image. It could be easily done using the connected-component labeling method [7]. After labeling, the maximum value in the labeling image was the number of cells in the detected image. It might also be a good idea to make cell-count processing a loss layer and achieve cell detection and counting from the network directly. We only used a post-processing method to automatically count cells from the algorithm output in this study. Figure 17.3 shows the procedure for cell detection and counting using our U-Net model.

17.2.3 Data Preparation

Holograms of RBCs were obtained from DHM. Although our model was based on the diffraction pattern with a prediction that did not need reconstructed phase images, we reconstructed phase images from the holograms for the purpose of U-Net training and evaluation. The numerical reconstruction method was used to reconstruct phase images with a size of 700×700 pixels, which corresponded

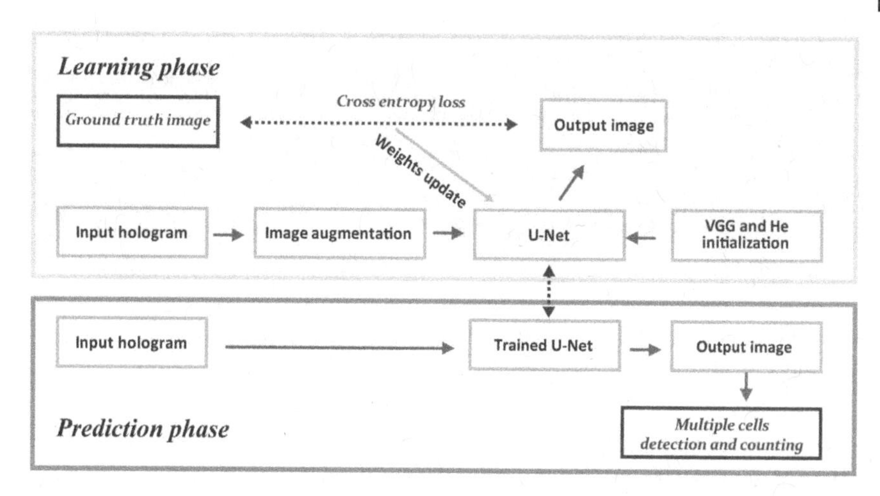

Figure 17.3 Cell detection and counting procedure at the single-cell level using U-Net.

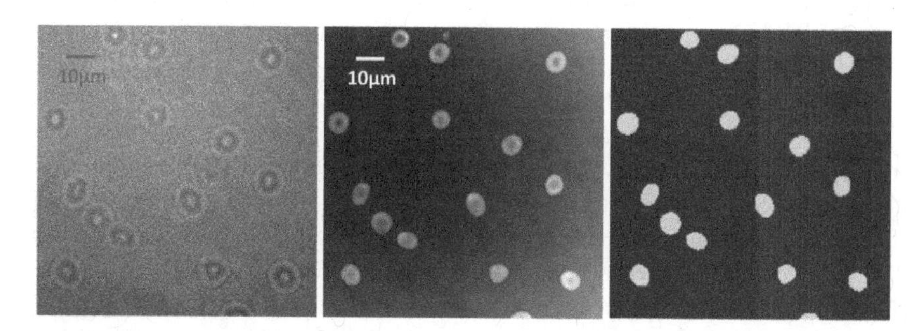

Figure 17.4 Illustration of (left) hologram, (center) reconstructed phase, and (right) labeled ground-truth images [1] / SPIE / CC BY 4.0.

to a field of view (FOV) of 100 μm × 100 μm. A biologist manually labeled 150 images for our model development. The labeling processing was based on reconstructed phase images. Figure 17.4 shows a hologram of RBCs, a phase image of RBCs, and the labeled image (ground truth).

Since deep-learning models have several hyper-parameters for training, such as learning rate and weight decay, we split our dataset into training and validation datasets for the purpose of hyper-parameter tuning and model evaluation. The training dataset had 125 holograms and the validation dataset had 25 holograms. For the performance evaluation, the fivefold cross-validation method was used based on the training dataset. All holograms (n = 150) were obtained with DHM at 40× magnification. We also had 50 holograms that

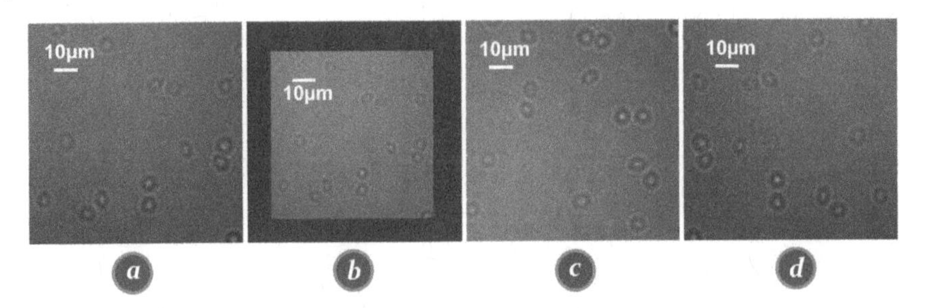

Figure 17.5 Illustration of image augmentation. (a) Original image, (b) image resizing, (c) image rotation, and (d) image flipping [1] / SPIE / CC BY 4.0.

included correspondingly labeled ground-truth images. These images were obtained with DHM at 20× magnification. In addition, 70 bead holograms were captured at 40× magnification. The original size of the diffraction pattern obtained with DHM at 20× magnification was 700 × 700 pixels with a FOV of 200 μm × 200 μm. We used these datasets to test the generalization of our U-Net model and compare prediction outputs using the training dataset at different magnifications.

Image normalization and augmentation were applied to each hologram before feeding it to our U-Net model. Image normalization was done by subtracting the mean value then dividing by the standard deviation. Both mean and standard deviation were computed from the training dataset. Image augmentation used during training consisted of image cropping, resizing, translation, rotation, and flipping. These augmentation methods were from a python package called *imgaug* [8]. Figure 17.5 shows some image augmentation examples. The hologram was padded with a zero value after resizing when it was smaller than 384 × 384 pixels. which was the input size to our U-Net model.

17.2.4 Hardware and Hyper-parameter Configuration

Our U-Net model was trained on a server with 48 CPUs (Intel(R) Xeon(R) CPU E5-2650), one P100 Nvidia graphics processing unit (GPU), and an Ubuntu 16.04 operating system. Both training and prediction were based on GPU parallel computing. The algorithm was implemented based on the Pytorch deep-learning framework [9]. The batch size used in the learning step was four and the loss function, which was a combination of a soft Dice coefficient and cross-entropy loss, was optimized using a gradient-descent algorithm with momentum set at 0.9. The learning rate was initialized as 0.01 and decreased by a factor of 10 every 30 epochs. L2 regularization was used to reduce the over-fitting problem. Its value was set at 0.0001. The number of epochs was set at 60.

17.2.5 Evaluation Metrics

Several metrics were used to evaluate cell detection and counting. First, we used accuracy, sensitivity, and Dice score coefficient (DSC) metrics. Sensitivity measures the proportion of positive cases correctly identified as such. A high sensitivity rarely overlooks an actual positive. The sensitivity within a predicted image is expressed as

$$Sensitivity = \frac{TP}{TP + FN},$$

where TP is the true positive and FN is false negative pixels within the image. The metric accuracy is the evaluation of all pixels within the image. It is defined as

$$Accuracy = \frac{TP + TN}{TP + TN + FP + FN},$$

where true negative (TN) is the number of non-cells pixels that are correctly classified, and FP is the false positive. Accuracy is the percentage of correctly classified pixels (both cells and background) out of the total number of pixels. The DSC is defined as

$$DSC = \frac{2TP}{2TP + FP + FN}. \tag{17.2}$$

The DSC is also used to evaluate cell segmentation results. The performance evaluation included two additional metrics: the Hausdorff distance (HD) and the ninety-fifth percentile Hausdorff distance (95th HD) [10, 11]. They were calculated by comparing binary objects in two images: the ground-truth image and the predicted image. It was an indicator of the largest segmentation error. HD was defined as the maximum surface distance between objects. It was calculated between boundaries of the predicted segmentation and the ground-truth segmentation. For two point sets X and Y (see Figure 17.6), the $HD(X,Y)$ was the longest distance one had to travel from a point in one of the two sets to its closest point in the other set. It was computed using the equation:

$$HD(X, Y) = max(hd(X, Y), hd(Y,X)),$$

where $hd(X, Y) = \max_{x \in X} \min_{y \in Y} \|x - y\|_2$, and $hd(Y,X) = \max_{y \in Y} \min_{x \in X} \|x - y\|_2$. $\|x - y\|_2$ was the Euclidean distance between point x and y. Compared to the Hausdorff distance, the ninety-fifth percentile Hausdorff distance is slightly more stable to small outliers and is commonly used in image segmentation. The smaller the value for both HD and 95th HD, the better performance is indicated.

We additionally defined metrics for correctly counted and over-counted cell metrics for to evaluate cell counting. If the centroid point of an isolated region in the predicted image from U-Net was within the cell of the ground-truth image, the cell was correctly counted. Otherwise, the cell was considered over-counted. When centroid points of more than two isolated regions were found in the same

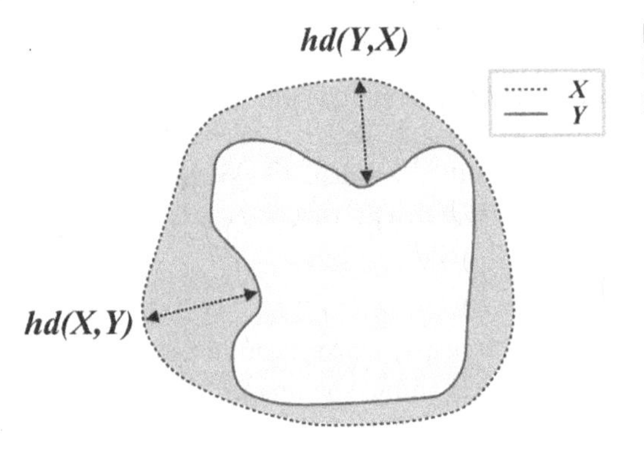

Figure 17.6 Illustration of the Hausdorff distance between point sets X and Y.

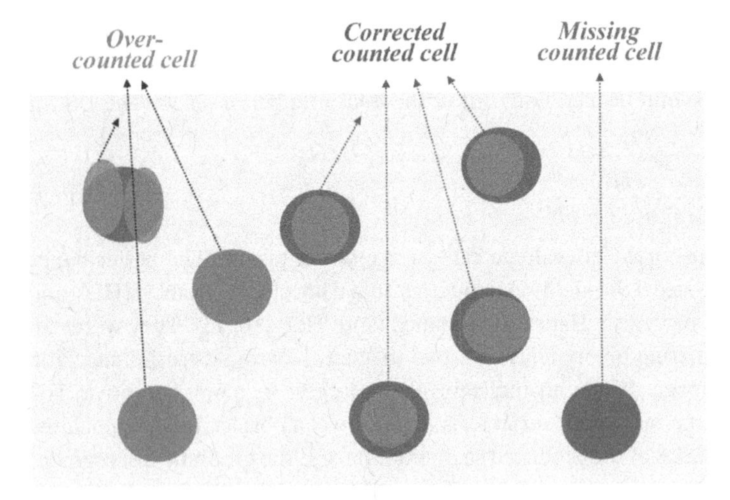

Figure 17.7 Illustration of different counted cell types. Dark grey = ground-truth cells; grey circles = predicted cells Adapted from [1].

cell in the ground-truth image, only one predicted isolated region was regarded as correctly counted cell while others were considered over-counted cells (see Figure 17.7). The throughput rate (the number of cells counted per second) was also given. The higher the number, the better cell-counting performance. All these metrics were measured using the fivefold cross-validation technique.

17.3 Experimental Results

Our U-Net model was trained using the fivefold cross-validation scheme based on training data. Hyper-parameters were tuned based on the validation dataset. DSC as defined in Eq. (17.2) was used to evaluate the U-Net model and tune

Table 17.1 DSC values from varied learning rates and weight decay.

	Learning rate			
Weight decay	**0.01**	**0.001**	**0.0001**	**0.00001**
0.01	0.9112	0.8788	0.8376	0.4053
0.001	0.9303	0.8626	0.8432	0.4244
0.0001	0.9321	0.8542	0.8468	0.3648
0.00001	0.9190	0.8800	0.8439	0.3325

hyper-parameters (learning rate and weight decay). We used the grid search method to tune our model. Candidate values for learning rate and weight decay during grid search were these empirical values. Table 17.1 shows varied values of learning rate and weight decay with their corresponding best DSC values on the validation dataset during U-Net learning.

As shown in Table 17.1, our U-Net model had the best Dice score value on the validation dataset when the learning rate was 0.01 and the weight decay value was 0.0001. Therefore, the final U-Net model was trained based on learning rate and weight decay with values of 0.01 and 0.0001, respectively. The fivefold cross validation was used for the performance evaluation. Hyper-parameters were tuned in the first round of fivefold cross validation while the other four rounds used tuned hyper-parameters in the first round. Figure 17.8 shows learning curves between epoch numbers versus the loss and DSC values. As shown in Figure 17.8, the best DSC value was achieved when the epoch number was 50. Thus, we used trained parameters at epoch 50 to create the final model. The average execution time for

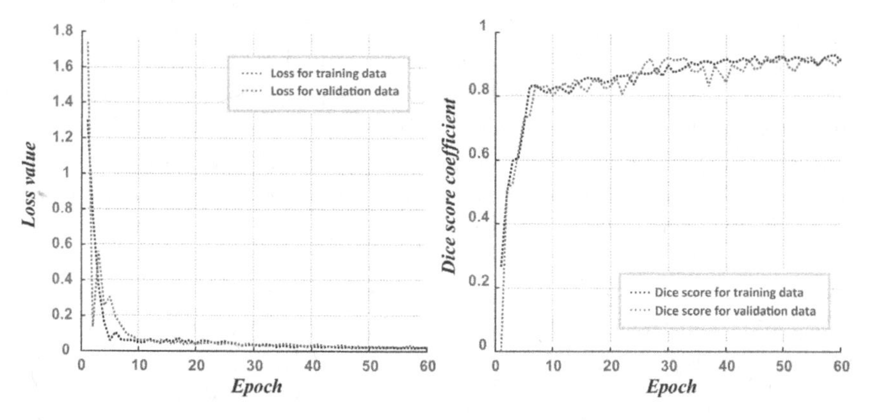

Figure 17.8 Loss and Dice score values during algorithm training. (Left) Loss values for training and validation datasets. (Right) Dice score values for training and validation datasets [1] / SPIE / CC BY 4.0.

the U-Net training with 50 epochs was 633.2311 seconds and the prediction time for a 384 × 384 image was 0.1050 seconds. Computation time for both U-Net training and prediction are highly promising.

Our trained U-Net model was then evaluated using the fivefold cross-validation technique. All holograms for model training were obtained from DHM at 40× magnification. To test the performance of our trained model on holograms obtained from different conditions, we also evaluated the trained U-Net model on holograms obtained from DHM at 20× magnification and bead holograms at 40× magnification. Table 17.2 shows some metric values for holograms obtained from 40× magnifications (125 RBC holograms in fivefold cross validation) or 20× magnifications (50 RBC holograms in fivefold cross validation) and bead holograms (70 bead hologram images in fivefold cross validation). The average and standard deviation values from the fivefold cross validation are given in Table 17.2. The total number of cells is the sum of correctly counted cells and missing counted cells. It is equal to the number of cells in ground-truth images. To compare the performances of different methods, the CNN algorithm in [12] for cell counting from holograms was adopted. PSPNet, another famous algorithm, was also included for the performance evaluation. All algorithm hyper-parameters were tuned using the validation dataset. Figure 17.9 shows the loss values for both CNN and PSPNet algorithms with one training and validation dataset. The initial learning rate was 0.01 for the CNN and 0.001 for the PSPNet. The epoch number was 60 for both CNN and PSPNet. The weight decay value was 0.0001 for both CNN and PSPNet. The input image size for CNN algorithm was 80 × 80 and the architecture was the same as that in [12] (see Figure 17.10). The mirroring method was used for the patch extraction along the image border. The PSPNet architecture is given in Figure 17.11. The input image size for PSPNet was 384 × 384. Resnet34 [13] was used as the encoder. All configurations about POOL, CONV, and UPSAMPLE can be found in [2]. Results from CNN and PSPNet algorithms are also included in Table 17.2. The average training time was 8819.83 seconds for CNN and 576.8960 seconds for PSPNet. For a 384 × 384 image, the average prediction time was 36.8764 seconds for CNN and 0.0872 seconds for PSPNet. As expected, FCN-based algorithms such as U-Net and PSPNet achieved much faster prediction time than CNN algorithms for semantic segmentation because the FCN method could automatically finish all pixel predictions in parallel. Throughput rates for CNN and PSPNet algorithms for different datasets are also given in Table 17.2.

As shown in Table 17.2, the trained model worked well for holograms obtained with DHM at 40× magnification where these holograms were identical to those used in the U-Net training. However, the U-Net model trained from holograms obtained with DHM at 40× magnification did not achieve good prediction results for holograms obtained with DHM at 20× magnification or bead holograms because the DSC value was low, and the number of over-counted cell was high.

Table 17.2 Evaluation of models trained on holograms at 40× magnification.

	RBC hologram (40×)(mean/std)			RBC hologram (20×)(mean/std)			Bead hologram (40×)(mean/std)		
Metrics	**CNN**	**U-Net**	**PSPNet**	**CNN**	**U-Net**	**PSPNet**	**CNN**	**U-Net**	**PSPNet**
DSC	0.85/0.003	0.91/0.002	0.88/0.003	0.32/0.009	0.42/0.02	0.35/0.01	0.21/0.007	0.26/0.01	0.31/0.01
HD	20.90/6.25	14.88/5.51	17.37/4.35	50.48/3.36	43.55/6.93	35.74/7.42	105.17/11.13	79.55/7.4	105.8/5.9
Sensitivity	0.92/0.006	0.96/0.003	0.92/0.003	0.81/0.01	0.83/0.02	0.84/0.02	0.51/0.03	0.76/0.04	0.46/0.04
Accuracy	0.97/0.0004	0.99/0.0003	0.98/0.0005	0.90/0.003	0.92/0.003	0.94/0.002	0.93/0.0004	0.95/0.001	0.95/0.0006
Throughput rate	0.37/0.01	125.4/4.65	148.71/6.6	1.04/0.08	430/27.14	405.73/33.69	0.93/0.04	422.47/47.05	305.96/22.83

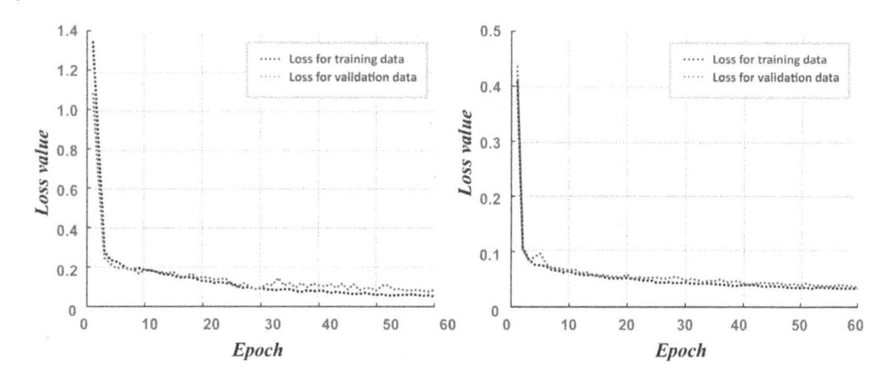

Figure 17.9 Loss values during training for (left) PSPNet and (right) CNN algorithms [1] / SPIE / CC BY 4.0.

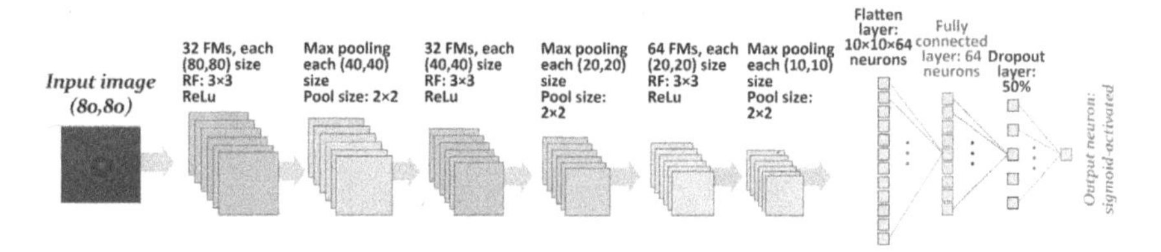

Figure 17.10 Illustration of the CNN architecture. FMs = feature map; RF = receptive field.

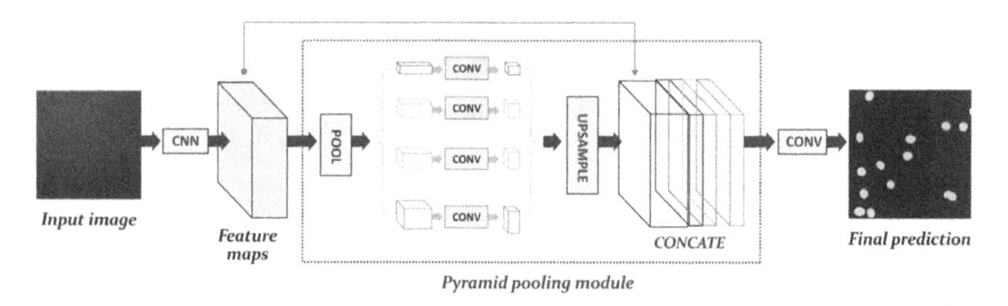

Figure 17.11 Illustration of the PSPNet architecture.

The model's performance decreased when the testing dataset was very different from the training dataset. Results from Table 17.2 also revealed that U-Net and PSPNet models produced very similar results. Both outperformed the CNN model. Figure 17.12 shows some prediction results from the CNN, U-Net, and PSPNet models for holograms obtained with DHM at 40× magnification, DHM at 20× magnification, and bead holograms, respectively.

To further investigate models, we re-trained them. We added 50 holograms obtained with DHM at 20× magnification and 70 bead holograms at 40× magnification into the training dataset. The fivefold cross validation was used.

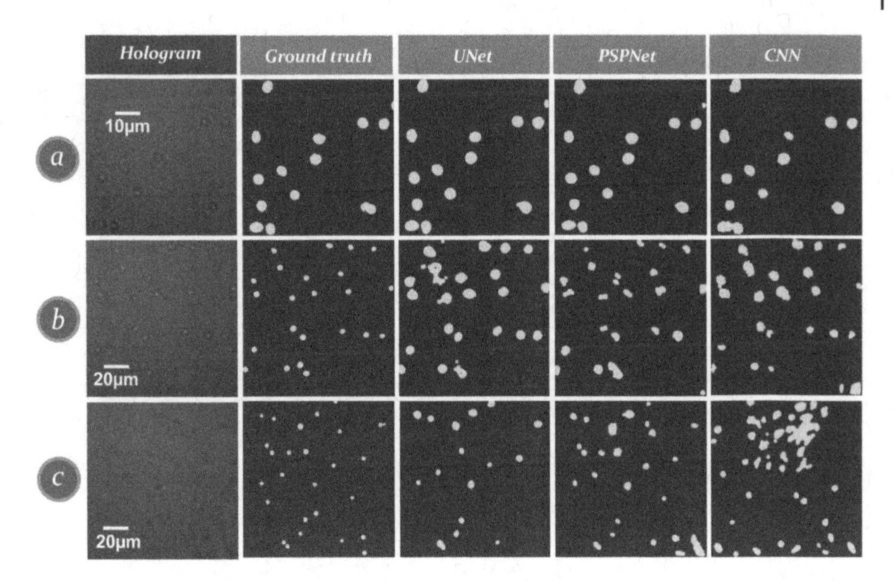

Figure 17.12 Illustration of prediction results. (a) RBCs at 40× magnification, (b) RBCs at 20× magnification, (c) bead holograms at 40× magnification. U-Net, PSPNet and CNN algorithms were trained with data at 40× magnification [1] / SPIE / CC BY 4.0.

The hyper-parameter was tuned using the validation dataset. Table 17.3 shows some evaluation results. We found that model performances for holograms obtained with DHM at 20× magnification and bead holograms greatly increased. The performance for holograms obtained with DHM at 40× magnification was still very good. Note that the performance of U-Net with holograms at 20× magnification was worse than that with holograms at 40× magnification. The reason could be a lack of holograms at 20× magnification in the training dataset because there were 125 holograms at 40× magnification. However, there were only 50 holograms at 20× magnification in the training dataset. The prediction performance for holograms at 40× magnification was also better than that for bead holograms. One reason could be a shortage of bead holograms (70 bead hologram images) used in the training. Another reason could be target complexity in bead holograms because the bead hologram included both focused and defused objects. Some beads were located at a different focus distance than other beads. These focused and defocused beads are shown in reconstructed phase images in Figure 17.13.

Figures 17.14 shows prediction results for input holograms at 40× and 20× magnifications and bead holograms for CNN, U-Net, and PSPNet algorithms. Prediction results in Figures 17.12a and 17.14a are very similar. However, results in Figures 17.14b and 17.14c are much better than results in Figures 17.12b and 17.12c. Furthermore, both focused and defocused beads in the bead holograms

Table 17.3 Evaluation of models trained on holograms from different sources.

	RBC hologram (40×)(mean/std)			RBC hologram (20×)(mean/std)			Bead hologram (40×)(mean/std)		
Metrics	**CNN**	**U-Net**	**PSPNet**	**CNN**	**U-Net**	**PSPNet**	**CNN**	**U-Net**	**PSPNet**
DSC	0.86/0.003	0.91/0.002	0.88/0.003	0.76/0.01	0.88/0.01	0.88/0.02	0.57/0.03	0.72/0.02	0.70/0.02
HD	17.85/4.27	13.17/4.32	15.44/3.85	25.85/9.43	23.08/8.96	22.32/8.17	78.76/6.4	43.51/6.81	51.09/9.38
Sensitivity	0.90/0.004	0.95/0.004	0.93/0.005	0.85/0.01	0.90/0.02	0.93/0.01	0.48/0.03	0.72/0.03	0.68/0.03
Accuracy	0.97/0.0003	0.99/0.0003	0.98/0.0007	0.96/0.0005	0.98/0.0004	0.98/0.001	0.94/0.0004	0.98/0.0003	0.95/0.0003
Throughput rate	0.36/0.01	125.18/4.42	149.90/6.06	0.77/0.08	288.41/26.9	355.73/35.37	0.50/0.04	221.58/17.42	267.01/17.21

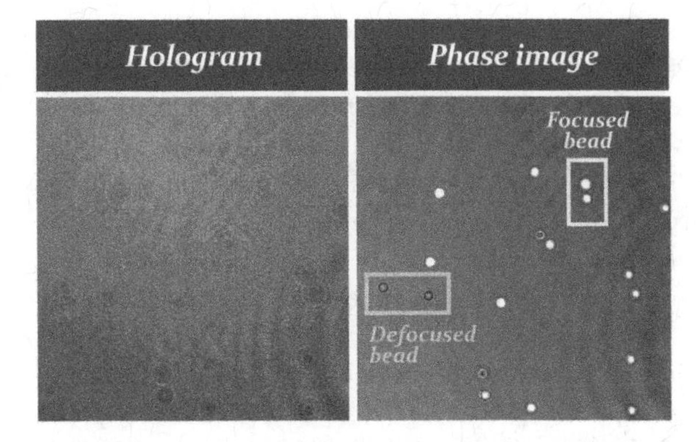

Figure 17.13 Illustration of focused and defocused beads in bead holograms. (Left) Bead hologram and (right) the corresponding reconstructed phase image with some focused and defocused beads indicated [1] / SPIE / CC BY 4.0.

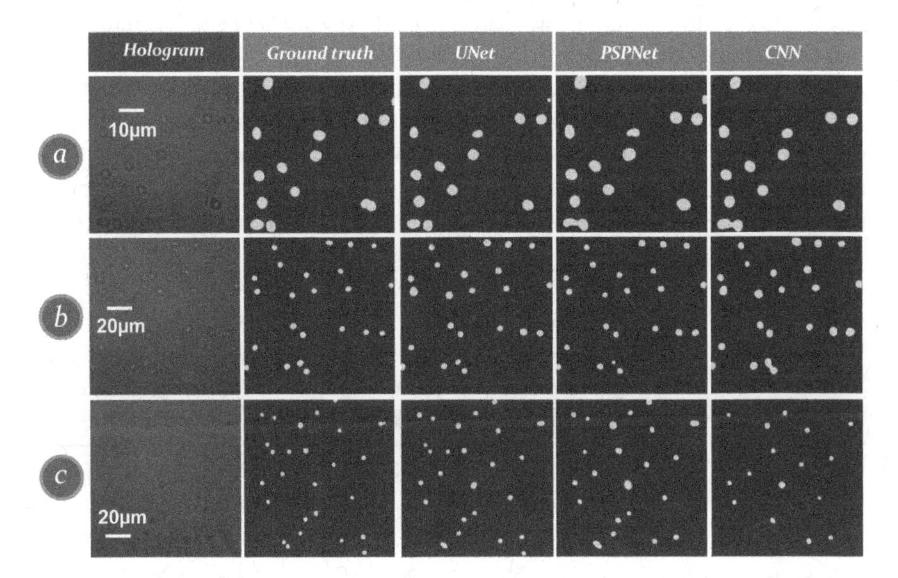

Figure 17.14 Illustration of prediction results. (a) RBCs at 40× magnification, (b) RBCs at 20× magnification, and (c) beads at 40× magnification. U-Net, PSPNet, and CNN algorithms were trained with holograms at 40× and 20× magnifications [1] / SPIE / CC BY 4.0.

were detected when the algorithm was trained on hologram images obtained from different sources. Regarding algorithms, both U-Net and PSPNet outperformed CNN. These findings also imply that it is better to train a deep-learning model using a training dataset that is similar to those used in the testing space. Our simulation results verified that deep FCNs could be used to detect and count cells from holograms directly without needing phase image reconstruction. Therefore, our algorithm can improve the efficiency of label-free detection and counting of cells at the single-cell level. Additionally, our high-throughput cell counting schemes at the single cell level may be further enhanced by increasing the image FOV. Future studies are needed to see if deep FCNs may detect, count, and classify biological cells in lower resolution holograms.

Cell detection and counting from diffraction patterns using deep FCNs can avoid numerical reconstruction, which is complex and inefficient. We also expect that results from diffraction patterns are comparable with those from reconstructed phase images. Consequently, we tested our U-Net model using phase images reconstructed from holograms obtained with DHM at 40× magnification. The training procedure was identical to that used for diffraction patterns. Table 17.4 shows fivefold cross-validation evaluation results for our U-Net model trained with reconstructed phase images; results for our U-Net trained with reconstructed phase images and diffraction patterns are very similar, indicating that cell detection and counting from diffraction patterns can be feasible. In our algorithm training, we applied an image augmentation method to increase the training dataset and reduce the over-fitting problem. To show the importance of image augmentation, we conducted U-Net model training without using any augmentation. Figure 17.15 is a learning curve between epoch number versus loss and DSC value for our U-Net model without any image augmentation.

Table 17.4 Evaluation of UNet based on DHM at 40× magnification.

Metrics	RBC hologram (with image augmentation)	RBC phase image (with image augmentation)	RBC hologram (without image augmentation)
DSC	0.91/0.002	0.92/0.003	0.86/0.003
HD	14.88/5.51	12.17/3.34	21.09/4.91
95th HD	4.30/1.24	3.00/0.75	10.21/2.23
Sensitivity	0.96/0.003	0.96/0.002	0.89/0.007
Accuracy	0.99/0.0003	0.99/0.0002	0.92/0.002
Correctly counted cells	323.4/10.03	323.6/8.16	258.5/12.42
Over-counted cells	5.8/2.18	4.3/1.28	14.1/3.57
Ground-truth cell number	328/11.84	328/11.84	328/11.84

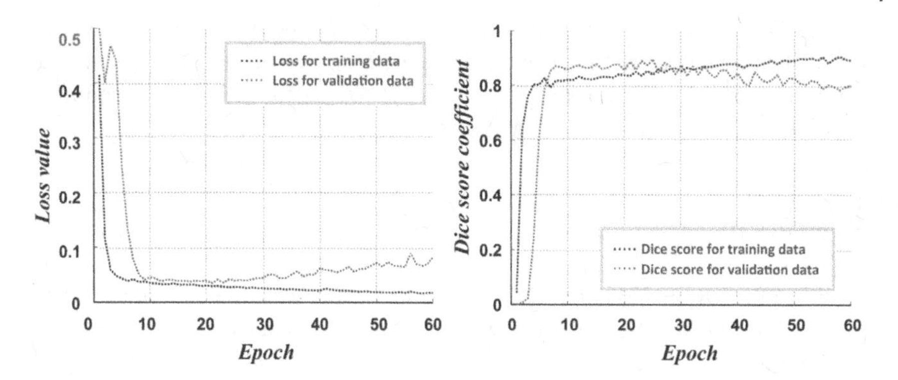

Figure 17.15 U-Net learning curve between (left) epoch numbers versus loss value and (right) DSC values without image augmentation [1] / SPIE / CC BY 4.0.

We noted that the performance for the training dataset became better with increasing epochs while the performance for the validation dataset peaked and then declined. It was a typical example of over-fitting where the model learned the training dataset too well to be generalized to new data. Corresponding evaluation results of our U-Net model without image augmentation are also given in Table 17.4. Results showed that the performance was not as good as that of U-Net with image augmentation during training.

17.4 Conclusions

We trained deep FCNs to detect and count cells directly from the diffraction pattern recorded by DHM, a label-free imaging technique. Simulation results showed that our model was capable of rapidly detecting and counting cells in holograms at the single cell level. Our method can reduce the phase reconstruction step and greatly improve the efficiency and convenience of analytical processing. Furthermore, our model is an end-to-end training and prediction technique that can be applied to an entire image and adapted to images of different sizes. Therefore, it avoids the drawbacks of convolutional neural networks and enables a high-throughput capability with a counting rate of more than 288 cells per second and a FOV of 200 μm × 200 μm. For cell detection and counting from the diffraction pattern, our results also indicated that the U-Net algorithm could achieve very good performance when we used holograms at 40× magnification. Our U-Net model outperformed the CNN algorithm in terms of accuracy and throughput rate. The PSPNet achieved similar results to U-Net, implying that other deep FCNs can also work for cell detection and counting of DHM diffraction patterns. In the future, some explainable deep-learning models could be used to address the black box feature of deep-learning models. We can also apply explainable machine-learning algorithms such as decision trees to features extracted from deep-learning models for cell detection and counting. We believe our deep-learning model provides a promising tool for rapid cell counting and applications in the field of hematology.

References

1 Yi, F., Park, S., and Moon, I. (2021). High-throughput label-free cell detection and counting from diffraction patterns with deep fully convolutional neural networks. *J. Biomed. Opt.* 26 (3): 036001.

2 Zhao, H., Shi, J., Qi, X. et al. (2017). Pyramid scene parsing network. In: *2017 IEEE Conference on Computer Vision and Pattern Recognition (CVPR)*, Honolulu, HI, USA (21–26 July 2017), 6230–6239. New York (NY): IEEE.

3 LeCun, Y., Bengio, Y., and Hinton, G. (2015). Deep learning. *Nature* 521 (7553): 436–444.

4 Deng, L. and Yu, D. (2014). *Deep Learning: Methods and Applications*. Hanover, MA: Now Publishers.

5 Liu, W., Wang, Z., Liu, X. et al. (2016). A survey of deep neural network architectures and their applications. *Neurocomputing* 234: 11–26.

6 He, K., Zhang, X., Ren, S., and Jian, S. (2015). Delving deep into rectifiers: Surpassing human-level performance on ImageNet classification. In: *2015 IEEE International Conference on Computer Vision (ICCV)*, Santiago, Chile (7–13 December 2015), 1026–1034. New York (NY): IEEE.

7 Umbaugh, S. (2017). *Digital Image Processing and Analysis: Applications with MATLAB and CVIPtools*. CRC Press.

8 Jung, A. (2019). *Imgaug Documentation*. Readthedocs.

9 A. Paszke, S. Gross, F. Massa et al. (2019). PyTorch: An imperative style, high-performance deep learning library. arXiv. https://doi.org/10.48550/ARXIV.1912.01703.

10 Simões, R., Wortel, G., Wiersma, T. et al. (2019). Geometrical and dosimetric evaluation of breast target volume auto-contouring. *Phys. Imaging Radiat. Oncol.* 12.

11 Rockafellar, R. and Wets, R. (2005). *Variational Analysis*. Springer-Verlag.

12 Trujillo, C. and Garcia-Sucerquia, J. (2019). Automatic detection and counting of phase objects in raw holograms of digital holographic microscopy via deep learning. *Opt. Lasers Eng.* 120: 13–20.

13 Ker, J., Wang, L., Rao, J., and Lim, T. (2017). Deep learning applications in medical image analysis. *IEEE Access* 6: 9375–9389.

18

Automated Tracking of Temporal Displacements of Red Blood Cells

18.1 Introduction

RBCs, particularly spherocytes, may exhibit temporal displacement or movement when their membranes are loosely attached to their substrates during sedimentation on a glass surface. In addition, RBCs investigated with time-lapse DHM may undergo lateral displacement due to microscope drift. Consequently, it is inevitable to develop a tracking algorithm to automatically investigate RBCs with temporal displacement or movement over time even though these RBCs are attached to a coverslip (they show temporal displacement when their membranes are loosely attached to a glass surface or due to microscope drift). By using a tracking scheme, it is possible to quantitatively analyze temporal displacement and 3D morphology of RBCs. This scheme may also be used in other studies like analyses of RBC fluctuation since it can localize the same RBC in time-lapse sequences. In this chapter, we will introduce a tracking algorithm to measure temporal movements of RBCs along x and y axes. We will focus on RBCs attached to a coverslip with 2D movements.

For cell tracking, there are several algorithms based on segmentation and tracking [1, 2]. These approaches first segment target objects and then establish the association of segmented cells between successive frames. Segmentation-based tracking algorithms need target objects to have a sharp border and the same object in adjacent frames to share an overlapping area. Otherwise, it is difficult to obtain accurate tracking results. Other types of cell-tracking algorithms are based on object model adjustment [3–6]. These approaches attempt to optimize a parameterized model shape to fit the model of the targeted object in the previous frame. Algorithms based on model adjustment such as snakes [3], level set [4], and mean-shift [5, 6] are reported to successfully track cells. However, tracking algorithms based on snakes and level set easily fail to provide an appropriate tracking outcome when the target cell does not share a partial overlap with adjacent frames. Although mean-shift-based algorithms do not have the aforementioned

Artificial Intelligence in Digital Holographic Imaging: Technical Basis and Biomedical Applications, First Edition. Inkyu Moon.

disadvantages, they take a long time to successfully search for the target cell when the tracked cell's positions in successive frames are widely separated. Furthermore, a mean-shift algorithm will fail to find the target cell in this situation due to the presence of several similar cells within the search window and when the scale or shape of the target cell changes quickly. To overcome these drawbacks, many modified mean-shift algorithms have been presented [7, 8]. Most of them aim to design robust kernels that are suitable for targets with a changing scale and shape. A Kalman filter [9] has also been applied to improve the efficiency of the mean-shift tracking algorithm. The Kalman filter can broadly reduce the search time and increase the tracking accuracy of the mean-shift tracking algorithm since it can predict the location of the target in the next frame. On the other hand, it is difficult to obtain the measurement value when we only use a Kalman filter to link the target cell.

In this study, we will combine the mean-shift method with the Kalman filter to achieve the goal of RBC tracking [10]. The mean-shift algorithm has a reasonable computational speed and a robust tracking performance when the image background is not complicated, and the target speed varies smoothly. It is also easy to implement computationally. On the other hand, a Kalman filter can predict the object position of future states and use a recursive method to estimate the target position, which can minimize the mean of the squared error when the process and measurement noise are independent with a normal probability distribution. For RBC tracking, the RBC target needs to be selected in the first frame. The Kalman filter is then applied to this RBC to predict the location of the RBC in the next frame. Mean-shift then searches the RBC from the predicted target position. Consequently, the search time can be dramatically decreased, and the tracking accuracy increased when the speed of the RBC is high or abruptly changes. Once the mean-shift algorithm has reached convergence, the center point of the search window is used as a measurement value for the Kalman filter to estimate the position of the tracked RBC. As a result, the velocity of the target RBC and the error covariance of the Kalman filter can be updated for the next prediction. Since the scale, shape, and direction of the target RBC may change during the capture stage of DHM, conventional kernels used in mean-shift such as Gaussian or Epanechnikov kernels [7, 11] may not be effective in modeling the RBC target, resulting in tracking failure. Accordingly, we designed a novel kernel adaptable to changes in RBC scale, shape, and direction. The designed kernel is an ellipse that resembles the shape of RBCs rather than a rectangle. Parameters such as lengths of major and minor axes used to represent the ellipse must be extracted at each frame to renew the kernel. Consequently, it is necessary to segment the tracked RBC to obtain values of these parameters as well as RBC properties such as average optical path difference (OPD) and projected surface area (PSA). To extract the target RBC after tracking in each frame, we used the marker-controlled watershed-segmentation algorithm (see Chapter 12 for more details) since the center point of the tracked RBC could be treated as an internal marker and the expanded ellipse surrounding the RBC could be regarded as an external marker. In other words, our tracking method contributes to the determination of internal and external markers required in the marker-controlled watershed algorithm. The marker-controlled watershed method has the advantage of obtaining an isolated target, which is helpful for RBC feature measurement. It also has an acceptable speed (see Chapter 12).

18.2 Mean-shift Tracking Algorithm

The conventional mean-shift tracking method based on a color histogram distribution has four basic steps [7, 11], which are described in this section.

18.2.1 Representation of Target Model

In the previous frame, the center point of the target is assumed to be x_0. There are n pixels within the target. The gray level of the target is divided into m bins. The color histogram distribution within the target model region is denoted as follows [11]:

$$q_b(x_0) = C\sum_{i=1}^{n} k\left(\left\|\frac{x_i - x_0}{h}\right\|^2\right)\delta(b(x_i) - b), b = 1...m, \tag{18.1}$$

where q_b is the probability of the quantized histogram bin b in the target model, $b(x_i)$ indicates the quantized histogram bin at point x_i, $k(x)$ means the kernel function, namely a weighted function in which greater weight is assigned when the pixel point is close to the center point, and h denotes the bandwidth of kernel function $k(x)$. Constant C is a normalized factor that satisfies $\sum_{i=1}^{m} q_b = 1$ and is defined [11] as

$$C = \frac{1}{\sum_{i=1}^{n} k\left(\left\|\frac{x_i - x_0}{h}\right\|\right)}$$

where $\delta(x)$ denotes a Kronecker delta function with the properties

$$\delta(x) = \begin{cases} 0 & \text{when } x \neq 0 \\ 1 & \text{when } x = 0 \end{cases}.$$

18.2.2 Representation of a Candidate Target Model

In the current frame, it is assumed that the center point of the candidate target is y and x_i is the pixel position in the candidate target. The color histogram is quantized into m bins, which should be the same as the target model at the previous frame. Similarly, the candidate target model is calculated [11] by

$$p_b(y) = C_2 \sum_{i=1}^{n} k\left(\left\|\frac{x_i - y}{h}\right\|^2\right) \delta(b(x_i) - b), b = 1...m, \tag{18.2}$$

where p_b is the probability of component bin b in the candidate target model. Definitions for other notations are the same as those in Eq. (18.1) [11]. Consequently, target and candidate target models are described as vectors $\overline{q}_b = \{q_1...q_m\}$, and $\overline{p}_b = \{p_1...p_m\}$, respectively.

18.2.3 Similarity Measurement Based on Bhattacharyya Coefficient

Once target and candidate target models are extracted, the similarity between the two models should be calculated. The purpose of target tracking is to find the best candidate target that can maximize the similarity. Usually, the similarity between these models is obtained by computing the distance as [11]

$$dist = \sqrt{1 - \rho[\overline{p}_b(y), \overline{q}_b(x_0)]}, \tag{18.3}$$

where $\rho[\ ]$ is the Bhattacharyya coefficient between target and candidate models, which is defined as [10]

$$\rho[\overline{p}_b(y), \overline{q}_b(x_0)] = \sum_{b=1}^{m} \sqrt{p_b(y) q_b(x_0)}. \tag{18.4}$$

The geometry meaning of the Bhattacharyya coefficient indicates the cosine value between two unit vectors. It can be used for comparing similarity between different vectors.

18.2.4 Target Localization

The distance between two models is minimized in Eq. (18.3) in the same manner that it is maximized in Eq. (18.4). First, the target search process in the current frame starts from the center point of the target in the previous frame. This point is denoted as point y_0. The candidate target model is computed as $\overline{p}_b = (y_0)$ in the current frame. Then $\rho[\overline{p}_b(y_0), \overline{q}_b(x_0)]$ is measured to achieve a maximum value. $\rho[\overline{p}_b(y_0), \overline{q}_b(x_0)]$ can be approximated according to the following equation by unwrapping a Taylor series to $\overline{p}_b(y_0)$ and omitting high-order items [11]:

$$\rho[\overline{p}_b(y), \overline{q}_b(x_0)] \approx \frac{1}{2} \sum_{b=1}^{m} \sqrt{\overline{p}_b(y_0) \overline{q}_b(x_0)} + \frac{1}{2} \sum_{b=1}^{m} \overline{p}_b(y_0) \sqrt{\frac{\overline{q}_b(x_0)}{\overline{p}_b(y_0)}}. \tag{18.5}$$

Using Eq. (18.2), we can rewrite Eq. (18.5):

$$\rho[\overline{p}_b(y), \overline{q}_b(x_0)] \approx \frac{1}{2}\sum_{b=1}^{m}\sqrt{\overline{p}_b(y_0)\overline{q}_b(x_0)} + \frac{C_2}{2}\sum_{i=1}^{N}\overline{w}_i k\left(\left\|\frac{x_i - y}{h}\right\|\right), \tag{18.6}$$

where $\overline{w}_i$ is defined by $\overline{w}_i = \sum_{b=1}^{m}\sqrt{\frac{\overline{q}_b(x_0)}{\overline{p}_b(y_0)}}\delta[b(x_i) - b]$.

Since the first term in Eq. (18.6) is not related to the center point of the potential candidate target, only the second term in Eq. (18.6) needs to be maximized to obtain the smallest distance in Eq. (18.3). In fact, the second term in Eq. (18.6) is the probability density estimation in the current frame. The new center point y_{j+1} of the candidate target can be acquired by a recursion calculation from position y_j [11]:

$$y_{j+1} = \frac{\sum_{i=1}^{n} x_i w_i g\left[\left\|\frac{x_i - y_j}{h}\right\|^2\right]}{\sum_{i=1}^{n} w_i g\left[\left\|\frac{x_i - y_0}{h}\right\|^2\right]},$$

where $g(x) = -k'(x)$. When $y_{j+1} - y_j$ is smaller than a predefined threshold, the recursion can be stopped, and the final position y_{j+1} can be used as the center point of the candidate target in the current frame.

18.3 Kalman Filter

The Kalman filter is a mathematical method to combine the predicted value obtained using data from the previous time step and the measurement value of the current time step to obtain an optimal a posteriori estimated value that can minimize the a posteriori error covariance. The Kalman filter is successfully used in many applications such as automation, military technology, space, and assisted navigation [9, 12].

18.3.1 State Space and Measurement Model

The state space and the measurement model for the discrete Kalman filter are defined [9] by

$$x_k = Ax_{k-1} + Bu_{k-1} + w_{k-1}, \tag{18.7}$$

and $z_k = Hx_k + v_k$, respectively. In the above state space and measurement model equations, $x \in \Re^n$ indicates the state variable vector, subscripts k and $k-1$

represent the time of the vector, $z \in \Re^m$ denotes the measurement variable vector, $u \in \Re^l$ indicates the input vector in the system, A denotes an $n \times n$ matrix, which can form a relation between the state variable at the previous time instant and that of the current time instant, B refers to an $n \times l$ matrix relating the input vector to the state variable, H denotes an $m \times n$ matrix, which relates the state variable to the measurement variable, while w and v represent process and measurement noise, respectively. They are assumed to be independent. In addition, w and v follow a normal probability distribution [9]: $p(w) \sim N[0, Q]$, $p(v) \sim N[0, R]$, where Q and R denote process and measurement noise covariance, respectively. These noises may vary over time. They are usually assumed to be constant. In this study, Q and R were also assumed to be constant. After the state space and measurement are successfully modeled, the Kalman filter algorithm can be applied to them. In this study, the position and velocity information of the target RBC were reflected in the state variable vector.

18.3.2 Kalman Filter Algorithm

The Kalman filter algorithm has two main steps: a prediction step (time update) and a correction step (measurement update). The prediction step can be expressed with the equations [9]:

$$\hat{\hat{x}}_k = A\hat{x}_{k-1} + Bu_{k-1}, \tag{18.8}$$

and

$$p_k^- = Ap_{k-1}A^T + Q, \tag{18.9}$$

where $p_k^- = E\left[\left(x_k - \hat{\hat{x}}_k\right)\left(x_k - \hat{\hat{x}}_k\right)^T\right]$ is an a priori estimate error covariance and $p_k = E\left[(x_k - \hat{x}_k)(x_k - \hat{x}_k)^T\right]$ is a posteriori estimate error covariance. At this step, the a priori estimate of the state variable vector $\hat{\hat{x}}_k$ and the a priori estimate error covariance p_k^- are computed from time step $k-1$ to step k. These estimates are then used at the second step-correction step, which is expressed with the equations [9]

$$K_k = p_k^- H^T\left(HP_k^- H^T + R\right)^{-1}, \tag{18.10}$$

$$\hat{x}_k = \hat{\hat{x}}_k + K_k\left(z_k - H\,\hat{\hat{x}}_k\right), \tag{18.11}$$

and

$$P_k = (I - K_k H)P_k^-, \tag{18.12}$$

where K_k is the Kalman gain used to minimize the a posteriori estimate error covariance p_k. At this step, the Kalman gain [Eq. (18.10)] is calculated with the

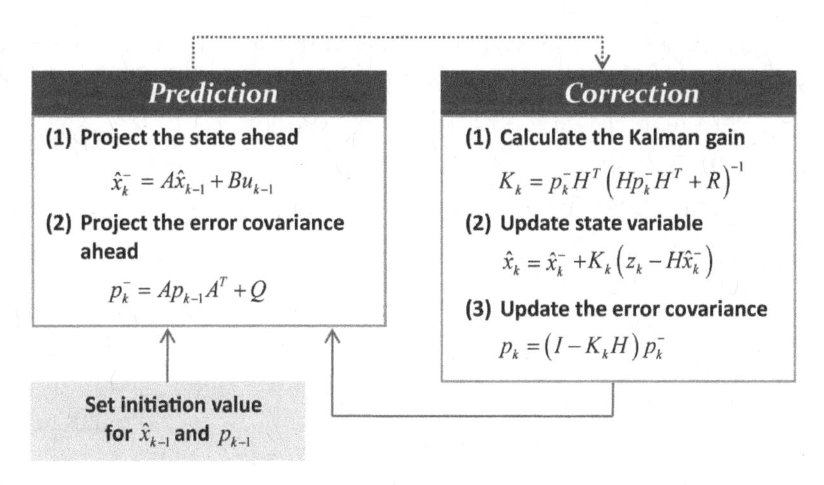

Figure 18.1 Flow diagram of the Kalman filter algorithm.

predicted value obtained in the previous step. The a posteriori state estimate of the state variable $\hat{x}_k$ is then corrected with the Kalman gain. Finally, the a posteriori error covariance P_k is generated for use at the prediction step in the next time instant. The equations within these two steps [Eq. (18.8) to Eq. (18.12)] are recursively called. As a result, the Kalman filter can track a target effectively because it only needs to store the data from the previous step. The process of the Kalman filter algorithm is illustrated in Figure 18.1 [9].

18.4 Procedure for Single RBC Tracking

18.4.1 System State Modeling

Before RBC tracking, we need to model the state of the system. The input vector u_{k-1} in Eq. (18.7) is set at 0 since there are no input data. The center point and velocity of the RBC are used as tracking parameters in the Kalman filter. The state variable vector in this RBC tracking system is expressed by $x_k = \left[c_x^k c_y^k v_x^k v_y^k\right]$, where $\left(c_x^k c_y^k\right)$ is the center point of the tracked RBC at time k along x and y directions; v_x^k and v_y^k denote the velocity of the tracked RBC in x and y directions at time k, respectively. In each time step, the velocity is different. v_x^k and v_y^k are defined by

$$v_x^k = \frac{C_x^k - C_x^{k-1}}{\Delta t}, v_y^k = \frac{C_y^k - C_y^{k-1}}{\Delta t}, \tag{18.13}$$

where variable Δt means the interval time between two frames. Here, the interval time Δt is 0.05 seconds. Other parameters in this system are modeled by

$$H = \begin{bmatrix} 1 & 0 & 0 & 0 \\ 0 & 1 & 0 & 0 \end{bmatrix}, \tag{18.14}$$

$$A = \begin{bmatrix} 1 & 0 & 0.05 & 0 \\ 0 & 1 & 0 & 0.05 \\ 0 & 0 & 1 & 0 \\ 0 & 0 & 0 & 1 \end{bmatrix}, \tag{18.15}$$

$$R = \begin{bmatrix} r & 0 & 0 & 0 \\ 0 & r & 0 & 0 \\ 0 & 0 & r & 0 \\ 0 & 0 & 0 & r \end{bmatrix}, \tag{18.16}$$

and

$$Q = \begin{bmatrix} q & 0 & 0 & 0 \\ 0 & q & 0 & 0 \\ 0 & 0 & q & 0 \\ 0 & 0 & 0 & q \end{bmatrix}. \tag{18.17}$$

After modeling the system, the next step is to use these matrices in the Kalman filter and combine them with the mean-shift algorithm to predict and track the target RBC.

18.4.2 Kernel Design in Mean-shift Algorithm

Because the shape, direction, and scale of RBCs can change over time, the kernel used in the conventional mean-shift algorithm is not robust enough to track the changed target without an update operation. In this case, it is easy to lose the target RBC during tracking since the target histogram obtained by this kernel is not correct. Therefore, we designed a novel kernel that is adaptive to changes in RBC scale, shape, and direction to better cover the target and establish its intensity histogram. A single RBC is approximately elliptical in shape and tends to be a circle, so an ellipse-like kernel can generally cover the target RBC. We can define the kernel for RBC tracking in the mean-shift algorithm as

$$k(x, y, a, b, \theta) = \frac{1}{2\pi ab} \times exp\left[-\left(\frac{(xcos\theta + ysin\theta)^2}{2a^2} + \frac{(ycos\theta - xsin\theta)^2}{2b^2}\right)\right], \tag{18.18}$$

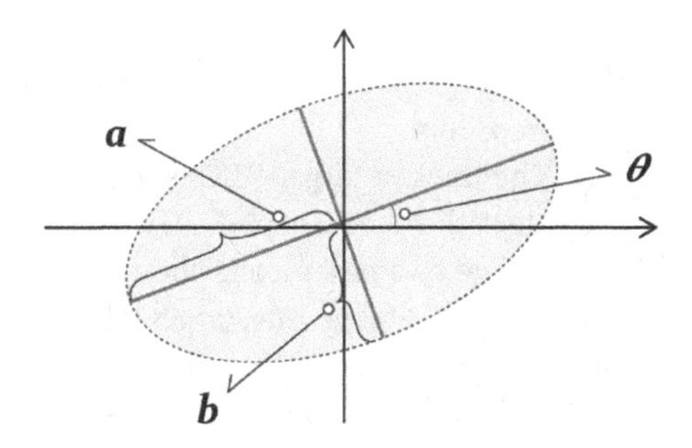

Figure 18.2 Illustration of ellipse kernel parameters.

where (x, y) is a point within the ellipse kernel, a and b are lengths of major and minor axes, respectively; θ is the angle between the x-axis and the major axis of the ellipse kernel. These parameters of the ellipse kernel are shown in Figure 18.2.

18.4.3 Single RBC Tracking with Mean-shift Algorithm and Kalman Filter

The process of RBC tracking with mean-shift algorithm and Kalman filter is described as follows:

Step 1: Select a single target RBC in the first frame. It can be manually or automatically obtained using motion detection methods [13, 14]. The target RBC is then segmented to calculate its center point and kernel parameters for tracking. Once the region of the target RBC is determined, extract the isolated RBC using the marker-controlled watershed algorithm since both internal and external markers can be easily obtained.

Step 2: With the center point of the target RBC extracted in the current frame, the predicted center point in the next frame is obtained using the Kalman prediction process. Before we use the Kalman filter, the process noise covariance (Q) and the measurement error covariance (R) should be initiated. These values can be calculated by tracking a series of RBC images manually. Q and R indicate the covariance matrix of prediction and observation errors, respectively. The velocity of the target RBC is also initiated. It is initially set to 0 because it can be updated once the RBC is successfully tracked in the next frame. The estimate error covariance P_0 must be initiated with a value that does not equal 0 [9, 12].

Step 3: Mean-shift starts to search for the RBC from the center point predicted by the Kalman filter in Step 2. After the mean-shift reaches convergence, the converged point is considered the measurement value in the Kalman filter. The Kalman correction procedure is then executed to obtain the estimated center point of the RBC and the estimate error covariance is updated accordingly. The speed of the target RBC in Eq. (18.13) is updated and used in the next Kalman prediction procedure.

Step 4: The tracked target RBC is segmented using the marker-controlled watershed algorithm (see Chapter 12) with the estimated center point. Corresponding ellipse edge and parameters shown in Figure 18.2 are computed to update the ellipse kernel. Meanwhile, target RBC properties such as velocity, motion, mean OPD, and PSA are dynamically observed and measured.

Step 5: Repeat the operation from Steps 2 to 4 until the tracking is complete.

18.4.4 Segmentation of a Single RBC

After the target RBC is tracked, it is segmented to obtain the parameters of the kernel and compute properties of the target RBC. The center point of the RBC is treated as an internal marker and the expanded ellipse edge is considered an external marker. Correct determinations of internal and external markers depend on the success of RBC tracking. Once internal and external markers are appropriately identified, the marker-controlled watershed method is applied to segment the target RBC. Kernel parameters such as a, b, and θ in Eq. (18.18) are calculated and used to update the kernel for RBC tracking in the next frame. Furthermore, RBC properties including OPD, PSA, MCH, MCH surface density, volume, and RBC fluctuation rate can be measured with the segmented RBC. In this study, we demonstrate that our tracking method can be used for the dynamic analysis of an RBC image sequence obtained with time-lapse DHM. The procedure for our RBC tracking algorithm based on the combined mean-shift and Kalman filter models is illustrated in Figure 18.3.

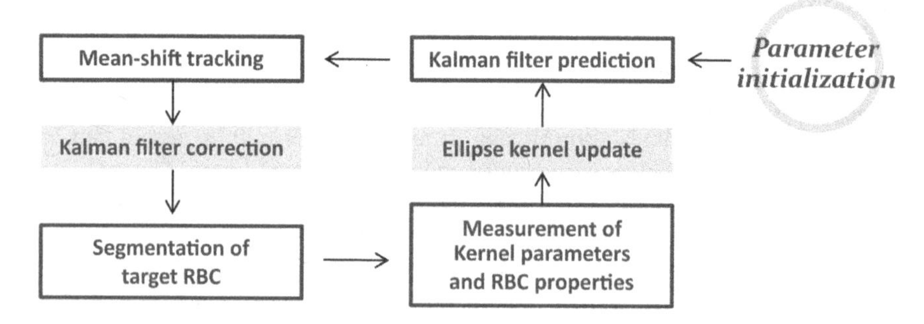

Figure 18.3 Procedure for single RBC tracking.

18.5 Experimental Results

OPD images of RBCs are reconstructed from the holograms recorded by DHM using digital reconstruction methods (see Chapter 5). Some OPD images of RBCs are shown in Figure 18.4.

The RBC target is manually identified in the first frame. The selected target RBC is then tracked using our tracking method, which combines the Kalman filter with mean-shift using the ellipse-like kernel defined in Section 18.4. Experimental results show that our method could successfully track a target RBC. Figure 18.5 shows the tracked target RBC in multiple frames. Consequently, the velocity of the target RBC is updated with Eq. (18.13) using the estimated center point of the target RBC in each frame. Figure 18.6 shows the velocity of the tracked RBC. We noted that there was a lateral movement of the target RBC, possibly due to a loose substrate attachment or microscope drift in the RBC OPD image sequence. The trajectory of the target RBC after tracking is also shown in Figure 18.7. Note that the total travel distance of the tracked RBC was approximately 23.30 μm in 10 seconds.

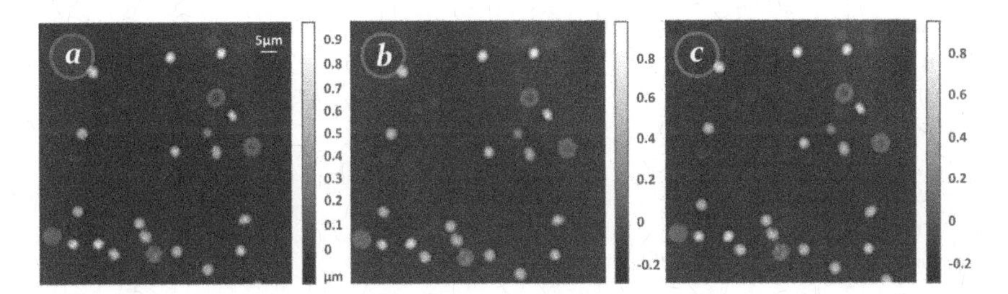

Figure 18.4 RBC's OPD images in (a) frame #1, (b) frame #100, and (c) frame #200 [10] / with permission of Optical Publishing Group.

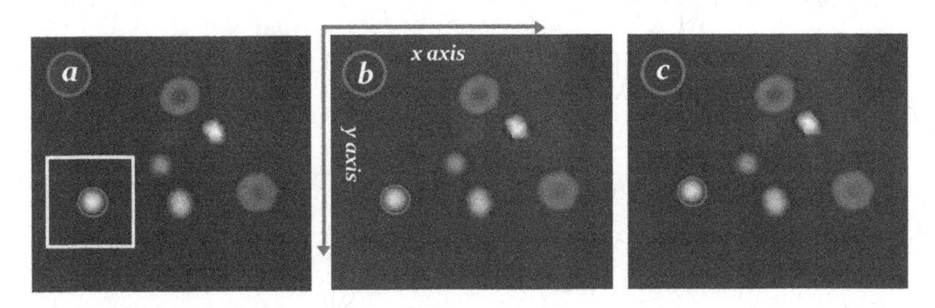

Figure 18.5 Tracked target RBC in (a) frame #1, (b) frame #100, and (c) frame #200 [10] / with permission of Optical Publishing Group.

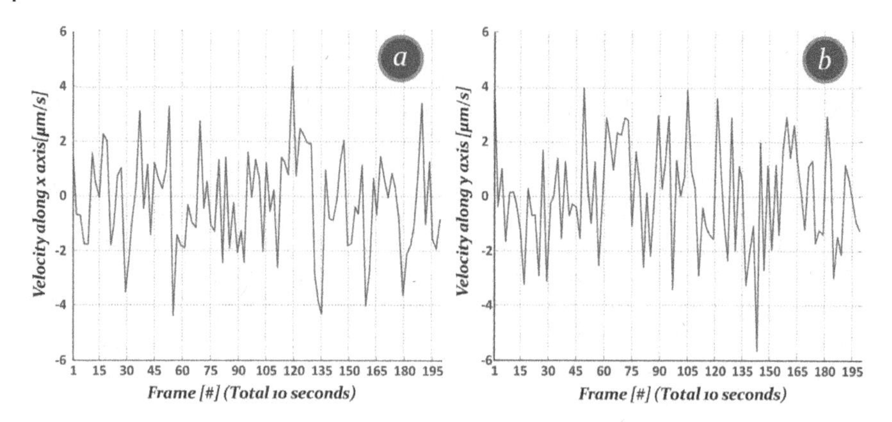

Figure 18.6 Velocity of target RBC along the (a) x-axis and (b) y-axis within 10 seconds.

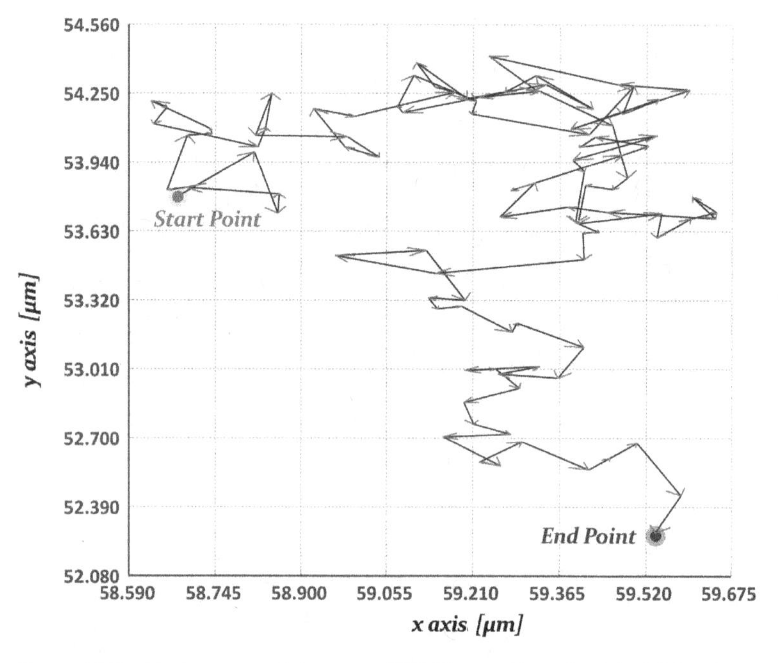

Figure 18.7 Motion curve of the target RBC. The total length of the RBC trajectory is about 23.30 μm in 10 seconds.

The next step is to segment the tracked RBC with the marker-controlled watershed algorithm to obtain RBC properties and update ellipse-like kernel parameters. The segmented target RBC in a different frame is shown in Figure 18.8. The target RBC's biophysical properties are computed with the segmented RBC after tracking. This method allows clinical parameters of RBCs, particularly its membrane fluctuations, to be dynamically analyzed, which could be beneficial for testing potential therapies and analyzing RBC-related diseases. Computed features, including OPD and PSA, based on tracked RBC are shown in Figure 18.9. Parameters of the ellipse-like kernel are updated simultaneously.

Previous simulation results showed that our method could successfully track spherocyte RBCs. To illustrate that our algorithm could also work for other RBC types, Figure 18.10 shows tracking results for a discocyte RBC. The trajectory of the discocyte RBC is given in Figure 18.11. Note that the movement direction or behavior of a spherocyte RBC (see Figure 18.7) is different from that of a discocyte RBC (see Figure 18.11). In our tracking algorithm, each target has its own model

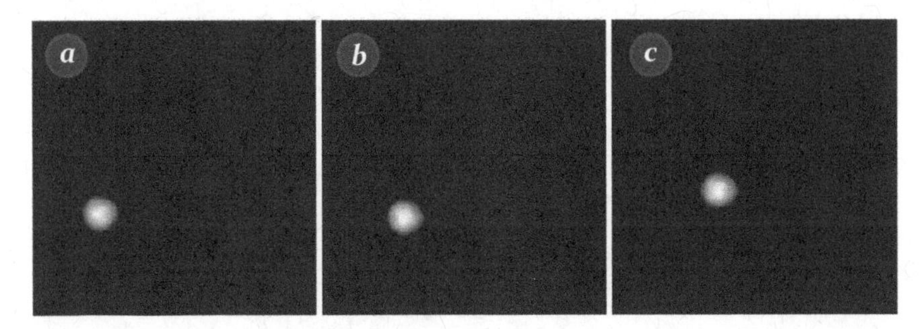

Figure 18.8 Segmented target RBC in (a) frame #1, (b) frame #100, and (c) frame #200 [10] / with permission of Optical Publishing Group.

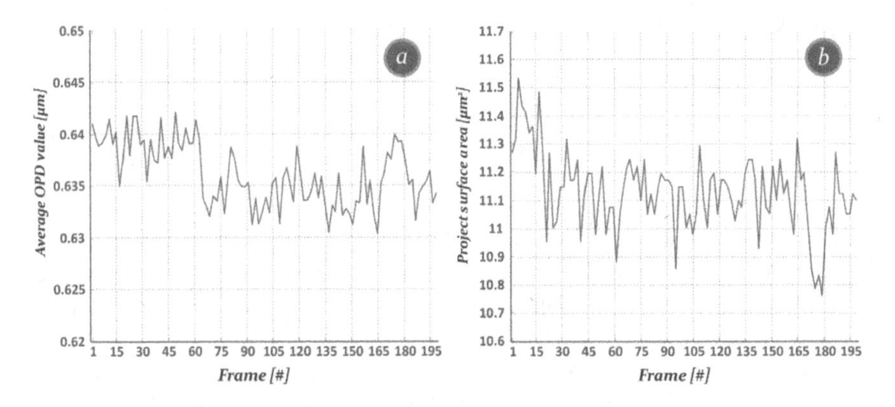

Figure 18.9 Properties of the target RBC: (a) Mean OPD and (b) PSA.

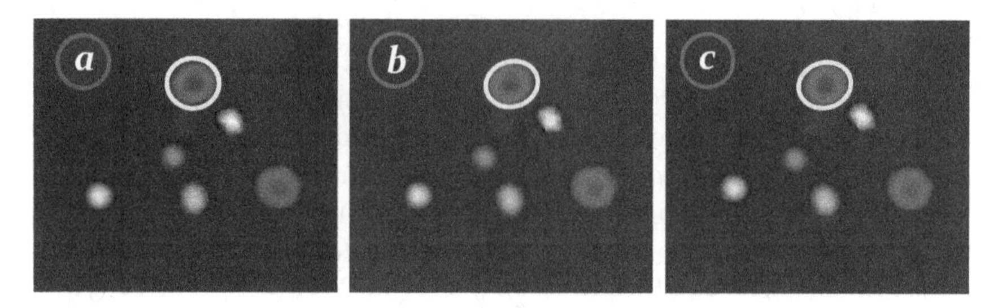

Figure 18.10 Tracked target discocyte RBC in (a) frame #1, (b) frame #100, and (c) frame #200 [10] / with permission of Optical Publishing Group.

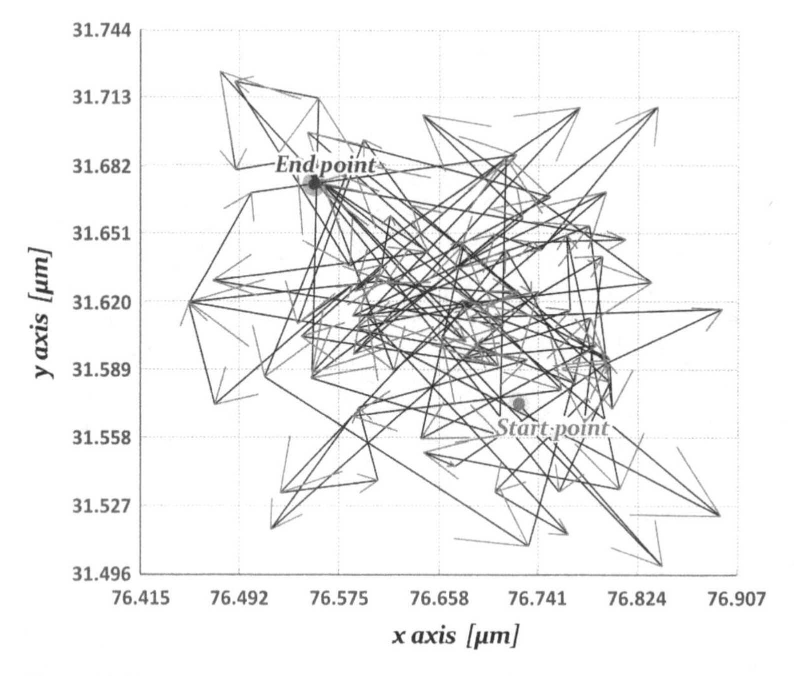

Figure 18.11 Motion curve of the target discocyte RBC. Total length of the trajectory of this tracked discocyte RBC is about 16.22 μm in 10 seconds.

which is presented by Eq. (18.1) so that each object is tracked using its own model. We can also track multiple RBCs when multiple models are designed before tracking. Therefore, our tracking algorithm can be applied to all RBCs in the field of view.

To evaluate the accuracy of estimated target RBC positions with our tracking algorithm, centroid points of the target RBC were calculated in time-lapse holographic images. These centroid points were then used as ground-truth values for comparison. The MSE between estimated RBC positions and measured centroid points from segmented target RBC was calculated in both x and y axes. Four RBCs randomly chosen from time-lapse holographic images were used as tracking targets. The corresponding MSE was then computed. Estimated RBC positions and measured centroid points for four target RBCs are shown in Table 18.1. Here, 10 position values are presented, although there are a total of 200 frames in time-lapse OPD images. MSEs between estimated RBC positions and calculated centroid points in x and y axes are also given in Table 18.1. We found that the

Table 18.1 Comparison of RBC position measurements.

	Cell 1	Cell 2	Cell 3	Cell 4
Estimated RBC positions with our tracking algorithm	(12.33,35.33)	(58.75,53.86)	(26.29,27.64)	(56.76,14.76)
	(12.35,35.75)	(58.70,54.27)	(26.02,27.46)	(56.82,14.69)
	(12.67,35.84)	(58.96,54.20)	(25.70,27.12)	(56.91,14.71)
	(12.75,35.77)	(59.10,54.24)	(25.60,26.85)	(56.77,14.73)
	(13.11,35.79)	(59.44,54.28)	(25.89,27.15)	(56.85,14.73)
	(13.32,35.29)	(59.65,53.62)	(25.26,27.80)	(56.93,14.85)
	(12.90,35.38)	(59.29,53.89)	(24.75,27.66)	(56.98,14.86)
	(13.10,35.20)	(59.39,53.59)	(24.96,27.86)	(56.90,14.85)
	(13.23,34.74)	(59.42,53.07)	(24.47,28.07)	(56.74,14.81)
	(13.52,34.34)	(59.22,52.72)	(24.57,28.17)	(56.75,14.90)
Calculated centroid points of the target RBCs	(12.09,35.34)	(58.74,53.70)	(25.96,27.28)	(57.11,14.33)
	(12.31,36.27)	(58.75,54.34)	(25.66,27.60)	(57.14,14.37)
	(12.55,35.65)	(58.92,54.00)	(25.69,26.70)	(56.94,14.37)
	(12.59,35.72)	(59.01,54.02)	(25.69,26.78)	(56.84,14.35)
	(12.92,36.12)	(59.48,54.32)	(25.79,27.32)	(56.87,14.69)
	(13.01,35.03)	(59.26,53.51)	(25.39,27.34)	(57.11,14.79)
	(12.63,35.54)	(58.94,53.58)	(24.98,27.67)	(57.10,14.31)
	(12.96,34.86)	(59.33,53.41)	(25.64,27.62)	(56.92,14.79)
	(13.25,34.53)	(59.46,53.02)	(24.81,28.19)	(57.13,14.74)
	(13.56,34.40)	(59.20,52.61)	(24.18,28.09)	(56.81,14.77)
MSE	0.0587 μm^2 (in x-axis) 0.0543 μm^2 (in y-axis)	0.0454 μm^2 (in x-axis) 0.0375 μm^2 (in y-axis)	0.0546 μm^2 (in x-axis) 0.0496 μm^2 (in y-axis)	0.0545 μm^2 (in x-axis) 0.0624 μm^2 (in y-axis)

MSE for each of the four target RBCs was very small. The average MSE was 0.05330 μm^2 along the x-axis and 0.05095 μm^2 along the y-axis. Root MSEs were 0.230 μm and 0.226 μm along the x- and y-axis, respectively. Considering that the pixel resolution was 0.155 μm, our tracking algorithm's inaccuracy was less than 2 pixels along both x and y axes. These experimental results demonstrate that our tracking method can measure the temporal displacements of RBCs with high precision and accuracy.

18.6 Conclusions

We presented a method to track a single RBC with temporal displacements in OPD images obtained by time-lapse DHM using the integration of a Kalman filter and the mean-shift algorithm. The Kalman filter was used to predict the position of the target RBC in the next frame to increase the efficiency of the mean-shift method. Moreover, an ellipse-like kernel adaptive to changes in RBC scale, shape, and direction was designed to establish the target model for mean-shift tracking. It greatly reduced the effect of background during mean-shift tracking. The tracked RBC was segmented using a marker-controlled watershed algorithm to obtain the isolated target RBC. After segmentation, parameters of the ellipse-like kernel were recursively updated. Moreover, we demonstrated that the OPD value of RBCs with temporal displacements could be dynamically computed. Our tracking method could be used for the quantitative analysis of RBC membrane fluctuation or dynamics to understand RBC-related diseases.

References

1 Chen, X., Zhou, X., and Wong, S. (2006). Automated segmentation, classification, and tracking of cancer cell nuclei in time-lapse microscopy. *IEEE Trans. Biomed. Eng.* 53: 762–766.

2 Yang, X., Li, H., Zhou, X., and Wong, S. (2005). Automated segmentation and tracking of cells in time-lapse microscopy using watershed and mean shift. In: *2005 International Symposium on Intelligent Signal Processing and Communication Systems*, Hong Kong, China (13–16 December 2005), 533–536. New York (NY): IEEE.

3 Ray, N., Acton, S., and Ley, K. (2002). Tracking leukocytes in vivo with shape and size constrained active contours. *IEEE Trans. Med. Imaging* 21: 1222–1235.

4 Mukherjee, D., Ray, N., and Acton, S. (2004). Level set analysis for leukocyte detection and tracking. *IEEE Trans. Image Process.* 13: 562–572.

5 Debeir, O., Ham, P., Kiss, R., and Decaestecker, C. (2005). Tracking of migrating cells under phase-contrast video microscopy with combined mean-shift processes. *IEEE Trans. Med. Imaging* 24: 697–711.

6 Tang, C., Dang, H., and Su, X. (2008). Multi-thread, increment-bandwidth and weighted mann-shift algorithm for neural stem cells tracking. In: *2008 Second International Symposium on Intelligent Information Technology Application*, Shanghai, China (20–22 December 2008), 576–579. New York (NY): IEEE.

7 Comaniciu, D., Visvanathan, R., and Peter, M. (2000). Real-time tracking of non-rigid objects using mean shift. In: *2000 Proceedings IEEE Conference on Computer Vision and Pattern Recognition (CVPR)*, Hilton Head, South Carolina, USA (13–15 June 2000), 142–149. New York (NY): IEEE.

8 Yilmaz, A. (2007). Object tracking by asymmetric kernel mean shift with automatic scale and orientation selection. In: *2007 IEEE Conference on Computer Vision and Pattern Recognition (CVPR)*, Minneapolis, MN, USA (17–22 June 2007). New York (NY): IEEE.

9 Welch, G. and Bishop, G. (2006). *An Introduction to the Kalman Filter*. Chapel Hill.

10 Moon, I., Yi, F., and Rappaz, B. (2016). Automated tracking of temporal displacements of a red blood cell obtained by time-lapse digital holographic microscopy. *Appl. Opt.* 55: A86–A94.

11 Comaniciu, D., Ramesh, V., and Meer, P. (2003). Kernel-based object tracking. *IEEE Trans. Pattern Anal. Mach. Intell.* 25: 564–577.

12 Grewal, M. and Anderws, A. (2001). *Kalman Filtering: Theory and Practice Using Matlab*. Wiley.

13 Gonzalez, R. and Woods, R. (2002). *Digital Imaging Processing*. Prentice Hall.

14 Jones, G., Paragios, N., and Regazzoni, C. (2012). *Video-Based Surveillance Systems: Computer Vision and Distributed Processing*. Springer.

19

Automated Quantitative Analysis of Red Blood Cell Dynamics

19.1 Introduction

Mature erythrocytes, or discocytes, are the main type of RBCs in blood circulation. Their biconcave shape and associated deformability are essential functional features. Their exceptional ability to deform, particularly when passing through narrow capillaries during microcirculation, is caused by this structure, which corresponds to the highest surface area for a given volume. Cell-membrane fluctuations (CMFs) exhibited by RBCs reflect their ability to deform. However, the exact mechanism(s) of CMFs remain unclear. Considering that CMFs reflect the biomechanical properties of RBCs, it is important to study the evolution of CMFs over time within the discocyte subpopulation of RBCs. Indeed, it was reported that during their transformation into temporary echinocytes and finally spherocytes, RBCs show a significant CMF decrease [1, 2].

In previous chapters, we showed that DHM systems could measure the morphological properties of discocyte RBCs, including PSA, surface area, sphericity coefficient, and two clinical cell parameters (MCH and mean corpuscular volume [MCV]) at the single cell level. In this chapter, we will introduce automated methods to measure quantitative fluctuation rates at the single RBC level as a function of their storage time using time-lapse DHM [3]. We will also analyze both the ring and dimple areas of RBC separately. Quantitative phase images were acquired every few days over a 71-day period to systematically analyze alterations in RBC parameters over their storage time.

Our quantitative analysis revealed some interesting findings. First, we found that older discocytes (stored for 71 days) exhibited more apparent stiffness than younger ones (stored for 4 days). In particular, the fluctuation rate in the dimple area was greater than that in the ring section of younger RBCs. Moreover, the MCV, MCH, PSA, and surface area did not change significantly during these time intervals. Interestingly, we found that the CMFs of a whole cell (RBCs stored for

Artificial Intelligence in Digital Holographic Imaging: Technical Basis and Biomedical Applications, First Edition. Inkyu Moon.

4 days) showed a significant negative correlation with the sphericity coefficient, indicating that the more that RBCs transition to a spherocyte shape, the less fluctuation they exhibit. In contrast, dimple fluctuations showed a significant positive correlation with the sphericity coefficient.

19.2 RBC Parameters

19.2.1 Cell Thickness

RBC thickness images can be computed using DHM phase images of RBCs using the equation

$$h(x,y) = \frac{\lambda \times \phi(x,y)}{2\pi(n_{RBC} - n_m)}, \tag{19.1}$$

where $h(x, y)$ is the RBC thickness, and n_{RBC} and n_m are refractive indices of the RBC and HEPA buffer, respectively. The value for n_{RBC} is obtained using the decoupling technique [4, 5], which was reported to be 1.418 ± 0.012 for a spherocyte population. The refractive index of HEPA (1.3334 ± 0.0002) is measured at room temperature with an Abbe-2WAJ refractometer.

19.2.2 Membrane Fluctuation Rates

To calculate the fluctuation rate, we defined a region of interest (ROI) and two independent variables: (i) $std(h_{cell} + h_{background})$, the temporal deviation within the RBC area (combining both the cell fluctuations and noise), and (ii) $std(h_{background})$, the mean temporal deviation calculated over all pixels located outside the RBC area (see Figure 19.1). Measured STDs for one pixel outside the cell (Figure 19.2 point "A"), on the ring (Figure 19.2 point "B"), and at the center (Figure 19.2 point "C") were 17, 42, and 29 nm, respectively, indicating that membrane fluctuation amplitudes were significantly larger than background noise level (see Figure 19.2). Accordingly, fluctuations at each pixel CMF(x, y) can be calculated using the following equation [6]:

$$CMF_{cell}(x,y) = \sqrt{\left(std\left(h_{cell} + h_{background}\right)(x,y)\right)^2 - \left(std\left(h_{background}\right)\right)^2}. \tag{19.2}$$

Chapter 18 showed that RBCs occasionally displayed significant lateral displacements in time-lapse sequences due to lose attachment of the cell to substrate (only a small part of the membrane is in contact with the substrate). Such cases also exhibited strong CMF changes. We removed these lateral displacements using *ImageJ* [7] software and the *StackReg* [8] plugin. Another point of concern was the

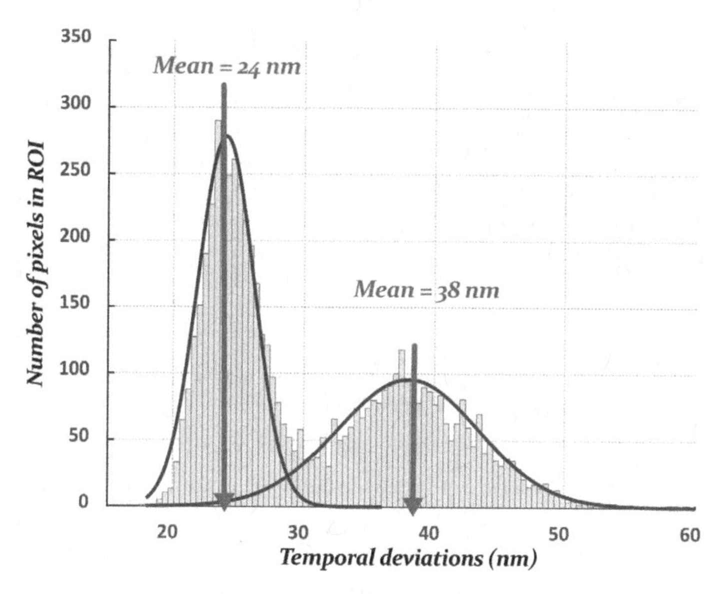

Figure 19.1 Distribution of temporal deviations within an ROI. The left-side distribution represents the background, and the right side corresponds to the RBC area (cell membrane and noise together).

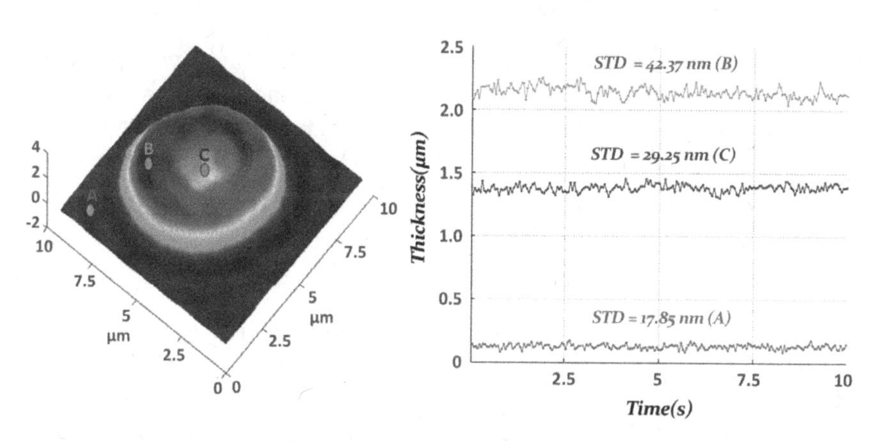

Figure 19.2 Thickness signals and standard deviations (STD) of the changes in three regions recorded at 20 Hz over a 10 s period. Standard deviations for the "A" (background location), "B" (on the cell ring), and "C" (in the dimple region) signals are 17, 42, and 29 nm, respectively [3] / with permission of Optical Publishing Group.

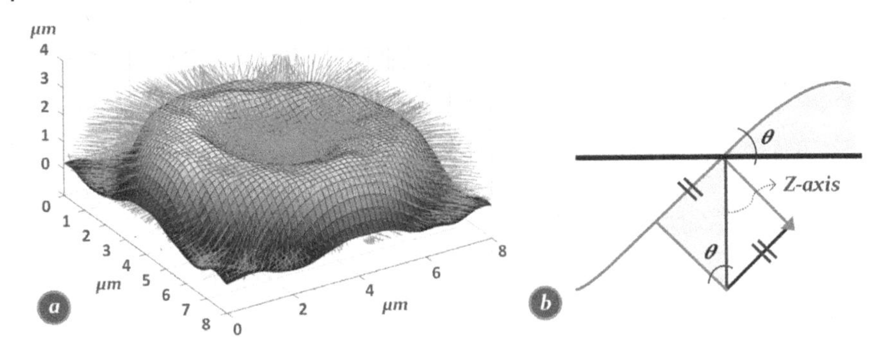

Figure 19.3 (a) RBC mesh with normal vectors. (b) *x*-*z* view of the membrane surface and its normal vector [3] / with permission of Optical Publishing Group.

steep membrane gradient at the RBC border and around the center, where a small lateral displacement of the RBC membrane had a strong impact on the thickness measurement because the thickness was along the *z*-axis (blue vector in Figure 19.3b). To avoid overestimation of cell fluctuations in such steep-slope areas, cell fluctuations should be evaluated in a direction perpendicular (normal) to the RBC membrane (green vectors in Figures 19.3a and 19.3b). Finally, normal fluctuations can be estimated as follows:

$$h_n(x,y) = h(x,y) \times cos\,(\theta(x,y)). \tag{19.3}$$

To calculate θ, normal vectors at each vertex of the RBC mesh were measured (see Figure 19.3a). To do so, a bicubic fit of data in x, y, and z axes was performed. Diagonal vectors were then calculated and crossed to form the normal at each vertex. Finally, θ is computed by:

$$\theta(i,j) = arctan\left(\frac{N_z(i,j)}{\sqrt{N_x^2(i,j) + N_y^2(i,j)}}\right), \tag{19.4}$$

where $N_x(i, j)$, $N_y(i, j)$, and $N_z(i, j)$ are the normal for pixel (i, j) in the x, y, and z directions, respectively.

The deviation map of the RBC is obtained by Eq. (19.2). $std(h_{cell} + h_{background})(x, y)$, the temporal deviation within the ROI that combines both RBC fluctuations and noise, is then calculated. We also computed the average value of $std(h_{background})$, which was the temporal deviation of all pixels outside the RBC's projected area. These two computed values were substituted into Eq. (19.2) and the resulting map evaluated. CMFs of the whole RBC, dimple, and ring are averages of CMF(x, y) over the projected area of the whole RBC, dimple, and ring, respectively. We used binary masks to analyze the ring and dimple parts of RBCs

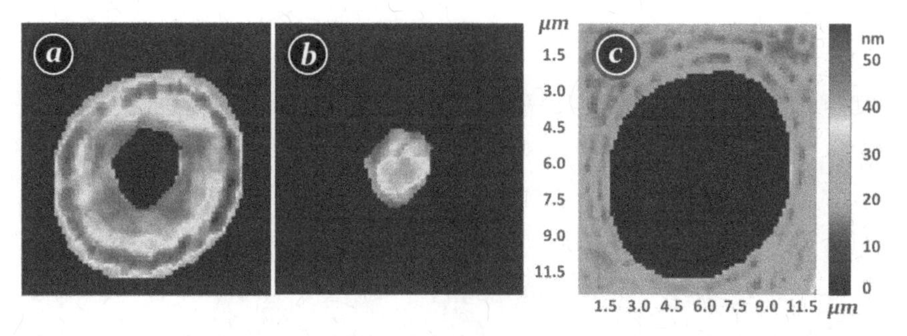

Figure 19.4 Deviation map of the (a) ring and (b) dimple sections of a 4-day-old RBC as well as (c) the background. Color-bar scale is in nanometers [3] / with permission of Optical Publishing Group.

separately as well as isolate the ROI background area. Masks were generated by applying thresholding and a binary operator to implement an erosion effect (of a few microns). The CMF amplitudes of the whole membrane, ring, and dimple were 35 ± 4.7 nm (in agreement with a previously reported value [6]), 35 ± 5.2 nm, and 36 ± 5.3 nm, respectively ($n = 33$, 4 days, see Figure 19.4).

Biconcave and spherocyte morphologies can be separated using sphericity coefficients (stomatocytes are removed from the biconcave set). The sphericity coefficient k is the ratio of the RBC thickness h_c at the cell center to the thickness h_t at a radius that is halfway to the cell perimeter [9]:

$$k = \frac{h_c}{h_t}. \tag{19.5}$$

A threshold value of $k_t = 0.97$ was assigned to distinguish biconcave cells from spherocytes and echinocytes. After making two masks to isolate the cell and background by applying a proper threshold and several image processing techniques, we calculated the PSA of the RBC. The RBC radius can then be evaluated with

$$r \simeq \sqrt{\frac{PSA}{\pi}}, \tag{19.6}$$

where PSA is defined by

$$PSA = Np^2, \tag{19.7}$$

where N is the total number of pixels that make up the RBC projected area resulting from the image segmentation algorithm, and p is the pixel size in the phase image (0.142 μm). Using r, we can choose candidate pixels that are likely to be close to the ring section (blue points in Figure 19.5). Since the above procedure could not identify points located precisely on the ring section, we sought the pixel with the greatest thickness value near each candidate pixel (within an area of size

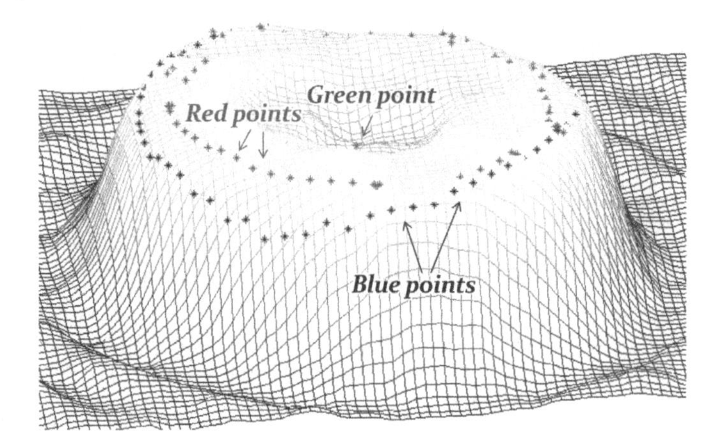

Figure 19.5 Blue points show those points obtained using r. Red points are the maximum values within a range of 3 × 3 pixels from blue points. The green point is the geometric center of the cell obtained by finding the geometrical center of the bounding box on the projected area or RBC [3] / with permission of Optical Publishing Group.

3 × 3 pixels). Finally, h_t was defined as the arithmetic average of the heights for all these updated points (red points in Figure 19.5). The single green point (h_c) was the center of the bounding box of the projected area of RBC on the x–y plane. After image segmentation and thresholding to binarize the RBC projected area, the binary result was bounded with a rectangle. The center of the bounding box was a quick estimation of the center of the RBC projected area.

19.2.3 Morphological- and Hemoglobin-related Parameters

For correlation analysis, we considered morphological parameters such as the MCV, PSA, surface area, sphericity coefficient, and MCH. The RBC volume at the single cell level in the thickness image can be expressed as

$$V \simeq p^2 \sum\nolimits_{(i,j)\in S_p} h(i,j), \tag{19.8}$$

where $h(i, j)$ is the thickness value at pixel (i, j) and S_p is the RBC projected area. The thickness summation is achieved over all pixels (i, j) of S_p. To measure the RBC surface area, we used the method described in Chapter 13.

The RBC MCH can be measured by calculating the dry mass of RBCs since RBCs are mainly composed of hemoglobin [4, 10]:

$$MCH = \frac{10\phi_{SP}\lambda PSA}{2\pi\alpha_{Hb}}, \tag{19.9}$$

where ϕ_{SP} is the mean phase value of the RBC projected area and $\alpha_{Hb} = 0.00196$ dl/g is a constant known as the specific refraction increment, which is related to protein concentration.

19.3 Quantitative Analysis of RBC Fluctuations

In our correlation analysis, we computed the Pearson product–moment correlation coefficient and conducted a *t*-test with a 95% confidence level [11]. Error bars shown in the plots represent twice the corresponding STDs. Numbers written after a ± sign are STDs.

The impact of the number of storage days was assessed by measuring fluctuation rates, morphological parameters, and MCH over time. In this study, more than 32 RBCs were extracted from each image (four blood samples) to measure these parameters. We also found a small increase in the sphericity coefficient, indicating that cells were becoming more spherical, less flexible, and stiffer [9], which could also affect the fluctuation rate. Specifically, the sphericity coefficient increased from 0.74 ± 0.09 for 4-day-old RBCs to 0.79 ± 0.12 for 71-day-old RBCs, consistent with reported values [9]. In the case of older RBCs, the STD of the sphericity coefficient was significantly larger (± 0.12). It illustrates the early stage of the gradual discocyte-spherocyte transformation process. In contrast, the MCH did not change significantly over the storage time. It only fluctuated around its average value of 32 ± 0.6 pg (see Figure 19.6), consistent with previous reports [5, 9]. Moreover, MCH was in agreement with the value obtained with Sysmex KX-21 (see Figure 19.6). This constancy demonstrates that while biconcave RBCs undergo morphological changes during storage, they do not leak their hemoglobin contents into the storage solution.

The CMF amplitude fluctuated over the storage time. However, the overall trend of CMF amplitudes as a function of storage time for the whole RBC, ring, and center decreased (F-statistics suggested that the linear regression line had a slope significantly different from zero, *p*-value <0.05), consistent with previous findings [12]. Our experimental results showed that RBC membranes stiffened as storage time increased, in agreement with reported decreases in flexibility [12, 13]. According to Figure 19.7, the fluctuation amplitude at the dimple area was generally larger than that in the ring region of 4-day-old RBCs ($p < 0.05$; two-sample Kolmogorov–Smirnov test), in agreement with previous findings [5].

Figure 19.8 shows the results of a correlation analysis of fluctuation amplitudes for the whole membrane and dimple of 4-day-old RBCs ($n = 33$) as functions of morphological and MCH parameters. The ring section exhibited the same trend as the whole membrane, so we excluded these results from our correlation analysis

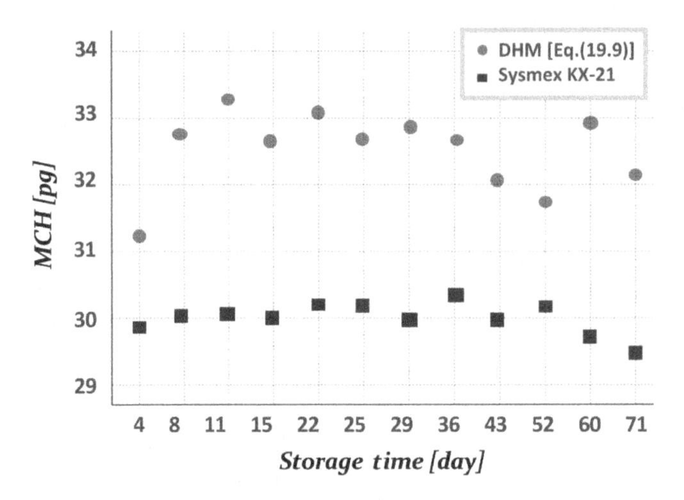

Figure 19.6 MCH level changes versus storage time. Dark grey points are MCH levels obtained with a Sysmex KX-21 hematology analyzer; grey points are MCH levels obtained using Eq. (19.9).

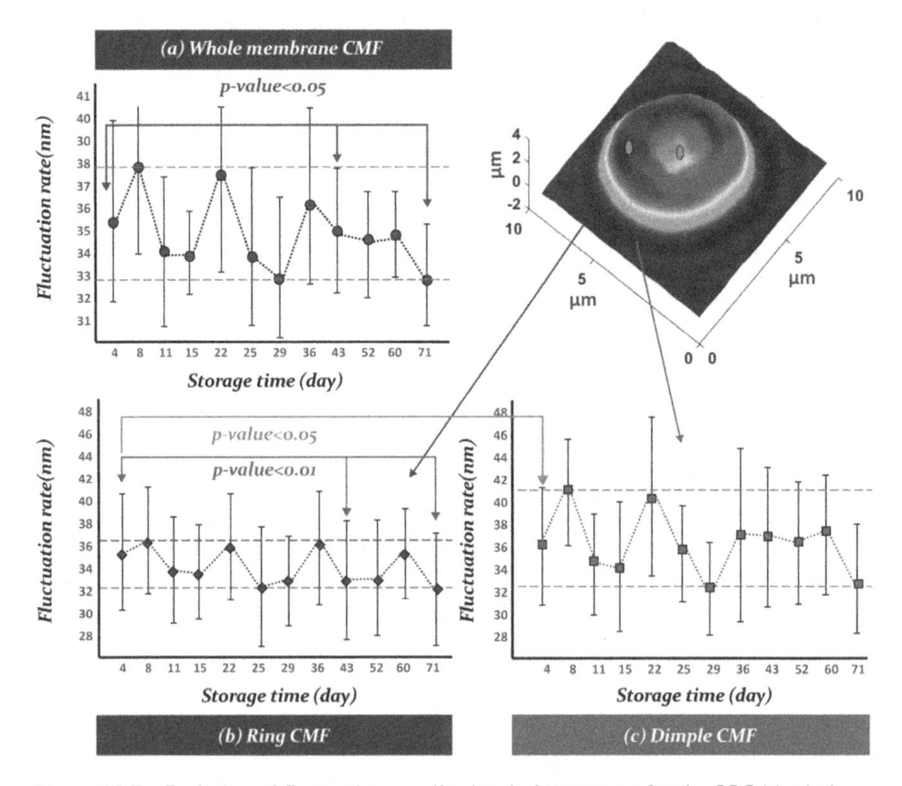

Figure 19.7 Evolution of fluctuation amplitudes during storage for the RBC (a) whole membrane, (b) ring, and (c) dimple. The length of the error bar measures two standard deviations (Statistical test is two-sample Kolmogorov–Smirnov test; $p < 0.05$).

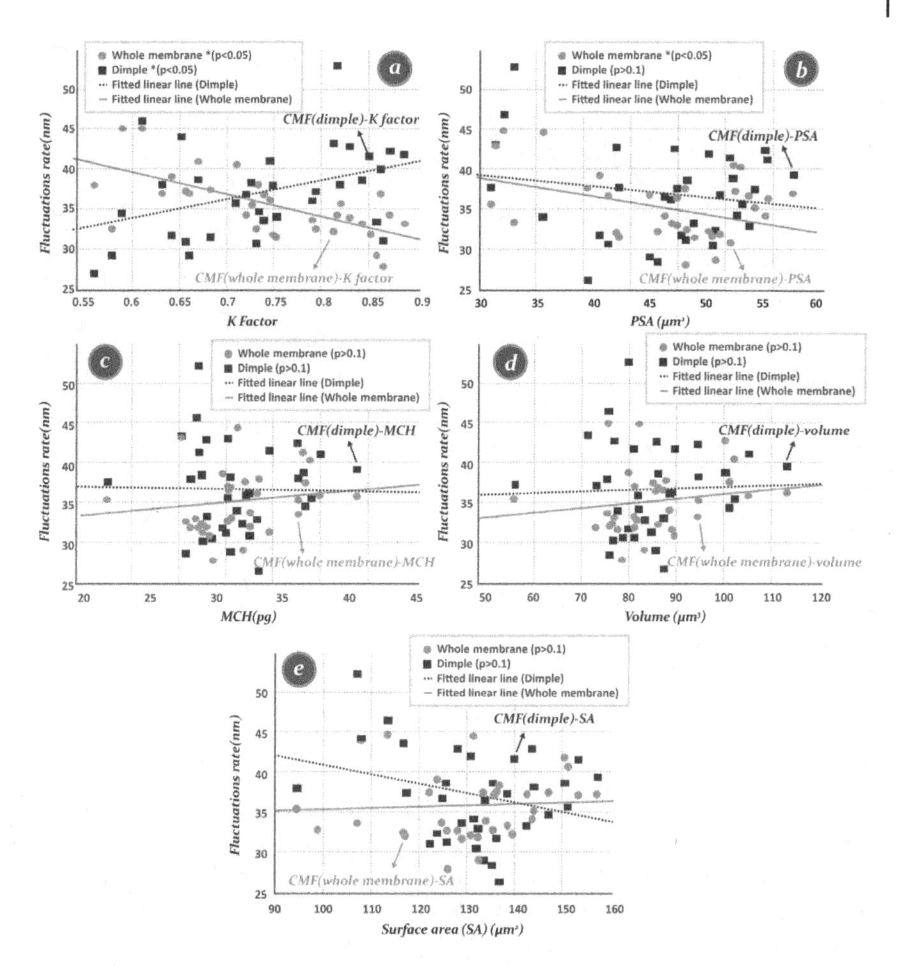

Figure 19.8 Results of a correlation analysis between the fluctuation rate and morphological or hemoglobin parameters for the entire membrane and the dimple region of 4-day-old discocyte RBCs: (a) sphericity coefficient (*k* factor), (b) PSA, (c) MCH, (d) MCV, and (e) surface area. ($n = 33$; $*$ indicates a significant linear correlation by Pearson correlation analysis, $p < 0.05$) [3] / with permission of Optical Publishing Group.

(Figure 19.8). Interestingly, the CMF amplitude of the whole membrane showed a strong negative correlation with the sphericity coefficient ($p < 0.05$; Pearson product–moment correlation test). The CMF amplitude of the whole RBC decreased as *k* increased. This could be due to the fact that RBCs become more spherical as they become stiffer [13]. On the other hand, there was a significant positive correlation between CMF amplitude in the dimple section and the *k* factor (Figure 19.8a). The lower the RBC PSA, the larger the observed CMF amplitude (Figure 19.8b). Fluctuation amplitudes of both the whole membrane and dimple (Figure 19.8c) did not

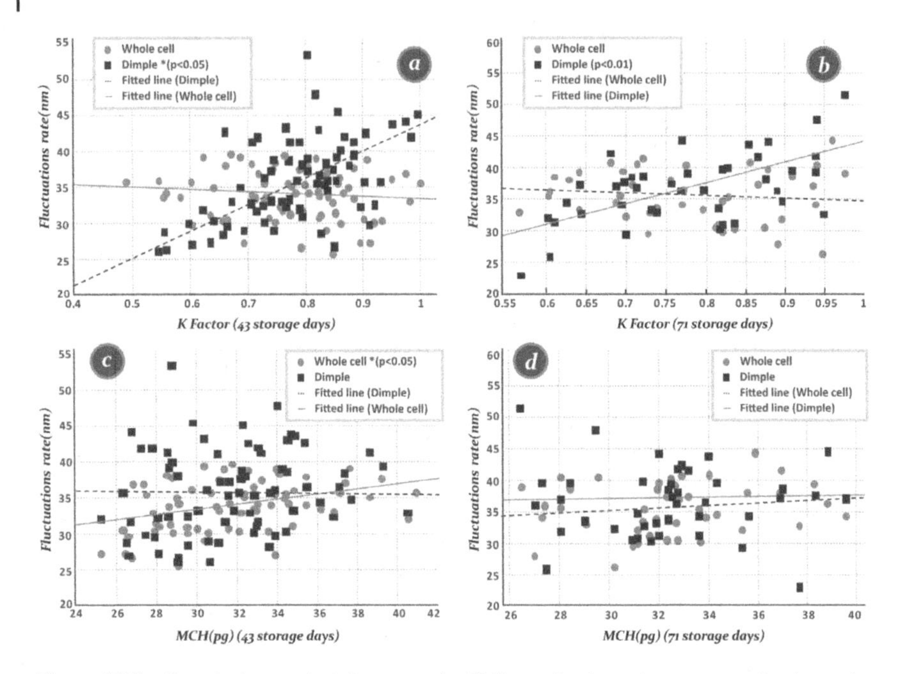

Figure 19.9 Correlation analysis between the CMF amplitude and parameters for the entire membrane and the dimple region of discocyte RBCs. The sphericity coefficient (*k* factor) for (a) 43- and (b) 71-day-old RBCs. The MCH values for (c) 43- and (d) 71-day-old RBCs. ($n = 33$; $*$ indicates a significant linear correlation by Pearson correlation analysis, $p < 0.05$) [3] / with permission of Optical Publishing Group.

significantly correlate with the MCH, also consistent with previous findings [13]. Figures 19.8d and 19.8e indicates that CMF amplitudes do not correlate with MCV or surface area values. We also computed the correlation between CMF amplitudes and both the *k* factor and MCH parameters of 43- and 71-day-old RBCs (see Figure 19.9). Our statistical model to evaluate CMF amplitudes indicated that both the ring and dimple areas differed between the younger and older discocyte RBCs. Moreover, our MCH results showed that the hemoglobin contents of RBCs remained constant over time.

Previous findings [12, 14] have indicated that RBCs become less flexible and much stiffer over longer storage days. Consequences of this flexibility loss by RBCs are significant to human organs since it is more difficult for stiff RBCs to traverse and oxygenate a microcapillary system. Figure 19.10 depicts two discocyte RBCs stored for different lengths of time. Figure 19.10a displays smaller sphericity coefficients and bigger amplitudes of fluctuation for younger RBCs. On the other hand, older RBCs have larger sphericity coefficients and smaller fluctuation amplitudes (see Figure 19.10b).

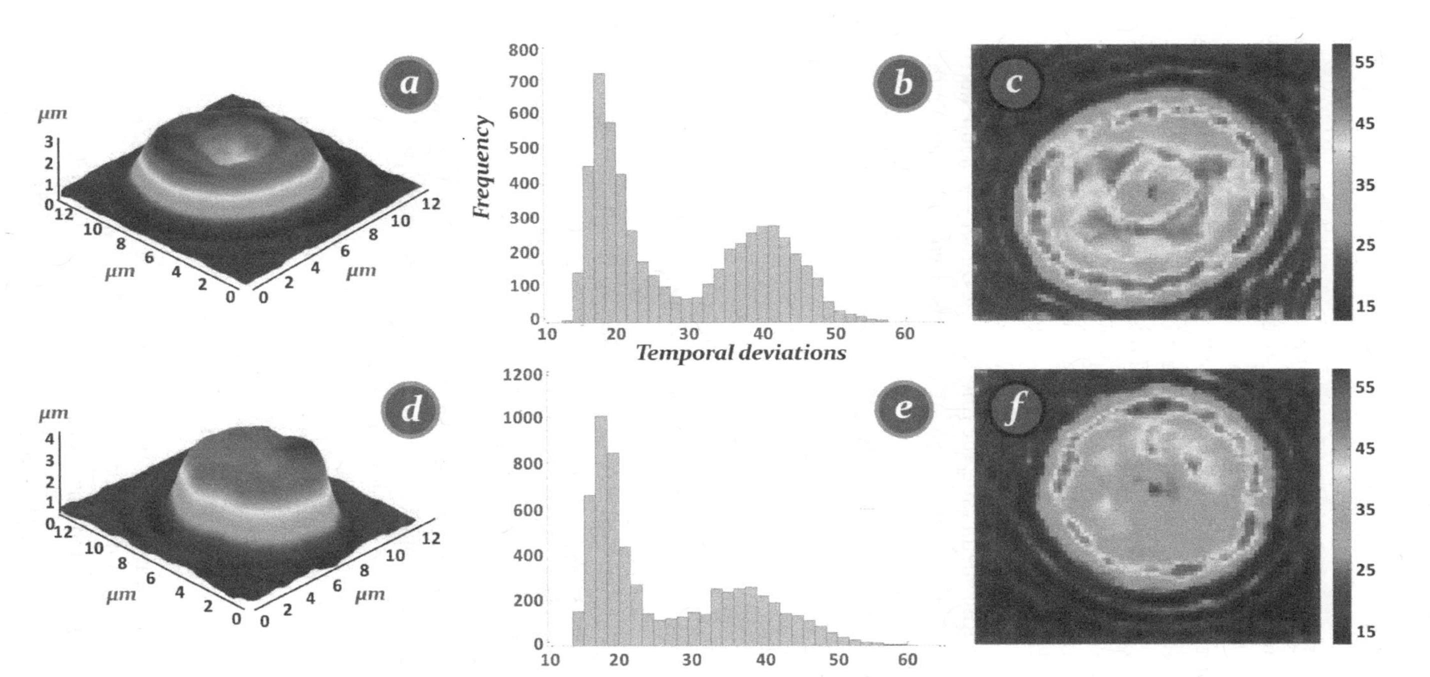

Figure 19.10 (a) 3D reconstruction of a 4-day-old RBC with $k = 0.85$ as well as the (b) temporal deviation distribution and (c) fluctuation map (surface average = 34.78 nm). (d) 3D reconstruction of a 71-day-old RBC with $k = 0.95$ as well as the (e) temporal deviation distribution and (f) fluctuation map (surface average = 26.33 nm). Color-bar scales are in nanometers [3] / with permission of Optical Publishing Group.

19.4 Conclusions

In this chapter, we quantified fluctuations in the membrane, dimple, and ring of discocyte RBCs with storage lesions using quantitative, DHM phase images. Since DHM could visualize RBCs at the single-cell level, we evaluated relationships between discocyte morphological features and CMFs as a function of the storage duration. Our measurements showed that membrane fluctuations were significantly correlated with the sphericity coefficient of observed discocytes. Increases in the discocyte sphericity coefficient were accompanied by a considerable reduction in the CMF, which corresponded to the RBC whole membrane, ring, and center (dimple), implying a loss of deformability. An increased sphericity coefficient was also observed as a function of the storage duration. The overall trend of CMF amplitudes – the entire RBC membrane, the ring, and the center – as a function of the storage duration was a decrease, implying that discocytes stiffened as they aged. We believe that this DHM discocyte flexibility study can help us evaluate the impact of storage duration on RBC quality and transfusion outcomes.

References

1 Korenstein, R., Tuvia, S., Mittelman, L., and Levin, S. (1994). Local bending fluctuations of the cell membrane. *Biomech. Active Movement Division Cells* 84: 415–423.
2 Bardyn, M., Rappaz, B., Jaferzadeh, K. et al. (2017). Red blood cells aging markers, a multi-parametric analysis. *Blood Transfus.* 15: 239–248.
3 Jaferzadeh, K., Moon, I., Bardyn, M. et al. (2018). Quantification of stored red blood cell fluctuations by time-lapse holographic cell imaging. *Biomed. Opt. Express* 9: 4714–4729.
4 Rappaz, B., Barbul, A., Emery, Y. et al. (2008). Comparative study of human erythrocytes by digital holographic microscopy, confocal microscopy, and impedance volume analyzer. *Cytometry A* 73: 895–903.
5 Rappaz, B., Marquet, P., Cuche, E. et al. (2005). Measurement of the integral refractive index and dynamic cell morphometry of living cells with digital holographic microscopy. *Opt. Express* 13: 9361–9373.
6 Rappaz, B., Barbul, A., Hoffmann, A. et al. (2009). Spatial analysis of erythrocyte membrane fluctuations by digital holographic microscopy. *Blood Cells Mol. Dis.* 42: 228–232.
7 Schneider, C., Rasband, W., and Eliceiri, K. (2012). NIH image to imageJ: 25 years of image analysis. *Nat. Methods* 9: 671–675.
8 Thévenaz, P., Ruttimann, U., and Unser, M. (1998). A pyramid approach to subpixel registration based on intensity. *IEEE Trans. Image Process.* 7: 27–41.

9 Jaferzadeh, K. and Moon, I. (2015). Quantitative investigation of red blood cell three-dimensional geometric and chemical changes in the storage lesion using digital holographic microscopy. *J. Biomed. Opt.* 20: 111218.
10 Barer, R. (1952). Interference microscopy and mass determination. *Nature* 169: 366–367.
11 Mukhopadhyay, N. (2000). *Probability and Statistical Inference*. CRC Press.
12 Bhaduri, B., Kandel, M., Brugnara, C. et al. (2014). Optical assay of erythrocyte function in banked blood. *Sci. Rep.* 4: 6211.
13 Kim, Y., Shim, H., Kim, K. et al. (2014). Profiling individual human red blood cells using common-path diffraction optical tomography. *Sci. Rep.* 4: 6659.
14 Evans, J., Gratzer, W., Mohandas, N. et al. (2008). Fluctuations of the red blood cell membrane: relation to mechanical properties and lack of ATP dependence. *Biophys. J.* 94: 4134–4144.

20

Quantitative Analysis of Red Blood Cells during Temperature Elevation

20.1 Introduction

There has been considerable scientific interest in studying the shape changes of erythrocytes induced by various conditions. Temperature, in particular, can affect steady-state volume, ion exchange rate, hemolysis rate, membrane dynamics, and cell deformation [1, 2]. Several studies suggest that red blood cell (RBC) membranes become less stable when they are exposed to temperatures above the normal body temperature. Another study showed that the uni-lamellar state of the RBC membrane is stable at 37 °C but changes to a multi-bilayer at a higher temperature [3]. Cell membrane fluctuations (CMFs) of the elastic membrane are based on the assumption that the driving force of fluctuations is purely thermal. Since membrane flickering is thermally dependent, temperature changes can directly affect membrane flickering. Thermal-induced changes in membrane profile can also affect RBC CMF maps and amplitude. In this chapter, we will introduce a quantitative method to monitor changes in the shape and CMF map of RBCs as a function of temperature at the single-cell level using label-free DHM [4]. We believe that thermal-induced changes can disrupt RBC membrane equilibrium in several ways that can be monitored quantitatively by DHM.

20.2 RBC Sample Preparations

RBCs were imaged at 17, 23, 37, and 41 °C with a sensitivity of ±0.1 °C for our RBC membrane fluctuation study. RBCs were imaged continually for all other parameters. Only RBCs with the discocyte shape ($n \geq 36$) were considered for the final analysis. All other shapes were excluded from the sample set.

Artificial Intelligence in Digital Holographic Imaging: Technical Basis and Biomedical Applications, First Edition. Inkyu Moon.

20.3 Experimental Results

20.3.1 Biochemical and Morphological Parameters

Several characteristics related to morphological features, RBC membrane fluctuations, and MCH, such as the PSA and sphericity coefficient, were investigated at the single-cell level. All these parameters were computed using the corresponding equations described in previous chapters (see Chapters 14 and 15). For this analysis, four temperatures were considered: 17, 23, 37, and 41 °C. At each temperature, we recorded 100 holograms with a sampling rate of 10 Hz. Holograms were numerically reconstructed after the experiment.

20.3.2 RBCs Trapped Between Cover Slip and Glass

Figure 20.1 shows RBC optical path difference (OPD) images at different temperatures as well as the profile and cross-section of one RBC at two temperatures: 17 and 41 °C. When RBCs are trapped between a cover slip and glass, the dimple section, or the central portion, of the RBC differs between the two temperatures.

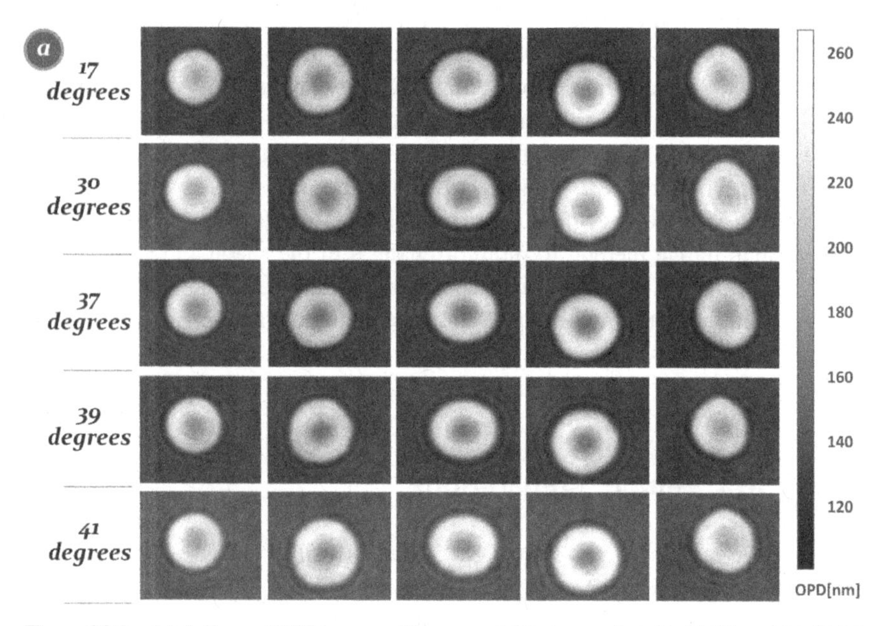

Figure 20.1 (a) Gallery of RBC images. The same RBCs imaged at (b) 17 °C and (c) 41 °C as well as (d) a cross-section of (b) and (c) drawn together. For (b) and (c), the PSAs were 60 and 64 μm^2, respectively; the sphericity coefficients were 0.65 and 0.45, respectively; and the MCH levels were 32.3 and 32.5 pg [4] / Springer Nature / CC BY 4.0.

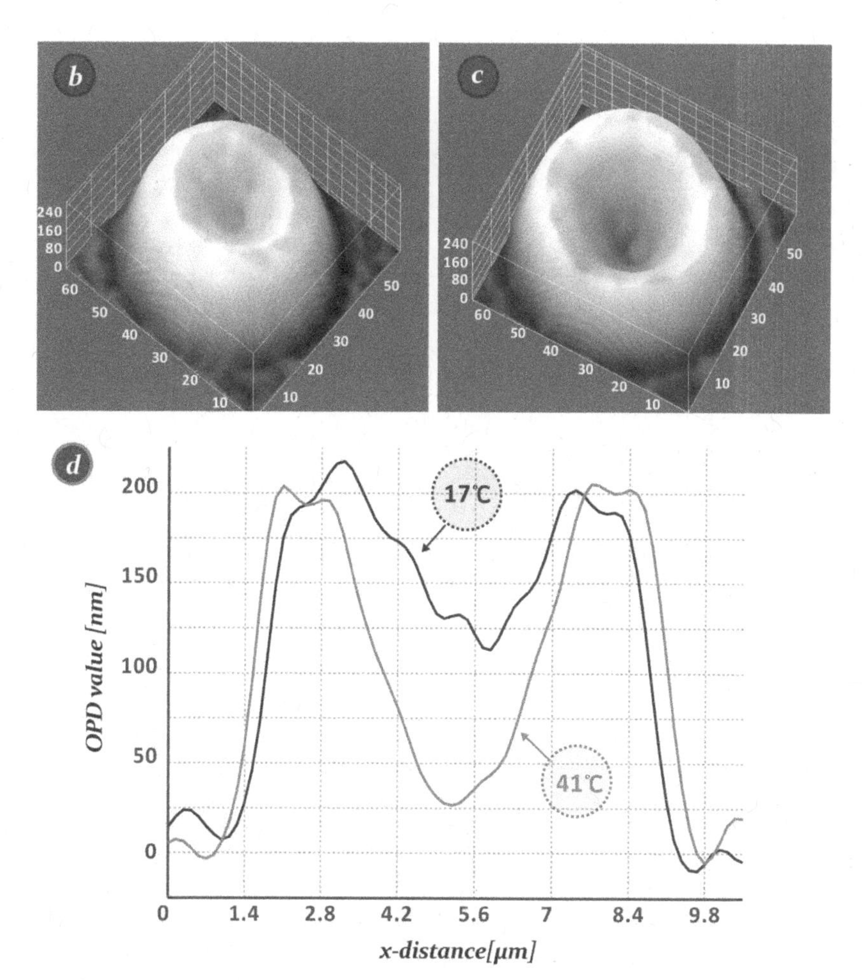

Figure 20.1 (Continued)

The impact of increasing temperature was evaluated by measuring morphological parameters, MCH, and magnitude of fluctuation rates. Figure 20.2a shows the time course of temperature elevation in this experiment. MCH remained unchanged, showing only small fluctuations around its average value of 30.76 pg (see Figure 20.2b). Increasing temperatures resulted in an increase in PSA but a decrease in sphericity coefficient (see Figures 20.2c and 20.2d). These results suggest that RBCs are losing their intracellular fluid, causing the RBC volume to drop. At the beginning of the experiment, the RBC and extracellular medium were isotonic. Thus, there was no water movement so that RBCs could maintain their shape. After some moments due to increasing temperature, the surrounding

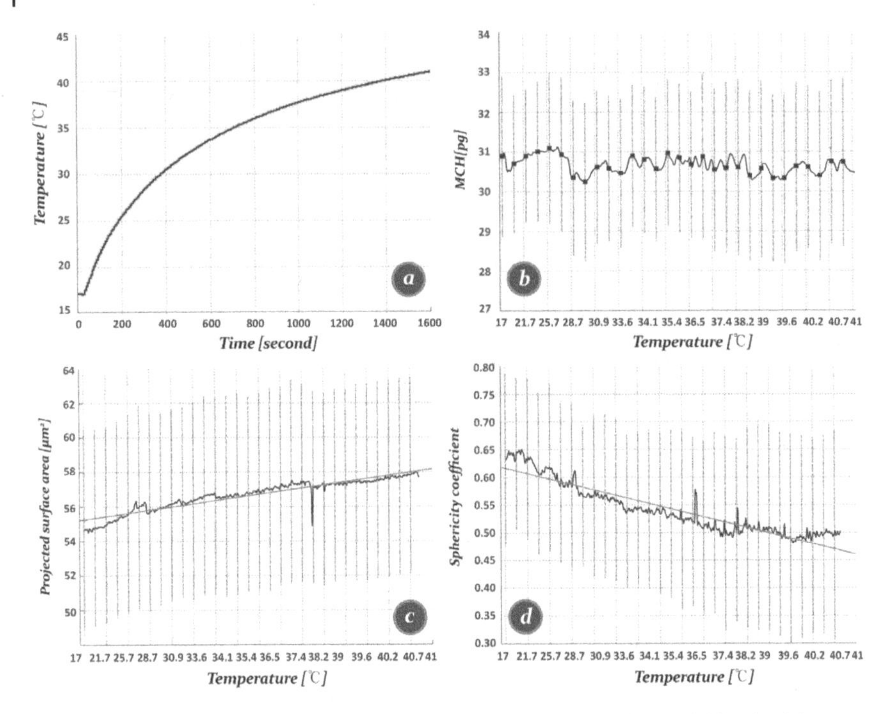

Figure 20.2 (a) Increasing temperature duration, (b) MCH (c), PSA, and (d) sphericity coefficient changes versus temperature ($n \geq 36$ cells). F-statistics performed on data shown in (c) and (d) suggested that the slope of the linear regression line was significantly different from zero with p-value <0.05. Error bars represent twice the corresponding standard deviations.

medium began to evaporate, creating an imbalance between intracellular and extracellular fluids. Intracellular fluidity increases because it is directly affected by the increasing temperature. As a result, a concentration gradient formed between the RBC membrane and the extracellular medium. Water left RBCs to diminish the gradient, leading to volume loss. This also increased concentrations inside the RBC to increase compared to lower temperatures due to the water loss at higher temperatures.

Figure 20.3 demonstrates the CMF maps of RBCs measured at four different temperatures. The CMF amplitude of the whole RBC is the average of the CMF (x, y) over the projected area of the whole RBC. The CMF amplitude for each temperature is also computed and the comparison given as a box plot (see Figure 20.3e). As shown in Figure 20.3, the CMF map at lower temperatures was usually present at the ring area of the RBC membrane. The RBC membrane fluctuations had a direct relationship with the Gaussian curvature of RBC [5].

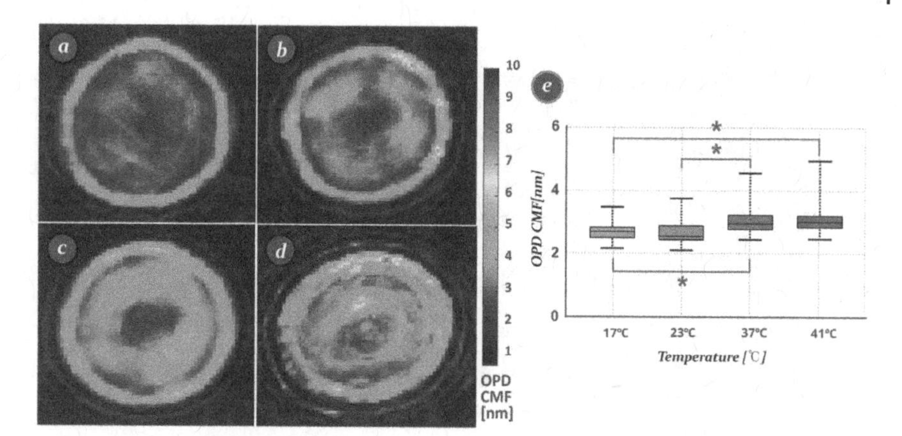

Figure 20.3 Fluctuation map for an RBC at: (a) 17 °C, (b) 23 °C, (c) 37 °C, and (d) 41 °C. (e) Box plot representation of the CMF amplitude for each temperature. * indicates a two-sample Kolmogorov–Smirnov test with p-value <0.05 [4] / Springer Nature / CC BY 4.0.

The deformation of the RBC membrane predominantly occurred at the area in which the Gaussian curvature approached zero. Higher temperatures caused loss of intracellular fluid, resulting in an RBC with a lower sphericity index. Thus, the Gaussian curvature was altered, and the fluctuation map differed in the case of higher temperatures. In Chapter 19, we reported a significant negative correlation between the CMF value and sphericity coefficient [6]. In addition, the Kolmogorov–Smirnov test showed that the ring CMF was greater than the dimple CMF for all temperatures (data not shown). We observed no significant changes in morphology, shape, or CMF when RBCs were kept at room temperature for one hour (data not shown).

20.3.3 RBCs Imaged on Chamber

In the second experiment, 200 μl of the suspension was dropped on the imaging glass of an 18 mm, round coverslip chamber. The temperature was raised and the RBCs imaged over time. The RBCs were imaged at 5 and 10 minutes after the temperature reached the desired level. Figure 20.4 depicts a gallery of RBC images at different temperatures as well as a profile of the same RBC at multiple temperatures.

Figure 20.5 shows the effects of temperature elevation on the RBC shape and membrane fluctuations. The sphericity coefficient of RBCs at a lower temperature is less than that at higher temperatures. There was no significant change in PSA. However, the CMF values at 17 and 37 °C are significantly different (p-value <0.005). We found no significant change in the volume or MCH at different

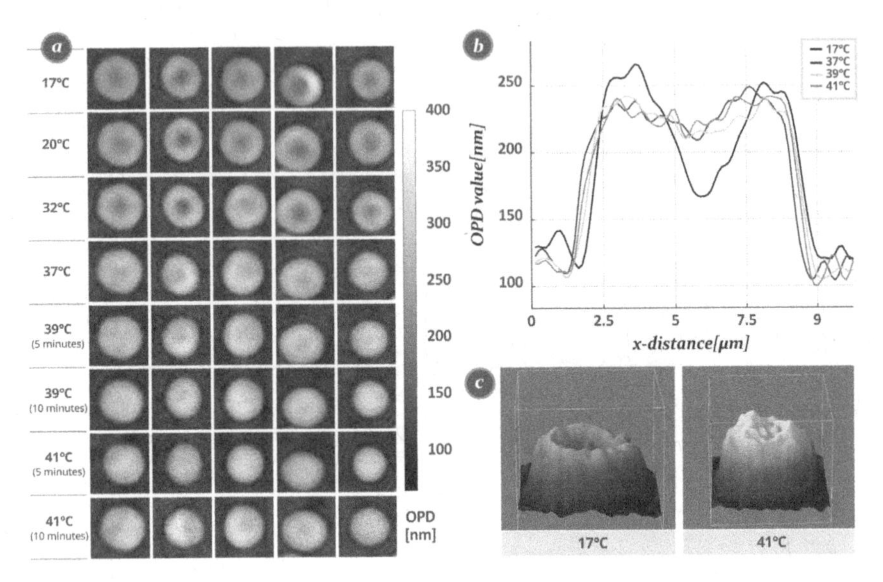

Figure 20.4 (a) Gallery of images of the same RBCs at multiple temperatures; (b) a cross-section of the same RBC at multiple temperatures; (c) a 3D representation of the RBC at 17 and 41 °C. For the RBC in (b) and (c) the PSAs were 49 and 50 μm^2, respectively; the sphericity coefficients were 0.86 and 0.91, respectively; and the MCH contents were 30.5 and 30.9 pg, respectively [4] / Springer Nature / CC BY 4.0.

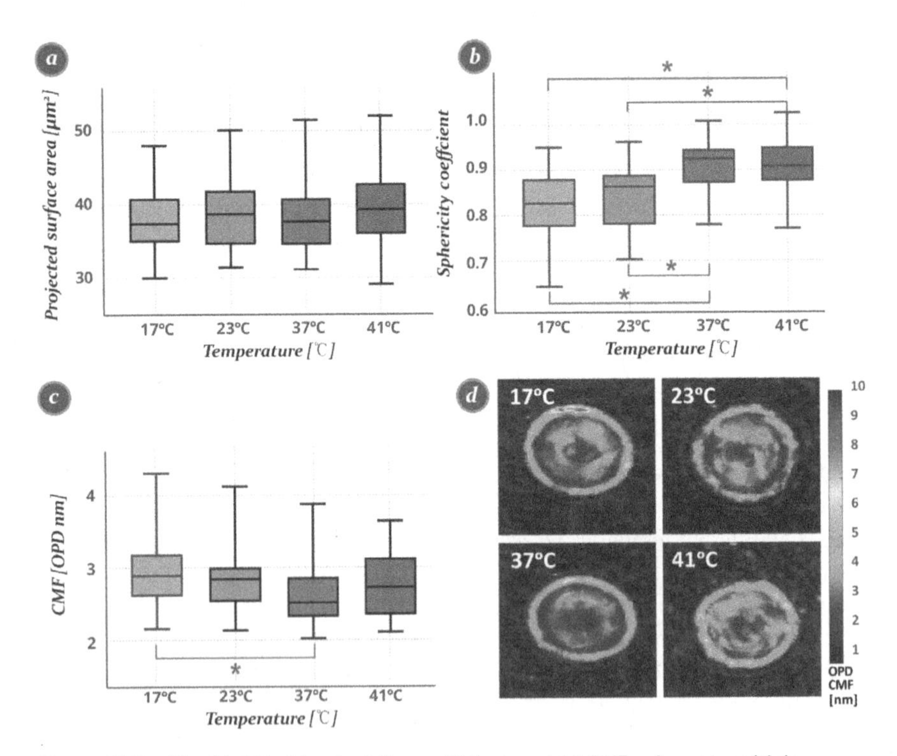

Figure 20.5 The (a) PSA, (b) sphericity coefficient, and (c) CMF values at multiple temperatures. (d) The CMF map at different temperatures. * indicates a two-sample Kolmogorov–Smirnov test, *p*-value <0.05; $n \geq 36$ cells) [4] / Springer Nature / CC BY 4.0.

temperatures (data not shown). We did not observe significant reversible changes in RBC membranes as they preserved their shape. Some RBCs gradually lost their membrane stability. Accordingly, RBCs with spiculated shapes (echinocytes) began to appear.

Cell membranes are made of phospholipids and more rigid at lower temperatures. They become softer at a higher temperatures, even very fluid and unstable at some temperatures above its physiological temperature (e.g. 37 °C for RBCs). Another hypothesis suggests that when the cell's physiological temperature is exceeded, the membrane bilayer will transform. As a result, the growth temperature of the cell is a key determinant of membrane bilayer stability.

20.4 Conclusions

Temperature changes may affect the mechanical properties and morphology of cells. In this chapter, we investigated the effects of brief (less than one hour) elevated temperatures on several RBC parameters using DHM. Our results indicated that RBCs retained their normal morphology, although there were changes in some parameters related to the RBC profile. For instance, the sphericity coefficient changed when RBCs were exposed to high temperatures, possibly because temperature can increase the fluidity of some membrane compounds.

References

1. Ben-Isaac, E., Park, Y., Popescu, G. et al. (2011). Effective temperature of red-blood-cell membrane fluctuations. *Phys. Rev. Lett.* 106: 238103.
2. Boss, D., Hoffmann, A., Rappaz, B. et al. (2012). Spatially-resolved eigenmode decomposition of red blood cells membrane fluctuations questions the role of ATP in flickering. *PLoS One* 7: e40667.
3. Gershfeld, N. and Murayama, M. (1988). Thermal instability of red blood cell membrane bilayers: temperature dependence of hemolysis. *J. Membr. Biol.* 101: 67–72.
4. Jaferzadeh, K., Sim, M., Kim, N., and Moon, I. (2019). Quantitative analysis of three-dimensional morphology and membrane dynamics of red blood cells during temperature elevation. *Sci. Rep.* 9: 1–9.
5. Popescu, G., Ikeda, T., Goda, K. et al. (2006). Optical measurement of cell membrane tension. *Phys. Rev. Lett.* 97: 218101.
6. Jaferzadeh, K., Moon, I., Bardyn, M. et al. (2018). Quantification of stored red blood cell fluctuations by time-lapse holographic cell imaging. *Biomed. Opt. Express* 9: 4714–4729.

21

Automated Measurement of Cardiomyocyte Dynamics with DHM

21.1 Introduction

It is crucial to improve the predictability of chemical toxicity through safety profiling assays during the lengthy drug discovery process. This should be done to detect potentially toxic compounds early in the process before considerable amounts of time and financial investments are made. Safety assessments are, therefore, performed in preclinical drug development to identify possible drug side effects, particularly those that may affect the electrical conduction and beating of the heart [1–3]. Consequently, it is critical to establish more informative tools for *in vitro* cardiotoxicity screens at early phases of drug development to prevent late-stage failure [4–6].

Cardiomyocytes, also known as myocardial cells, are main contractile elements of heart muscles. These cells can collaborate with each other to generate the human heartbeat and control blood flow in blood vessels of the circulatory system [3]. Like many other types of cells, cardiomyocytes are mostly transparent. As a result, traditional imaging systems based on bright-field intensity can only provide a low contrast image with limited informative details of the cell structure. Although some optical imaging techniques such as phase contrast and differential interference contrast microscopies can provide contrast for transparent cells, they cannot offer quantitative information on their thickness.

Various imaging systems have been used to analyze cardiomyocytes. For example, with fluorescence microscopy [7, 8], specific biological molecules are fluorescently stained, and the location of a protein traced, or the activity of specific ions can be monitored over time [9, 10]. However, fluorescence can fade or interfere with the molecule being analyzed [11, 12]. Atomic force microscopy [13] allows the measurement of advanced physical-mechanical parameters such as stiffness and elastic modulus. It also provides a high-resolution profile of the sample [14]. However, it has a limited spatial sampling speed. Phase contrast and

Artificial Intelligence in Digital Holographic Imaging: Technical Basis and Biomedical Applications, First Edition. Inkyu Moon.

differential interference contrast microscopies can monitor isolated cardiomyocyte contraction non-invasively. However, they require advanced image correlation analysis [15]. Although each of these techniques has its own advantages and drawbacks, when combined, they offer complementary information that can be exploited in multimodal setups [16].

In this chapter, we will introduce automated methods to quantitatively analyze dynamic phase profiles of beating cardiomyocytes reconstructed from holograms captured by DHM [17]. The contraction profile, relaxation profile, and beating activity of cardiomyocytes are determined from the reconstructed quantitative phase image. Other characteristic parameters used to categorize phenotypes [2, 18] such as the rising time, falling time, peak width, and frequency are also analyzed. These parameters are useful for determining the impact of medication candidates on cardiomyocytes. We analyzed the dynamic beating profile of cardiomyocytes obtained with DHM using two methods: monitoring the average or obtaining the variance information in optical path difference (OPD) images of cells. We further quantified contraction and relaxation movement by analyzing the difference between two successively acquired OPD images. From our experimental findings, we offer automated procedures for recording multiple parameters of cardiomyocyte dynamics captured by DHM for a new methodology in drug toxicity screens.

21.2 Cell Culture and Imaging

iCell cardiomyocytes (human-induced pluripotent stem cell–derived cardiomyocytes) obtained from Cellular Dynamics Int. (Madison, WI, USA) were cultured according to the manufacturer's instructions and grown for 14 days before recording. Measurements were taken using a Chamlide WP incubator system for 96-well plates (LCI, South Korea) set at 37 °C with 5% CO_2 and high humidity. OPD images were acquired with an off-axis DHM. Images were recorded using a Leica 20×/0.4 NA objective. Time-lapse images were acquired at 10 Hz for one minute.

21.3 Automated Analysis of Cardiomyocyte Dynamics

For the automated quantitative analysis of cardiomyocyte dynamics, we computed the beating activity using two methods: the averaged OPD images and variance of OPD images. The contraction and relaxation characteristics of cardiomyocytes were also measured using our automated procedure. Figure 21.1 shows two cardiomyocyte OPD images reconstructed from holograms. The high similarity

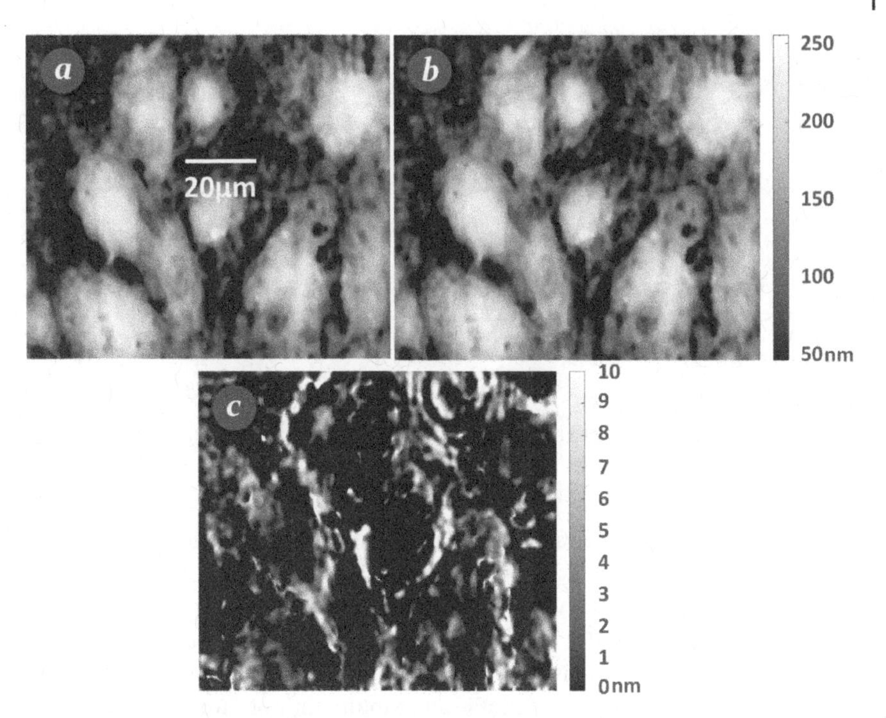

Figure 21.1 Optical path difference (OPD) images of cardiomyocytes captured at different times (a total of 540 frames). Cardiomyocyte OPD images at (a) maximum and (b) minimum peaks. (c) The difference between (a) and (b) [4] / with permission of American Chemical Society.

between these two OPD images highlights the need for extensive analysis to quantify the beating dynamics.

21.3.1 Cardiomyocyte-beating Profile Measurements Using Averaged OPD Images

The beating profile of cardiomyocytes was obtained by thresholding cardiomyocyte OPD images at 10% of the maximum OPD signal. Thresholded images were then averaged. The threshold value was used to restrain the effect of noise. This process is described by

$$\overline{opd}^{(i)} = average\left(opd\_thresh(x,y)^{(i)}\right),$$

$$with\ opd\_thresh\,(x,y)^{(i)} = \left\{\begin{array}{ll} opd(x,y)^{(i)} & if\ opd(x,y)^{(i)} \geq max\left(opd(x,y)^{(i)}\right) \times 0.1 \\ 0 & if\ opd(x,y)^{(i)} < max\left(opd(x,y)^{(i)}\right) \times 0.1 \end{array}\right\},$$

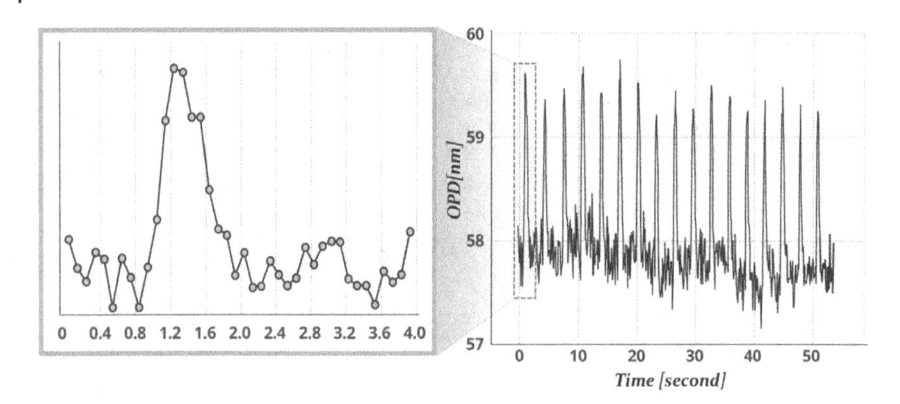

Figure 21.2 The beating activity of cardiomyocyte (inset shows a single beat).

where $\overline{opd}^{(i)}$ is the average value of ith OPD image after thresholding, $opd\_thresh(x, y)^{(i)}$ is the OPD value at (x, y) on the ith thresholded cardiomyocyte image, $opd(x, y)$ is the OPD value at (x, y) for the ith cardiomyocyte image in the optical path difference while $1 \leq x \leq M$, $1 \leq y \leq N$ (M and N are the size of cardiomyocyte OPD image) and $max(opd(x, y))^{(i)}$ is the maximum value of the ith cardiomyocyte image.

Figure 21.2 shows the beating profile and capture time of cardiomyocytes using our method, including a small inset with a zoom on a single beating pattern. The beating activity of cardiomyocytes is clearly visible (short peaks of high amplitude). Figure 21.3 shows the beating activity of cardiomyocytes with different threshold settings. Although the resulting OPD values are slightly different, the beating profiles of cardiomyocytes are approximately the same, as shown in Figure 21.3. Our final results were very similar since multiple parameters only depended on the beating profile of cardiomyocytes. Because these beating profiles are similar under different threshold settings, the values of multiple parameters will be comparable as well. In other words, changing the threshold value will not largely affect the final parameter data. Multiple parameters including the amplitude, rising time, falling time, IBD_{50}, IBD_{10}, rising/falling slope, beating rate, and beating period based on beating profile in Figure 21.2 are then derived. These parameters are described in Table 21.1 [1].

To measure the above-defined parameters, peaks were detected by applying the first derivative technique to the original data curve in Figure 21.2 and looking for locations where the first derivative values are zeros. Figure 21.4a shows detected peaks based on cardiomyocyte beating profiles. Figure 20.4a shows that several peaks, including false peaks, were detected. During the sorting process, we removed positive peaks with values below a given threshold and negative peaks

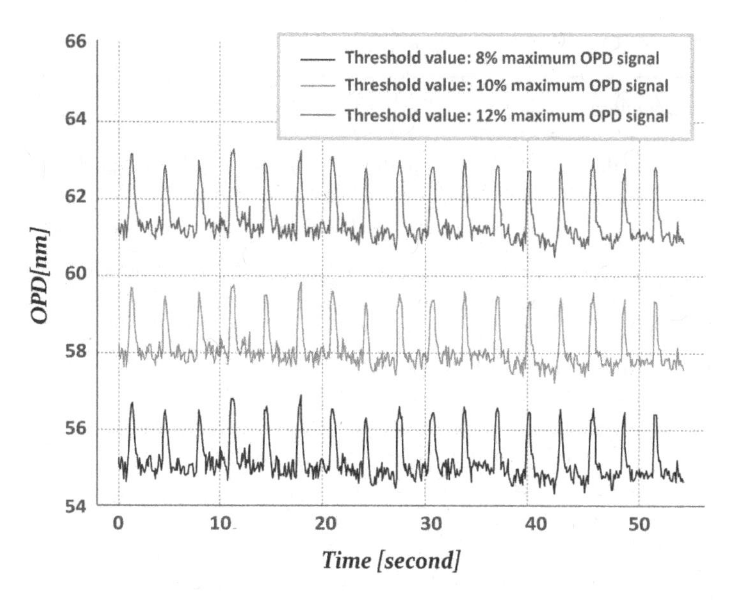

Figure 21.3 The beating activity of cardiomyocyte under different threshold values.

with values above a given threshold. This threshold was automatically determined with Otsu's method [19] using all positive peaks detected in Figure 21.4a. Furthermore, the minimum negative peak between two nearby positive peaks was extracted. The maximum positive peak between two neighboring negative peaks was then selected. This process reduced some incorrect peaks and resulted in the appropriate peaks for each beating period, as shown in Figure 21.4b.

Accordingly, we extracted the beating profile between two adjacent negative peaks, which we considered one beating period (see Figure 21.5). It was noted that beating periods computed between two negative peaks were almost equal to those computed between two positive peaks. At the same time, the extracted beating profile for each beating period was fitted with polynomials of degree 9 in a least-square criterion. Examining polynomials of up to 9 degrees (the data samples of some beating profiles were about 10) to fit the data based on the final fitting errors yielded a degree of 9. The fitted 9-degree polynomial is described by

$$f(x) = \sum_{i=1}^{10} a_i x^{10-i},$$

where x indicates the sample point on each beating period and a_i, which is the coefficient of the polynomial, is obtained with a least-squares criterion based on sample points. The average absolute error (absolute error between actual and fitted points) for each sample point was computed at 0.19. Figure 21.5 depicts one of the fitted polynomial curves with the measured parameters.

Table 21.1 Characteristic cardiomyocyte parameters.

Parameter	Definition
Amplitude	Value difference from each positive peak to the following negative peak (Amplitude = $Amp_{max} - Amp_{min}$) [see Figure 21.5]
Rising time	The time elapsed from Amp_{20} to Amp_{80} (=$T_3 - T_1$) [see Figure 21.5]
Falling time	The time elapsed from Amp_{80} to Amp_{20} (=$T_6 - T_4$) [see Figure 21.5]
IBD_{50}	The time elapsed for two points equal one Amp_{50} (=$T_5 - T_2$) [see Figure 21.5]
IBD_{10}	The time elapsed for two points equal one Amp_{10} (=$T_7 - T_0$) [see Figure 21.5]
Rising/falling slope	The change of increased/decreased amplitude between Amp_{80} and Amp_{20} (=$Amp_{80} - Amp_{20}$) [see Figure 21.5]
Beating rate	The total number of positive/negative peaks in 1 minute (= total number of positive peaks/total time)
Beating period	The time between two adjacent positive and/or negative peaks (= the time of i_{th} positive peak – the time of $(i-1)_{th}$ positive peak)
Frequency	The number of beats period per second (= total number of beats in a period/total time)

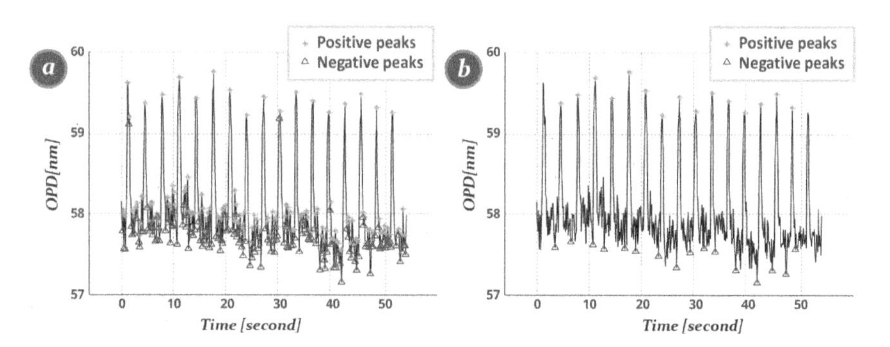

Figure 21.4 Detected peaks in the beating profile of cardiomyocytes. (a) Detected peaks on raw data. (b) Filtered peaks based on the results of (a) [17] / with permission of Optical Publishing Group.

The amplitude (Amp) value defined in Table 21.1 can be computed using the fitted polynomials curves by subtracting the minimum value from the maximum value on the fitted curve. The corresponding time (in seconds) on the x-axis for Amp_{10}, Amp_{20}, Amp_{50}, and Amp_{80} (see Figure 21.5) can also be calculated by solving the fitted polynomial equation. All parameters mentioned above were then

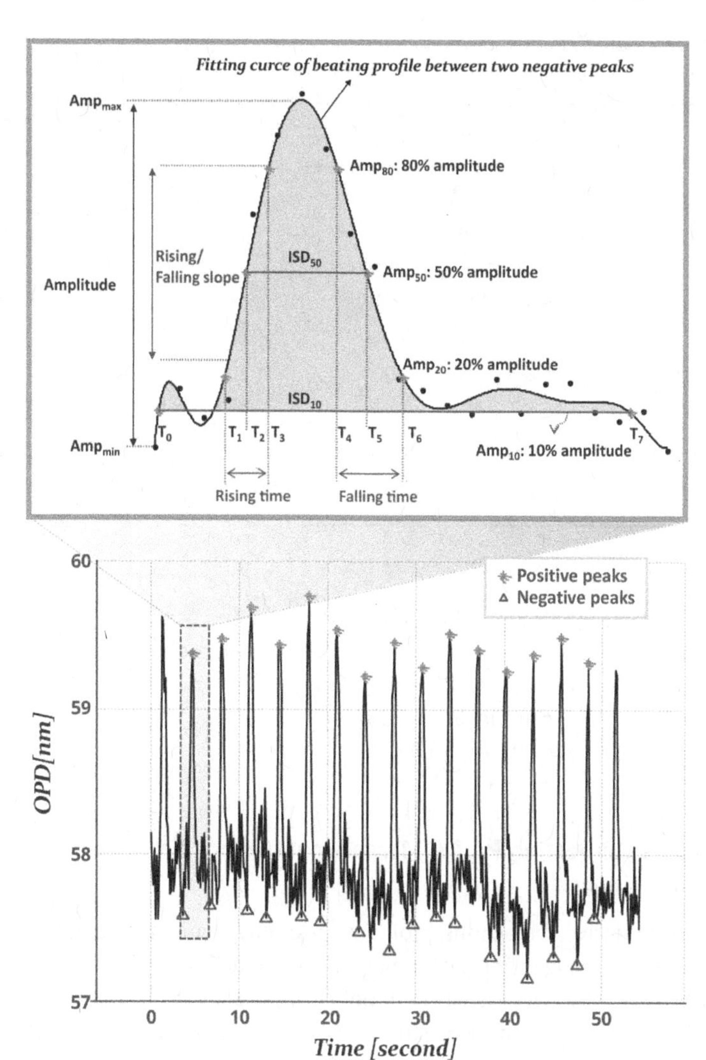

Figure 21.5 An illustration of fitted curves, parameters are within one beating period of the cardiomyocyte beating profile.

measured for each individual beating period. A population average with the coefficient of variation ($cv = \dfrac{standard\ deviation}{average\ value}$, a parameter often used in high-throughput screening) was then calculated (see Table 21.2). Multiple parameters were calculated by fitting a curve between two negative peaks, which allowed for the isolation of a full beating pattern and better fitting rising/falling time and

Table 21.2 Measured values for multiple parameters of cardiomyocyte-beating profiles.

Multi-parameter	Values (mean/cv)
1: Amplitude:	2.05/0.30
2: Rising time:	0.58/1.05 (seconds)
3: Falling time:	0.86/0.69 (seconds)
4: IBD_{50}:	0.79/0.33 (seconds)
5: IBD_{10}:	2.94/0.19 (seconds)
6: Rising/Falling slope:	1.19/0.30
7: Beating rate:	21.86/0
8: Beating period:	2.94/(sd = 0.10 seconds)
9: Frequency	0.34/0

amplitude. Furthermore, the IBD can only be quantified if a full beat is present in the analyzed period. However, as shown in Figure 21.4, positive peaks tended to be more stable and less noisy. Thus, we also calculated the beating period as defined by the time between two adjacent positive peaks and found similar results (2.93/sd = 0.10 seconds versus 2.94/0.10 seconds).

21.3.2 Cardiomyocyte-beating Profile Measurement Using the Variance of OPD Images

An alternative way to derive the beating profile of cardiomyocytes is to calculate the variance of each OPD image after the temporal mean of the image stack is subtracted. This alternative way is less sensitive to noise (originating from shot noise, speckle, and contribution of out-of-focus structures). However, it demands more computing resources. This method can be defined by

$$\delta_{opd}^{(i)} = variance\left[opd(x,y)^{(i)} - \overline{opd}_{temp}\right],$$

where $opd(x,y)^{(i)}$ is the ith OPD image with $1 \leq x \leq M$ and $1 \leq x, y \leq N$ (M and N are the size of cardiomyocyte OPD image), $\delta_{opd}^{(i)}$ denotes the variance of the ith cardiomyocyte image after temporal mean subtracted, and $\overline{opd}_{temp}$ is the temporal mean, which is calculated as the mean value of the image stack in the temporal dimension. Figure 21.6 presents the beating profile measured using this method. Compared to the previous analysis method (Figure 21.4), this method is more stable and less sensitive to noise (changes in the absolute value of the OPD signal).

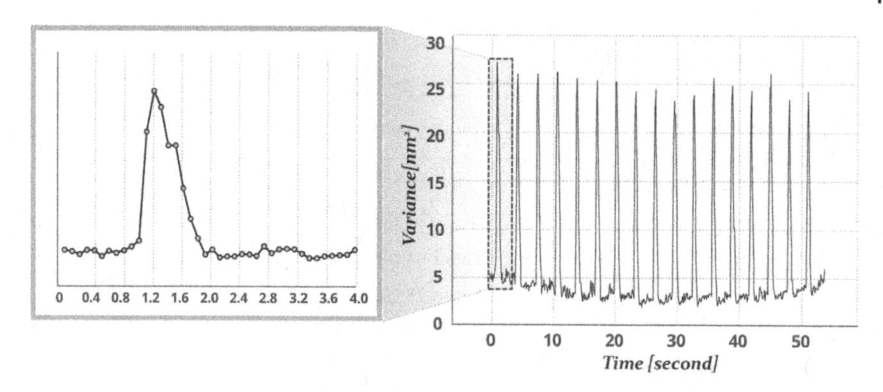

Figure 21.6 The measured beating profile with variance information (inset shows a single beat).

Similar to the previous method, peaks in Figure 21.6 can be detected using the first derivative property (see Figure 21.7a). In addition, positive and negative peaks are screened with a threshold value obtained using Otsu's method [19]. Consequently, the minimum negative peak between two neighboring positive peaks and the maximum positive peak between two neighboring negative peaks (see Figure 21.7b) were chosen.

The beating profile within one beating period (between two negative peaks) can be individually extracted and fitted using a polynomial equation of degree 9 in a least-square error, as shown in Figure 21.5. The average absolute error for each sample point is computed to be 2.0359. Consequently, the same parameters can be measured as in the previous method. The corresponding mean and coefficient of variation (cv) are given in Table 21.3. These measured parameters are in excellent agreement with the literature [12].

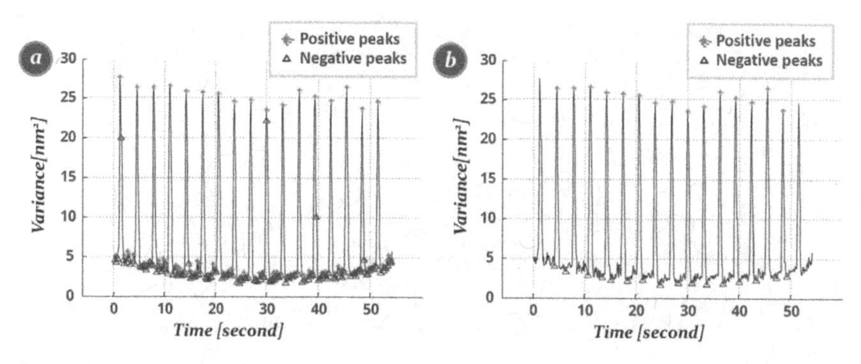

Figure 21.7 Detected peaks on a cardiomyocyte beating profile. (a) Multiple detected peaks. (b) Multiple detected peaks with some false peaks removed [17] / with permission of Optical Publishing Group.

Table 21.3 Multiple measured parameters of the cardiomyocyte beating profile.

Multi-parameter	Values (mean/cv)
1: Amplitude:	23.06/0.17
2: Rising time:	0.45/0.98 (seconds)
3: Falling time:	0.26/1.04 (seconds)
4: IBD50:	0.66/0.23 (seconds)
5: IBD10:	2.30/0.25(seconds)
6: Rising/Falling slope:	13.83/0.17
7: Beating rate:	21.91/0
8: Beating period:	2.93/(sd = 0.10)
9: Frequency	0.34/0

21.3.3 Cardiomyocyte Contraction and Relaxation Measurements

To observe the contraction and relaxation features of cardiomyocytes, each captured image in the temporal stack is subtracted from the next one. The spatial variance of the OPD is then measured to quantify the degree of spatial displacement between successive frames. The resulting image contains cardiomyocyte contraction and relaxation information (both indicated by an increase in the temporal variance signal). Two of the subtracted images are shown in Figure 21.8. Figures 21.8a,b are images at the minimum and maximum of a beat. Figure 21.9a shows the beating profile of cardiomyocytes with contraction and relaxation information, where one

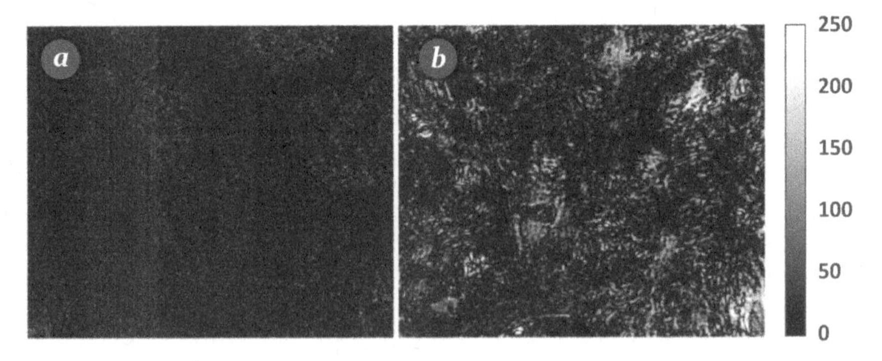

Figure 21.8 Illustrations of different images. (a) An image at the minimum peak of a beat. (b) A different image at the maximum peak of a beat [17] / with permission of Optical Publishing Group.

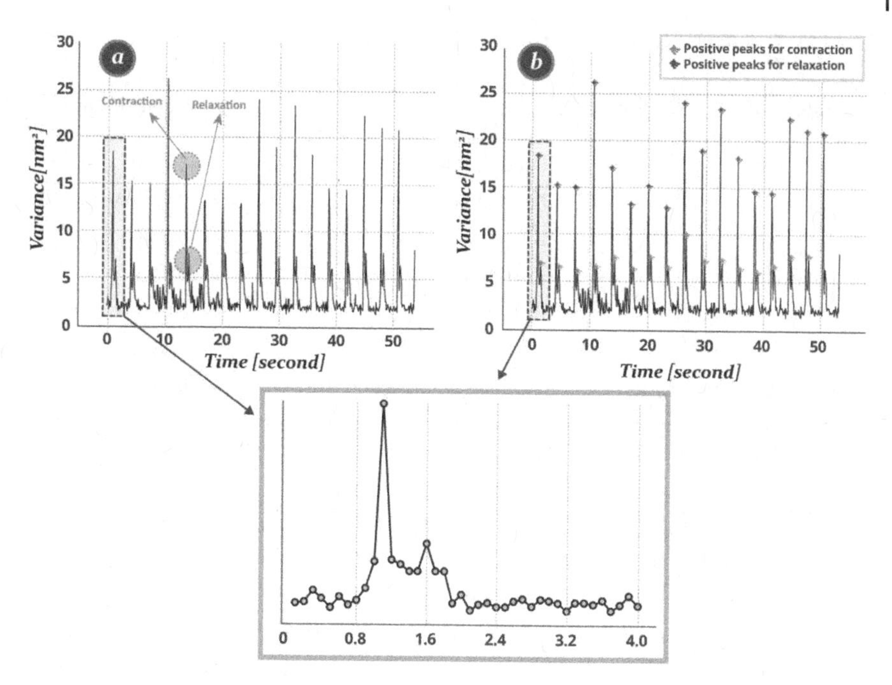

Figure 21.9 Cardiomyocyte beating profiles with contraction and relaxation information. (a) Raw data of a cardiomyocyte beating profile. (b) Cardiomyocyte beating profile with contraction and relaxation peaks indicated (inset shows a single beat with contraction and relaxation peaks).

higher peak represents contraction, and the neighboring lower peak represents relaxation. Peaks were then detected with the first derivative criterion (locations with a zero first-derivative values). Similarly, the positive peaks for contraction were extracted with Otsu's thresholding algorithm using all positive peaks detected. A maximum peak between two nearby contraction peaks was then picked as a positive peak for relaxation and the inappropriate peaks removed. Figure 21.9b shows the resultant curves with peaks indicated from Figure 21.9a. Finally, the beating rate, beating period, and frequency for cardiomyocytes contraction and relaxation were measured using the positive peaks detected, including contraction and relaxation peaks. The time between the cardiomyocyte contraction and the following relaxation can also be computed with the detected peaks shown in Figure 21.9b. These measured data are given in Table 21.4. Tables 21.1–21.3 show that our three methods can produce similar values for the beating rate, beating period, and frequency using the identical cardiomyocyte image sequence.

Four more cardiomyocyte image sequences were examined to show the generality of our methods. One of these image sequences was obtained under difficult

Table 21.4 Measured parameters for the cardiomyocyte contraction and relaxation curves.

Multi-parameters		Values (mean/cv)
Contraction	Beating rate:	21.71/0
	Beating period:	2.93/(sd = 0.08)
	Frequency	0.34/0
Relaxation	Beating rate:	20.84/0
	Beating period:	3.05/(sd = 0.49)
	Frequency	0.32/0
Time between contraction and the following relaxation		0.41/0.14

conditions (severe disturbances occurred due to out-of-focus debris that flowed through the field of view). Other sequences were comparable in quality to that previously presented. Four image sequences with peaks detected using each of our three methods are given in Figures 21.10, 21.11, and 21.12. Figures 21.10a,

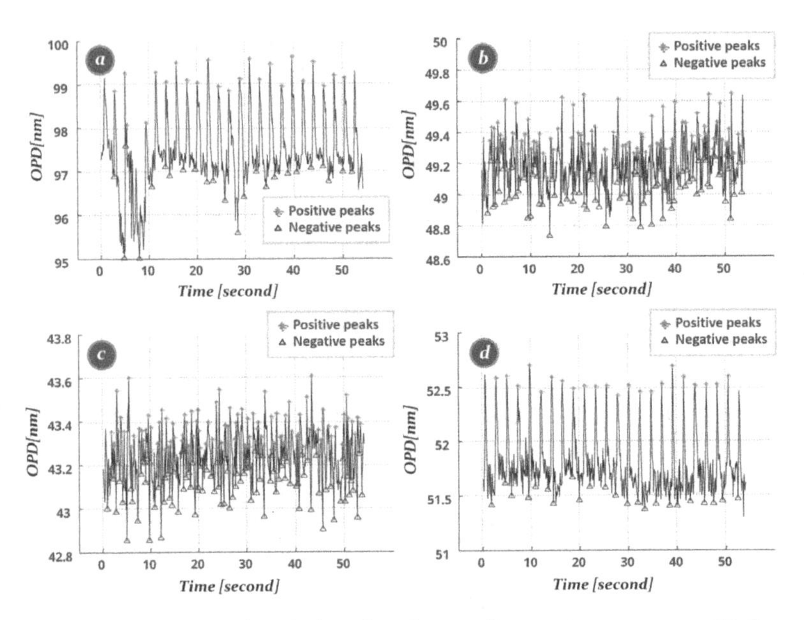

Figure 21.10 Detected peaks based on four cardiomyocyte sequences with the method shown in Section 21.3.1 (cardiomyocyte beating profile measurement using averaged OPD images). (a) Cardiomyocyte image sequences acquired in difficult conditions. (b), (c), and (d) "noise-free" recordings (i.e. no debris) similar to the previous recording [17] / with permission of Optical Publishing Group.

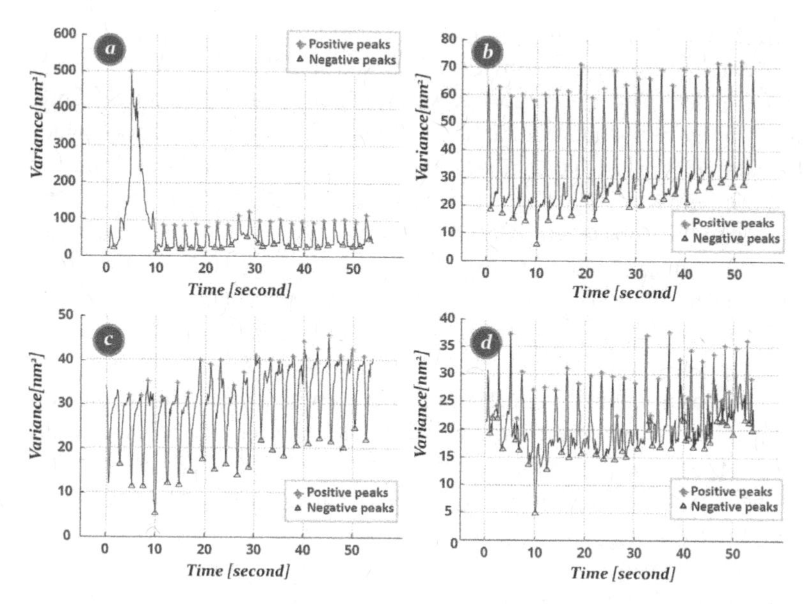

Figure 21.11 Detected peaks based on four cardiomyocyte sequences with the method shown in Section 21.3.2 (cardiomyocytes beating profile measurement using variance of OPD images). (a) Cardiomyocyte image sequences acquired in difficult conditions. (b), (c), and (d) "noise-free" recordings (i.e. no debris) similar to the previous recording [17] / with permission of Optical Publishing Group.

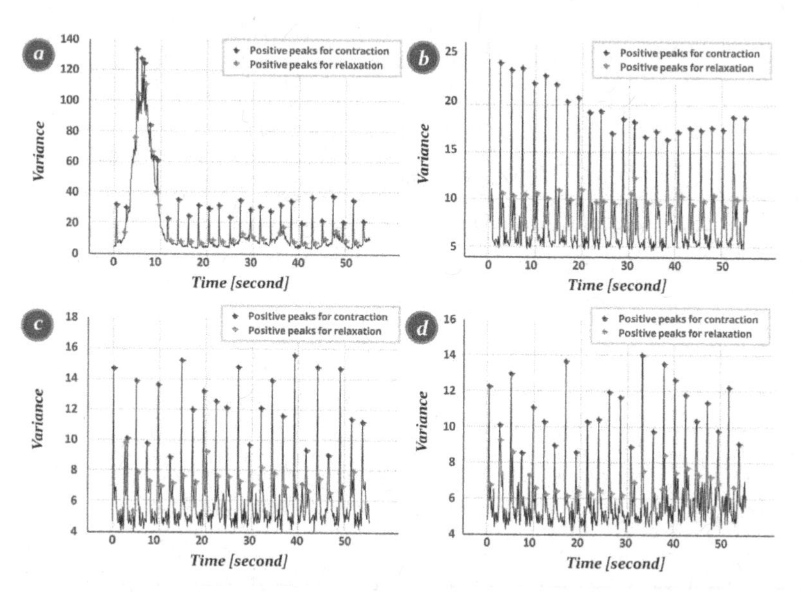

Figure 21.12 Detected peaks based on four cardiomyocyte sequences with the method in Section 21.3.3 (cardiomyocyte contraction and relaxation measurements). (a) Cardiomyocyte image sequences acquired in difficult conditions. (b), (c), and (d) "noise-free" recordings (i.e. no debris) similar to the previous recording [17] / with permission of Optical Publishing Group.

21.11a, and 21.12a are from image sequences under tough/difficult conditions. It should be emphasized that our method can detect all peaks, even in image sequences with debris interference. However, we found that our first method produced many noisy peaks under difficult conditions (first image sequence), which made the parameter measurement inaccurate. On the other hand, our second and third methods were more reliable at analyzing these image sequences, even under difficult conditions. Consequently, all the needed multi-parameters were measured using these images, demonstrating the robustness of our analysis algorithm. Finally, the mean value of each parameter described in each of our three methods was computed using all five sequences. The computed mean values for the three methods are given in Table 21.5, Table 21.6, and Table 21.7.

Combined measurements demonstrated the robustness of our methods and how they could quantify important cardiomyocyte dynamic characteristics, which might be used to screen the cytotoxic effects of compounds. The beating profile contains more information than what can be obtained by electrophysiology or fluorescence imaging because it integrates the effect of all ion channels involved, thus providing a signature that can be used to predict the effect of a specific compound. For instance, inhibitors of hERG channels (the main class of channels assessed in cardiac safety, these channels are involved in producing the repolarization current) all result in a similar profile [4]. Furthermore, due to the noninvasive aspect of measurements, both short-term and long-term effects of compounds can be monitored and analyzed with DHM.

Table 21.5 The measured values of multiple parameters for the cardiomyocyte-beating profile of five sequences (the first method).

Multi-parameter	Values (mean/cv)
1: Amplitude:	0.85/1.50
2: Rising time:	0.62/1.37 (seconds)
3: Falling time:	0.85/0.99 (seconds)
4: IBD_{50}:	0.44/0.50 (seconds)
5: IBD_{10}:	2.44/0.45 (seconds)
6: Rising/Falling slope:	0.51/1.52
7: Beating rate:	71.14/0.81
8: Beating period:	0.95/ (sd = 0.84 seconds)
9: Frequency	1.16/0.83

Table 21.6 The measured values of multiple parameters for the cardiomyocyte-beating profile of five sequences (the second method).

Multi-parameter	Values (mean/cv)
1: Amplitude:	40.01/1.21
2: Rising time:	0.57/0.86 (seconds)
3: Falling time:	0.43/1.13 (seconds)
4: IBD50:	0.79/0.65 (seconds)
5: IBD10:	2.35/0.38(seconds)
6: Rising/Falling slope:	24.00/1.21
7: Beating rate:	34.48/0.36
8: Beating period:	1.98/(sd = 0.83)
9: Frequency	0.55/0.37

Table 21.7 The measured parameters for the cardiomyocyte contraction and relaxation curve of five sequences (the third method).

Multi-parameters		Values (mean/cv)
Contraction	Beating rate:	28.52/0.16
	Beating period:	2.19/0.19
	Frequency	0.45/0.16
Relaxation	Beating rate:	28.52/0.17
	Beating period:	2.22/0.31
	Frequency:	0.45/0.18
Time between contraction and the following relaxation		0.59/0.62

21.4 Conclusions

The dynamics of human cardiac muscle cells and their spontaneous beating rates are quantitatively investigated through the fusion of DHM and information processing algorithms. We demonstrate the DHM is suitable for monitoring and quantifying the beating function of cardiomyocytes as well as automatically measuring multiple cardiomyocyte parameters based on quantitative phase profiles acquired with DHM. Our method is rapid, noninvasive, and effective. It allows for an automated analysis of normal cardiomyocyte dynamics and abnormal activities. Our automated, noninvasive measurement procedures open a new avenue for the cardiotoxicological screening or profiling of candidate molecules in preclinical drug discovery and safety testing programs.

References

1 Xi, B., Wang, T., and Li, N. (2011). Functional cardiotoxicity profiling and screening using the xCELLigence RTCA Cardio System. *J. Assoc. Lab. Autom.* 16 (6): 415–421.

2 Sirenko, O., Crittenden, C., and Callamaras, N. (2013). Multiparameter in vitro assessment of compound effects on cardiomyocyte physiology using iPSC cells. *J. Biomol. Screen.* 18 (1): 39–53.

3 Shaked, N., Satterwhite, L., and Bursac, N. (2010). Whole-cell-analysis of live cardiomyocytes using wide-field interferometric phase microscopy. *Biomed. Opt. Express* 1 (2): 706–719.

4 Abassi, Y., Xi, B., Li, N. et al. (2012). Dynamic monitoring of beating periodicity of stem cell-derived cardiomyocytes as a predictive tool for preclinical safety assessment. *Br. J. Pharmacol.* 165 (5): 1424–1441.

5 Carlson, C., Koonce, C., Aoyama, N. et al. (2013). Phenotypic screening with human iPS cell-derived cardiomyocytes HTS-compatible assays for interrogating cardiac hypertrophy. *J. Biomol. Screen.* 18 (10): 1203–1211.

6 Sirenko, O., Cromwell, E.F., Crittenden, C. et al. (2013). Assessment of beating parameters in human induced pluripotent stem cells enables quantitative in vitro screening for cardiotoxicity. *Toxicol. Appl. Pharmacol.* 273 (3): 500–507.

7 Yuste, R. (2005). Fluorescence microscopy today. *Nat. Methods* 2 (12): 902–904.

8 Stephens, D. and Allan, V. (2003). Light microscopy techniques for live cell imaging. *Science* 300 (5616): 82–86.

9 Salnikov, V., Lukyanenko, Y., Lederer, W., and Lukyanenko, V. (2009). Distribution of ryanodine receptors in rat ventricular myocytes. *J. Muscle Res. Cell Motil.* 30 (3–4): 161–170.

10 Hickson-Bick, D., Sparagna, G., Buja, L., and McMillin, J. (2002). Palmitate-induced apoptosis in neonatal cardiomyocytes is not dependent on the generation of ROS. *Am. J. Physiol. Heart Circ. Physiol.* 282 (2): H656–H664.

11 Sarah, M. and Antony, B. (2012). Preparation of plant cells for transmission electron microscopy to optimize immunogold labeling of carbohydrate and protein epitopes. *Nat. Protoc.* 7: 1716–1727.

12 Moerner, W. and Fromm, D. (2003). Methods of single-molecule fluorescence spectroscopy and microscopy. *Rev. Sci. Instrum.* 74 (8): 3597–3619.

13 Alsteens, D., Trabelsi, H., Soumillion, P., and Dufrêne, Y. (2013). Multiparametric atomic force microscopy imaging of single bacteriophages extruding from living bacteria. *Nat. Commun.* 4: 2926.

14 Azeloglu, E. and Costa, K. (2010). Cross-bridge cycling gives rise to spatiotemporal heterogeneity of dynamic subcellular mechanics in cardiac myocytes probed with atomic force microscopy. *Am. J. Physiol. Heart Circ. Physiol.* 298 (3): H853–H860.

15 Kamgoue, A., Ohayon, J., Usson, Y. et al. (2009). Quantification of cardiomyocyte contraction based on image correlation analysis. *Cytometry A* 75 (4): 298–308.

16 Pavillon, N., Benke, A., Boss, D. et al. (2010). Cell morphology and intracellular ionic homeostasis explored with a multimodal approach combining epifluorescence and digital holographic microscopy. *J. Biophotonics* 3 (7): 432–436.

17 Rappaz, B., Moon, I., Faliu, Y. et al. (2015). Automated multi-parameter measurement of cardiomyocytes dynamics with digital holographic microscopy. *Opt. Express* 23: 13333–13347.

18 Xi, B., Wang, T., Li, N. et al. (2011). Functional cardiotoxicity profiling and screening using the xCELLigence RTCA Cardio System. *J. Assoc. Lab. Autom.* 16 (6): 415–421.

19 Gonzalez, R. and Woods, R. (2002). *Digital Imaging Processing*. New York: Prentice Hall.

22

Automated Analysis of Cardiomyocytes with Deep Learning

22.1 Introduction

Many efforts have been made to develop complementary methods for human-induced pluripotent stem cell–derived cardiomyocytes (HiPSC-CMs) characterization to reduce drug development costs and cardiotoxicity-related drug attrition. CM characterization methods including patch clamping [1, 2], calcium imaging [3, 4], and image processing–based contraction-relaxation [5, 6] are reported. The mechanical probe is another widely used method for CM characterization [7]. There are drawbacks associated with each method that require either expertise and costly equipment or plating CMs onto specialized material, which makes the process difficult. Furthermore, these methods were only applied to the entire slide images of multiple CMs, so CM characterization methods at the single-cell level are lacking. As a result, it is important to exploit high-throughput and reliable methods for single-cell CM characterization.

In Chapter 21, we employed DHM for non-invasive, label-free studies of HiPSC-CMs and subsequent beating activity quantification [8]. We showed that the nucleus section of HiPSC-CMs from time-lapse DHM clearly reflects its rhythmic beating pattern, which might be less noisy and more informative for subsequent characterization. The CM beating activity at the single-cell level can thus be efficiently characterized if the dry-mass redistribution signal is observed only in the CM nucleus region.

To analyze the beating activity of a single CM, the region of interest (ROI) can be separated from the whole phase image of multiple CMs where the ROI is a region of the CM nucleus. The non-ROI includes the surrounding cytoplasm and membrane. Beating profiles in the non-ROI usually contain undesirable noisy peaks, which can make further dynamic activity characterization difficult. In general, CMs appear in any shape, size, and orientation by their very nature [9]. Therefore,

Artificial Intelligence in Digital Holographic Imaging: Technical Basis and Biomedical Applications, First Edition. Inkyu Moon.

it is a challenging task to extract the ROI from phase images of CMs. Recently, deep-learning methods have been applied to many image analysis tasks. Specifically, deep-learning models have shown great potential in medical image segmentation.

In this chapter, we will show how to use an FCN to extract nuclei from CM phase images to characterize single CMs [10, 11]. We will introduce an FCN-based network architecture consisting of parallel multi-pathways featuring concatenation for accurate CM nucleus extraction. We compared the performance of our method for CM nucleus extraction to that of the U-Net model. Our experimental results indicate that our model outperformed the U-Net model. Finally, multiple parameters related to the beating profile (contraction period, relaxation period, resting period, beating interval, and beat rate) of several single CMs were measured at the single-cell level from their phase images.

22.2 Region-of-interest Identification with Dynamic Beating Activity Analysis

Figure 22.1 shows one off-axis digital hologram of CMs and its corresponding reconstructed optical path difference (OPD) images of CMs. Note that the OPD and phase values are exchangeable with the following equation: $OPD = \frac{\lambda \times \varphi}{2\pi}$, where φ denotes the phase value. A dynamic beating activity comparison between the ROI and non-ROI was carried out to precisely identify the ROI as illustrated in Figure 22.1d. Beating results are shown in Figure 22.2. The dynamic of beating profile was obtained from the spatial variance between successive time-lapse OPD images as explained in Chapter 21. As stated in Chapter 21, OPD variance reflects the time course of cell dry-mass redistribution during the contraction-relaxation cycle of CMs. Since the OPD value redistribution of the ROI is dominant compared to that of non-ROI (see Figure 22.2), the OPD variance in the ROI might significantly reflect beating activity. The OPD variance in the non-ROI, on the other hand, shows no beating activity.

22.3 Deep Neural Network for Cardiomyocyte Image Segmentation

Deep-learning methods have been applied to a wide range of problems including medical image analysis [12]. The CNN is a type of deep-learning model that uses locally shared weights to capture hierarchical data characteristics. Multiple consecutive convolution kernels can capture the properties of input data, followed

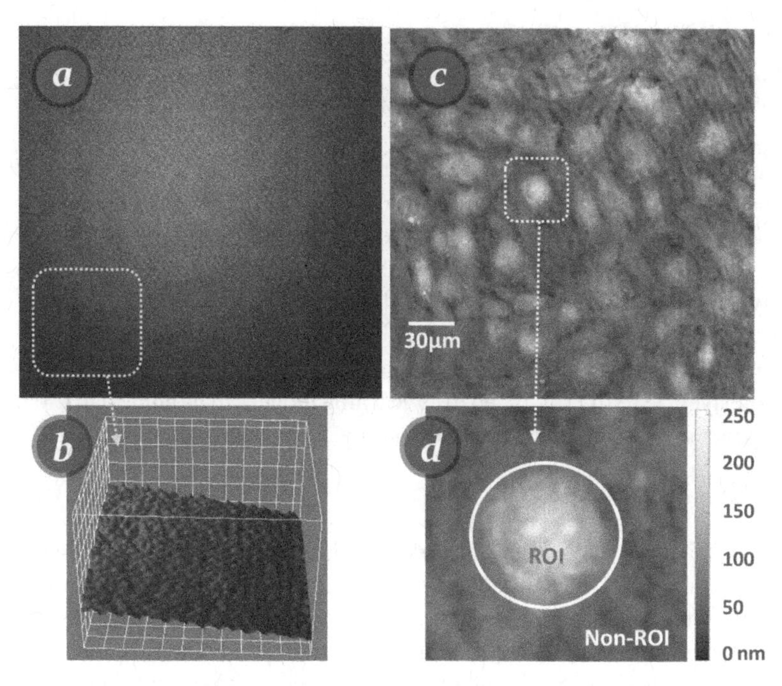

Figure 22.1 (a) A recorded hologram of cardiomyocytes and (b) an inset of the 3D portion of the hologram. (c) The OPD image after numerical reconstruction, which provides high contrast data for quantitative analysis. (d) A single cardiac cell with its nucleus section marked with a white line (the ROI) to be extracted for dynamic beating profile quantification. The OPD image was reconstructed from 540 images recorded at a sampling frequency of 10 Hz [10] / with permission of Optical Publishing Group.

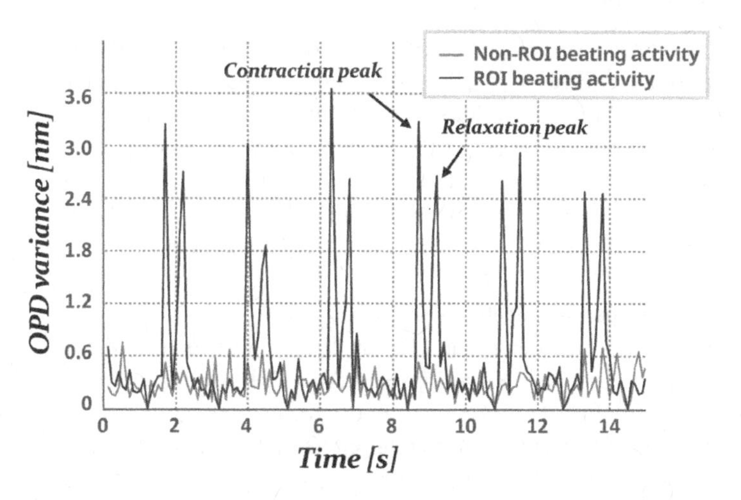

Figure 22.2 Beating activity comparison of ROI versus non-ROI for precise ROI identification.

by max-pooling layers for data dimension reduction. CNN mainly consists of the following elements: (i) a set of filters that can be trained to extract local features, (ii) a nonlinear function as an activation function, and (iii) a max-pooling layer that aggregates local feature specifications to reduce data dimensions. The max-pooling operation for down-sampling is used to obtain the maximum value of each filter in the convolution layer. The summation of each convolution layer is applied to a rectified linear unit (ReLU) as an activation function. The ReLU function is a nonlinear function that can increase the nonlinearity of CNN feature maps. Different network architectures can be designed depending on the task, which may be more efficient than simple CNN-based models. A fully convolutional network (FCN) is a type of CNN in which the fully connected layer is replaced with another convolution layer [13]. An FCN-based deep-learning model can be applied to CM's nucleus extraction for CM characterization at the single-cell level.

22.3.1 U-Net

U-Net is a widely used end-to-end network model for semantic segmentation. It consists of encoder and decoder pathways with skip connections between corresponding layers that offer good segmentation performance. The designed architecture is based on symmetric pathways for accurate localization. The first section of the U-Net extracts deep features while the second section is responsible for segmentation with extracted features. However, the U-Net architecture has some drawbacks such as a lack of flexibility and scalability. Deeper networks provide better segmentation while increasing parameter space and causing gradient vanishing [14].

22.3.2 Our Network Model

We will introduce a new FCN-based network architecture for accurate CM nucleus extraction. It takes advantage of parallel multi-pathway feature concatenation with dense connection blocks and residual connections [15]. The overall structure and building blocks of our network architecture are shown in Figure 22.3. The overall network structure leads to a better training performance compared to the U-Net model. It also improves the pixel classification accuracy. The building blocks of our FCN-based network model are briefly explained below.

22.3.2.1 Parallel Multi-pathway Feature Concatenation

In multi-pathway feature concatenation, different feature maps extracted by various kernel sizes are concatenated. Regarding the kernel size of convolution layers, it is difficult to decide which kernel size is more efficient for the task at hand because different kernel sizes will result in different features. The commonly used convolutional kernel size is 3×3. In our model, we used 1×1, 3×3, and 5×5

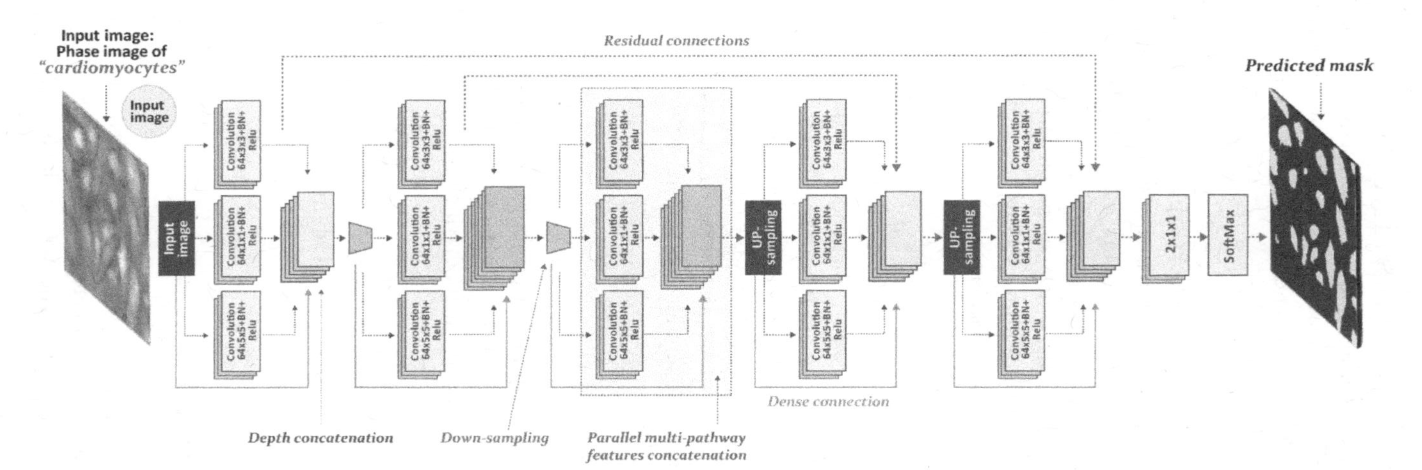

Figure 22.3 The proposed FCN-based network architecture for cardiac-cell ROI extraction. It is made of several modular blocks as follows. A parallel multi-pathway feature concatenation (yellow box) takes advantage of different kernel sizes: 1 × 1, 3 × 3, and 5 × 5. Each pathway is a composition of a convolutional layer, batch normalization layer, and rectified linear unit. Features are concatenated at the end. A 2 × 2 max-pooling layer with a stride of two is used for down-sampling data. A dense connection technique is used for efficient gradient propagation to prevent a vanishing gradient. Residual skip connections are denoted with dotted horizontal arrows [10] / with permission of Optical Publishing Group.

kernel sizes in different pathways in parallel followed by batch normalization (BN) and ReLU. Features from different pathways are concatenated at the end. A max-pooling layer with a size of 2×2 and a stride of 2 was used to down-sample feature maps from different pathways to reduce data dimension.

22.3.2.2 Dense Connection

We used the convolutional dense connection blocks explained in [16]. Within each dense block, layers were directly connected with their preceding layers, which was implemented through the concatenation of feature maps in subsequent layers. Dense connection blocks provide several advantages including efficient gradient propagation to avoid vanishing gradients, which commonly happens in deep networks. They can also reuse feature maps from previous layers instead of only the last layer, resulting in better network performance.

22.3.2.3 Residual Connection

Residual connection uses skip connections or short-cuts to jump over some layers to facilitate the training of deep networks [17, 18]. In addition, the shorter connection between layers close to the output and input offers better performance and reduces the number of parameters. Pooling operations can lead to the loss of some spatial information. These skip connections allow the network to recover such lost spatial information. In our network model, the output of the standard 3×3 convolutional layer prior to the pooling operation was transferred to the corresponding output of the up-sampling layer.

22.3.3 Patch Extraction

Deep-learning models, in general, require a large number of samples for training. There are several methods to increase the number of training samples such as data augmentation. We used a patch extraction method, which could increase the number of samples to a sufficient level, to train our FCN-based deep-learning model. We extracted patches from the OPD image containing multiple CMs using a sliding window and captured patches along with the corresponding ground truth (manually extracted) with a size of 32×32 pixels. Captured patches using a sliding window were not overlapped (see Figure 22.4).

22.4 Experimental Results

After manually annotated OPD images were reconstructed, we trained our model using the training dataset (OPD images generated by patch extraction) and corresponding ground-truth labels. The whole dataset contained 2500

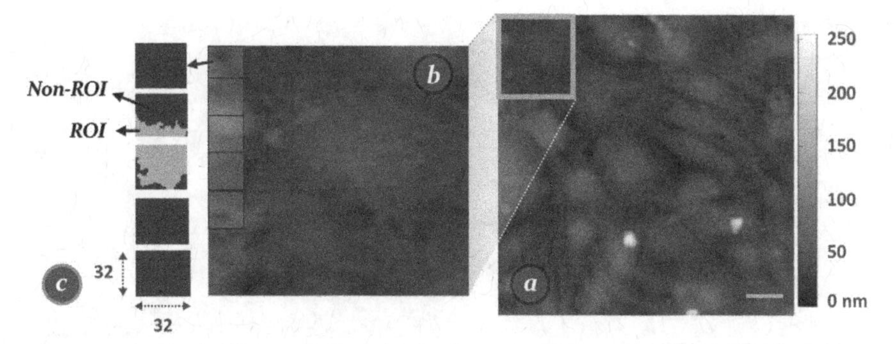

Figure 22.4 Sliding window patch extraction method for training data preparation. (a) Original OPD of multiple cardiomyocytes (grey bar = 20 μm). (b) Magnified portion of the original phase image with the indicated patches. (c) Corresponding ground-truth patches with the ROI (light grey) and non-ROI (dark grey) portions [10] / with permission of Optical Publishing Group.

extracted patches along with the corresponding ground truth. Of the whole dataset, 80% ($n = 2000$) were used for training and 20% ($n = 500$) were used for testing. We evaluated our trained model for ROI extraction using test images containing multiple CMs (see Figure 22.5a). Predicted masks from our model and the U-Net model are shown in Figures 22.5b,c. Figure 22.5d shows the ground-truth mask. The predicted mask by the U-Net model contained extra seeds and exhibited both over- and under-segmentation. However, our model effectively handled different edge directions and specific ambiguous attributes of the ROI and non-ROI, despite difficulties of ROI extraction including ROI size, shape, and orientations.

Figures 22.6 shows an example of beating activity comparison of the whole slide of OPD image with multiple CMs before and after ROI extraction (see Figures 22.6a,c) and their corresponding beating activity (see Figures 22.6b,d). As previously stated, the beating activity before ROI extraction was noisy, with extra incorrect peaks due to the OPD variance in the non-ROI section, which might make quantifying the beating profile difficult. However, the ROI section reflects a clean and less noisy beating activity with the wrong peaks removed. Hence, it is more relevant for CM dynamic characterization.

22.4.1 Single Cardiomyocyte Beating Profile Quantification

ROIs of CMs were extracted in the first step of our method. The resulting mask image was multiplied by each OPD image in the sequence. Finally, the spatial variance between successive images was computed to obtain the beating profile as explained in Chapter 21. The computed variance was sensitive to redistribution

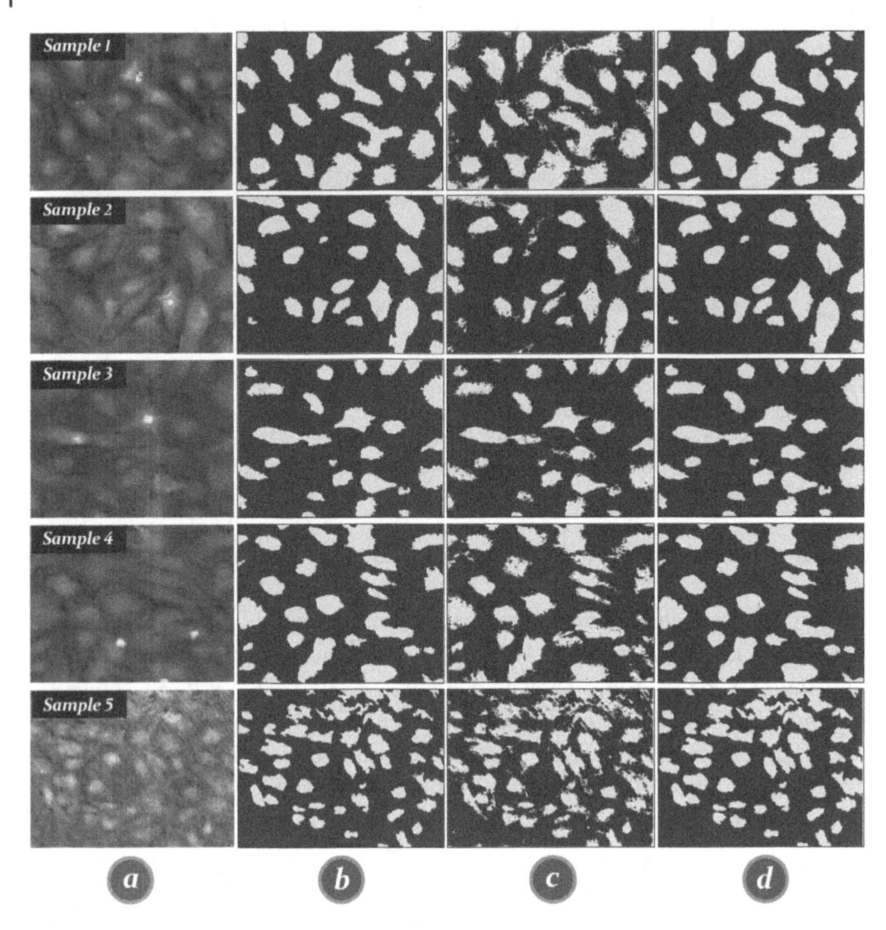

Figure 22.5 Results of ROI extraction using the proposed FCN-based method compared to the U-Net network model. (a) Original phase image of multiple cardiac cells obtained by DHM. (b) Predicted mask using our trained FCN-based model. (c) Predicted mask using the U-Net network model. (d) Ground-truth mask extracted manually [10] / with permission of Optical Publishing Group.

of dry mass within CMs to monitor characteristics of the cardio beating over time. The corresponding equation is shown as follows:

$$opd_{var} = var[opd_i - opd_{i-1}], \tag{22.1}$$

where opd_i and opd_{i-1} are the i_{th} and $i\text{-}1_{th}$ images. Specifically, opd_{var} (OPD variance) represents the time course of the cell dry-mass redistribution during CM's beating activity. The opd_{var} signal contains information about redistribution of dry mass within CMs, namely contraction and relaxation durations. After ROI extraction, we monitored CMs' beating activity and measure their dynamic parameters

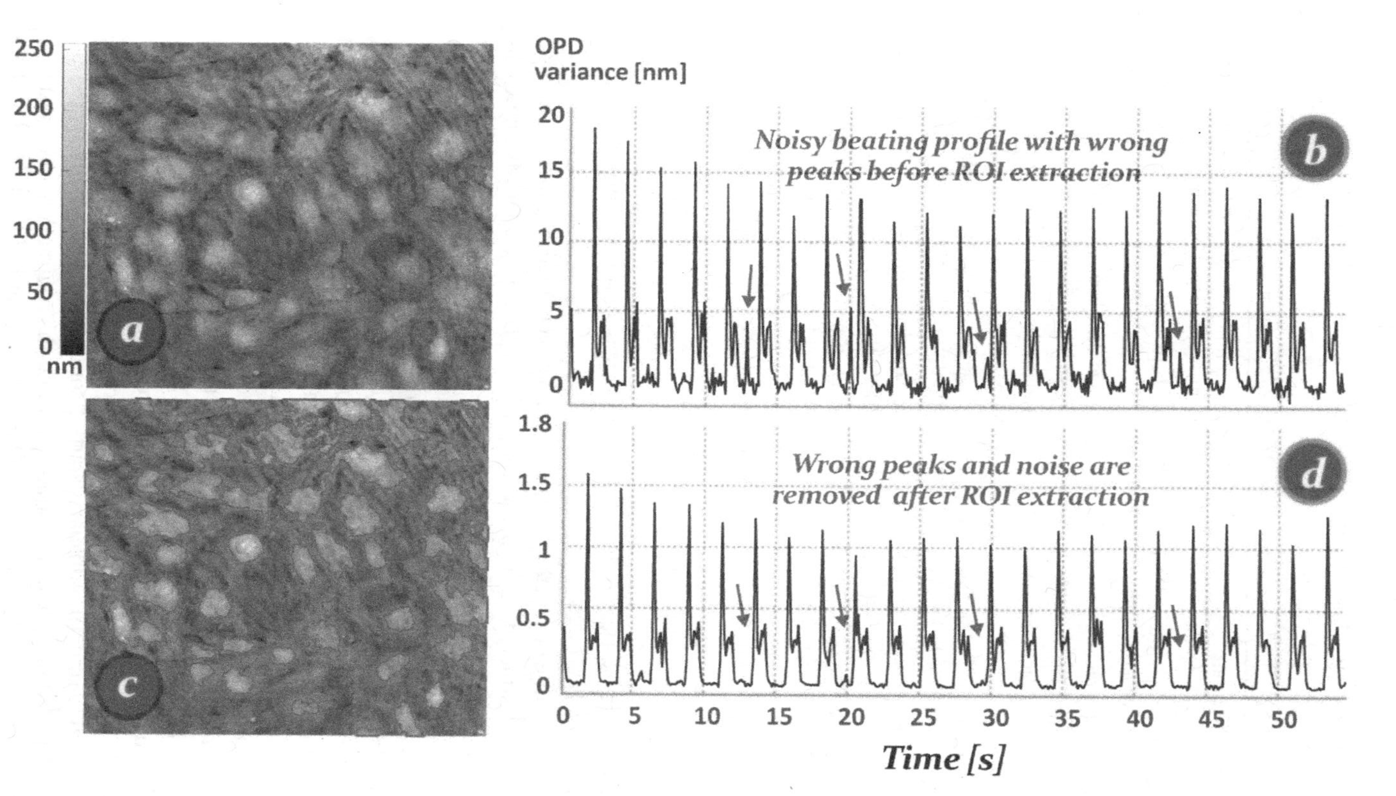

Figure 22.6 Example of a CM beating profile reconstruction before and after ROI extraction. An OPD image of multiple CMs (a) before ROI extraction and (b) the corresponding beating activity profile. An OPD (c) after ROI extraction using our method (grey outline) and (d) the corresponding beating activity profile [10] / with permission of Optical Publishing Group.

Table 22.1 Descriptions of parameters to quantify cardiomyocyte dynamics.

	Parameters	Description
Contraction	Beat rate	The total number of contraction peaks in 1 minute (number of red points in Figure 22.7b)
	Beating interval AVG [see #1 in Figure 22.7c]	The time between two adjacent contraction peaks.
	Contraction period AVG [see #2 in Figure 22.7c]	The average time between the start-of-contraction and end-of-contraction points.
	Contraction period STD	The standard deviation of the contraction beating period.
Relaxation	Relaxation period AVG [see #3 in Figure 22.7c]	The average time between the start-of-relaxation and end-of-relaxation points.
	Relaxation period STD	The standard deviation of the relaxation beating period.
Resting	Resting period AVG [see #4 in Figure 22.7c]	The average time between the end-of-relaxation and the next start-of-contraction points.
	Resting period STD	The standard deviation of the resting beating period.

at the single cell level. Descriptions of the parameters for quantifying CM dynamics are shown in Table 22.1.

To further examine our method for single-cell level characterization, six individual CMs were extracted from the whole slide OPD image of CMs shown in Figure 22.7a. The corresponding beating profile was calculated from Eq. (22.1) (see Figure 22.7b). We measured several parameters related to the beating activity profile for each individual CM. To characterize the dynamic beating profile activity of single CMs using ROI extraction, we detected two main peaks of contraction and relaxation using the Otsu thresholding method. The first peak with a larger amplitude value in most cases is the contraction, and the corresponding relaxation peak is represented by the second peak, which has a lower amplitude value. The contraction and relaxation peaks are then used to define three auxiliary points: (i) start-of-contraction, (ii) end-of-contraction, and (iii) end-of-relaxation showed in Figure 22.7c. The following steps are used to locate auxiliary points: (i) the end-of-contraction point (start-of-relaxation) (Figure 22.7c, blue points) is obtained by finding the smallest amplitude value between contraction and the corresponding relaxation peaks; and (ii) auxiliary start-of-contraction and end-of-relaxation points are detected using a search strategy around the contraction and relaxation peaks. The time difference between the end-of-contraction (start-of-relaxation)

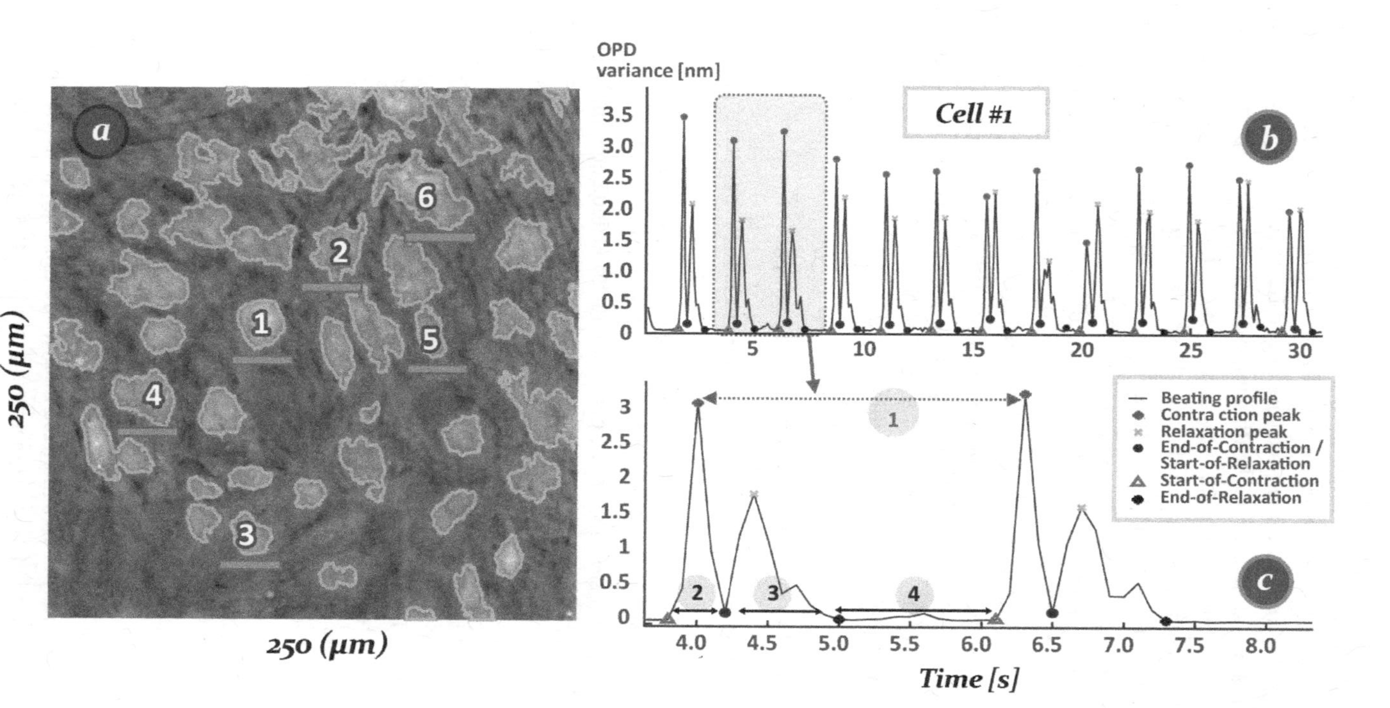

Figure 22.7 (a) Original OPD image of multiple cardiomyocytes in which single cells are denoted for further quantification. (b) Beating activity profile derived from cell #1 (at 30 s) and (c) details on quantification parameters explained in Table 22.1 [10] / with permission of Optical Publishing Group.

and the end-of-relaxation is considered as relaxation period and the time difference between the start-of-contraction and the corresponding end-of-contraction points specifies the contraction period. The time difference between two consecutive contraction peaks determines beating intervals. The resting period is the time difference between the end-of-relaxation and the next start-of-contraction points. Extracted features of the beating profile and their corresponding descriptions are listed in Table 22.1. Figure 22.8 shows a beating activity profile reconstructed from cell numbers 1, 2, 3, 4, 5, and 6 with detected contraction-relaxation peaks and auxiliary points for precise single-CM characterization. The dynamic beating profile quantification of single CMs demonstrates that the contraction period is shorter than the relaxation period due to the presence of different ion channels and transporters expressed in cardiomyocytes as well as the mechanisms by which their activities are sequentially orchestrated when CMs contract and relax (see Figure 22.9). As shown in Figure 22.9, all CMs have similar average beating rates, average beating intervals, and resting times.

22.4.2 Network Performance Accuracy Measurement

A set of evaluations were carried out to compare our model's performance against the U-Net model in terms of training process performance, pixel classification accuracy, and the Dice coefficient. The learning curve evaluation showed that our model convergence was faster than the U-Net model (see Figures 22.10a and 22.10b). The confusion matrix of pixel classification indicates the accuracy percentage of correctly classified pixels versus the misclassification rate. As shown in Figure 22.10c, the classification accuracy rates using our FCN-based model were 99.76% and 99.28%, respectively, while the same evaluation for the U-Net model showed classification accuracies of 93.69% and 91.25% (see Figure 22.10d). These experimental results indicate that our method is more robust with more accurate pixel classification. To statistically evaluate the segmentation performance of our model against the U-Net model for each CM sample, we performed the Dice coefficient analysis (see Table 22.2). The Dice coefficient was calculated using the equation

$$DSC = \frac{2TP}{FP + 2TP + FN}$$

where TP, FP, and FN were the numbers of true positive, false positive, and false negative detections, respectively.

To validate our method for single CM characterization using the nucleus section and see if the segmentation area included nuclei, one CM sample was stained with Hoechst dye. We applied our segmentation model prior to and after nuclei

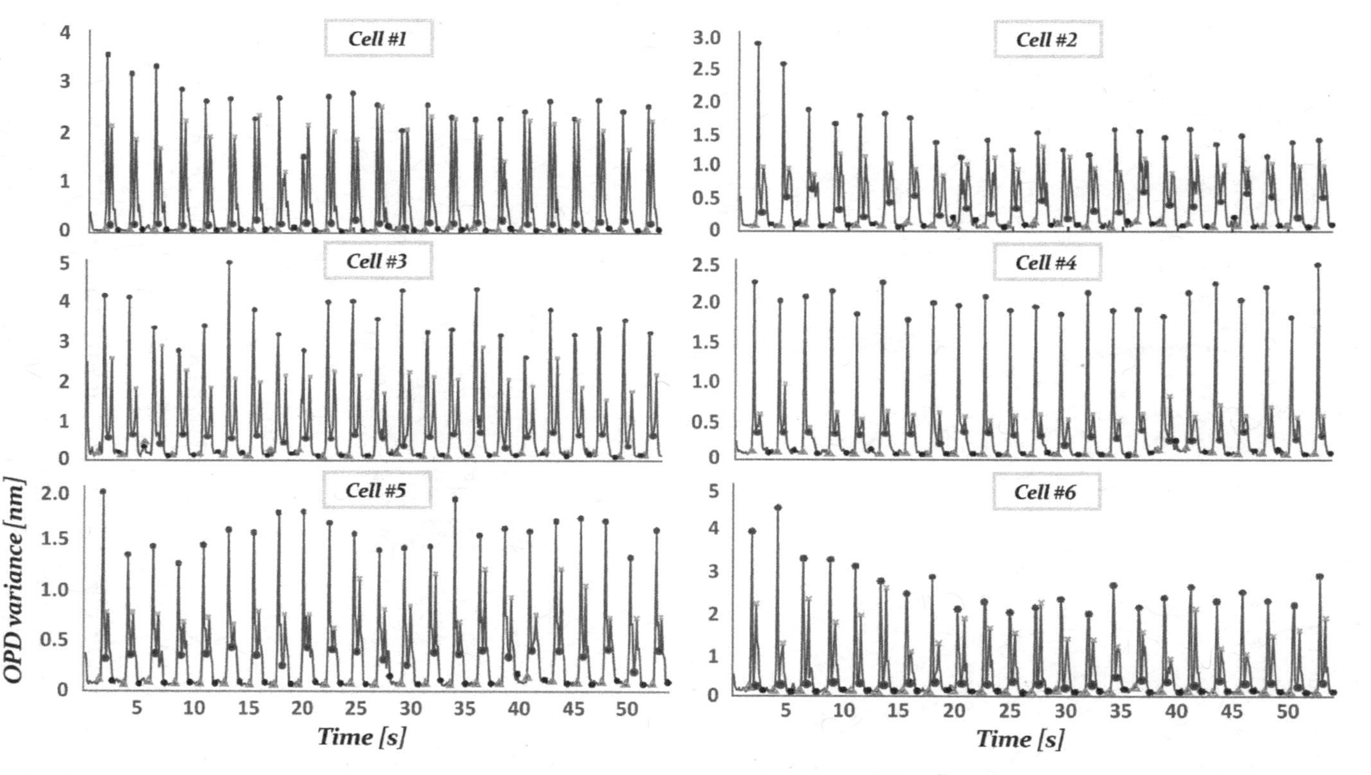

Figure 22.8 Beating profiles of single cells extracted from sample number five (cell #1, 2, 3, 4, 5, and 6). The sample was recorded with a 10 Hz sampling frequency for 54 seconds [10].

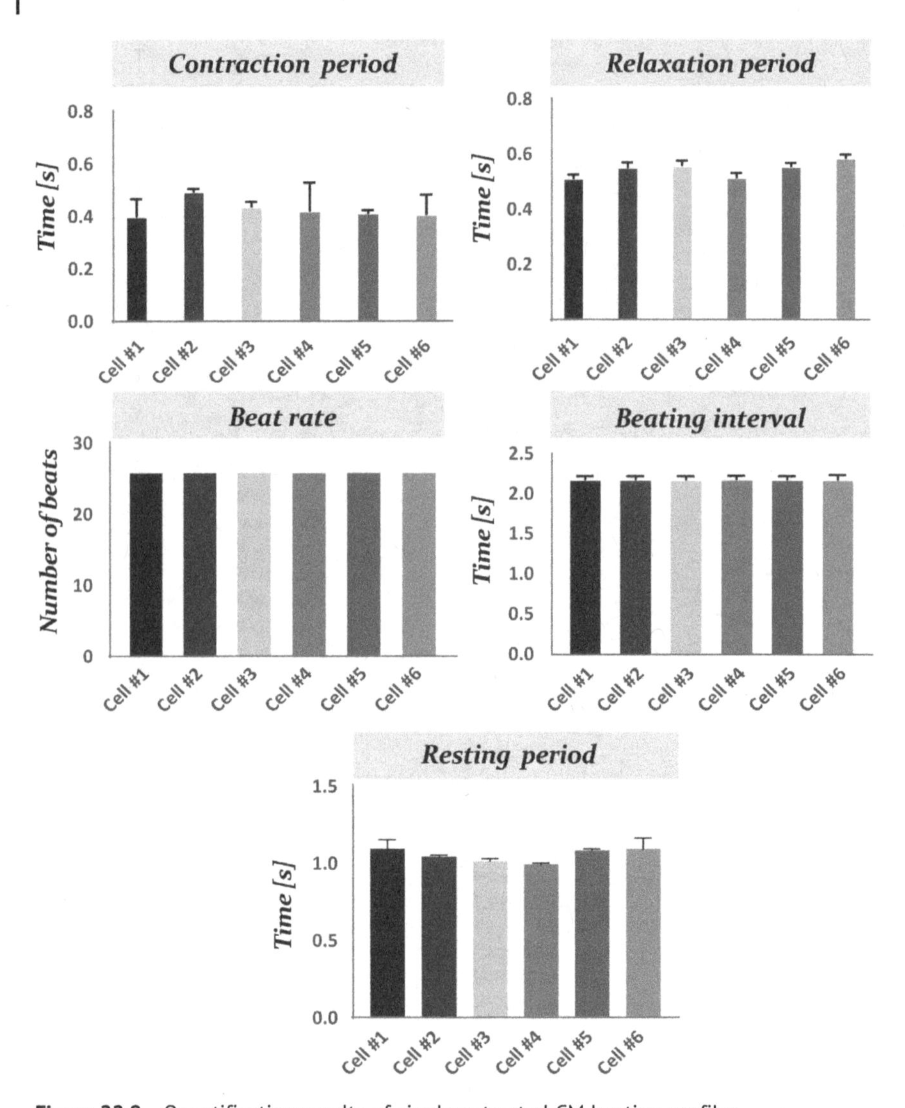

Figure 22.9 Quantification results of single extracted CM beating profiles.

staining. ROI and non-ROI sections were marked for visual comparison. As shown in Figure 22.11, the ROI section mainly included the nuclei (marked in blue).

22.4.3 Automated Quantification of Cardiomyocyte Synchronization

For optimal functionality of the cardiac muscle system, CMs must respond to the commands of mechanical contraction. The signal propagates through CMs, resulting in the myocardium pumping blood out of the ventricle chamber. It requires

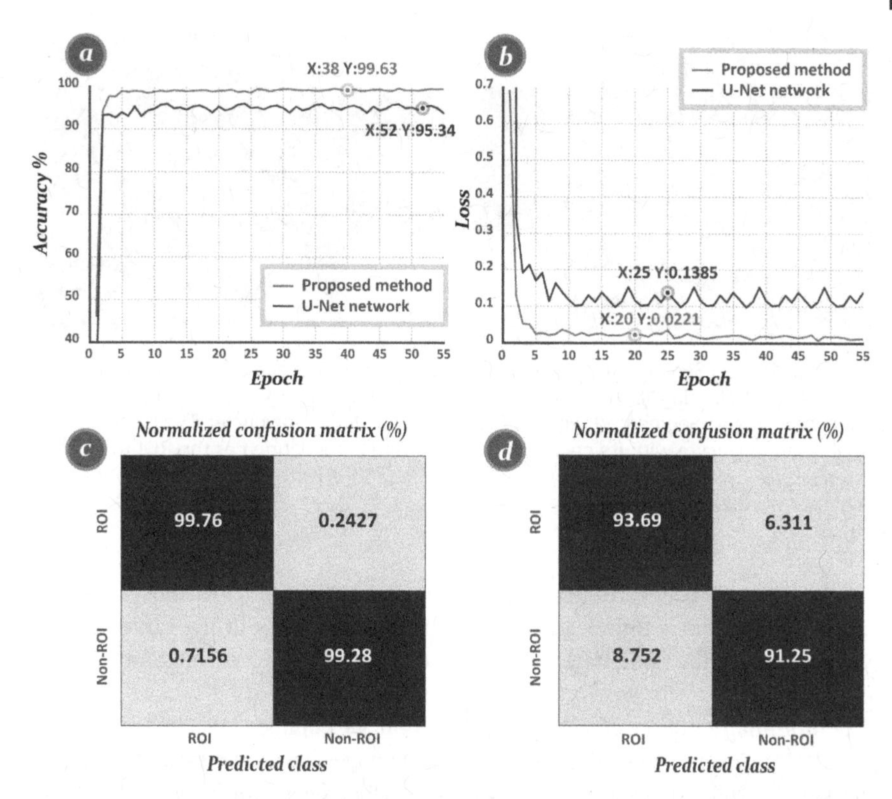

Figure 22.10 A comparison of learning curves and confusion matrix for pixel classification of the proposed FCN-based model versus the U-Net model. (a) Training process accuracy in 55 epochs and (b) the corresponding loss curves. (c) A confusion matrix of pixel classification prediction accuracy for the proposed model. (d) A confusion matrix of pixel classification prediction accuracy for the U-Net model.

Table 22.2 Segmentation performance evaluation using the Dice coefficient for the proposed method against the U-Net model.

	Dice (%)	
Cardiomyocyte samples	**Proposed method**	**U-Net**
Sample #1	96	88
Sample #2	95	89
Sample #3	94	88
Sample #4	96	87
Sample #5	97	88

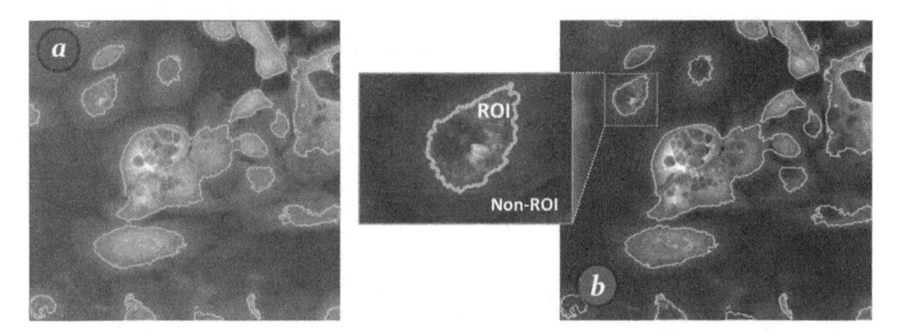

Figure 22.11 (a) Segmentation results of the original OPD image of multiple CMs before Hoechst nuclei staining. (b) An overlay of the original OPD image with nuclei staining on the segmented image. Segmented sections (ROI) mostly included the nuclei. The inset shows a single cardiomyocyte with its nuclei marked (grey line) and defined as the ROI to be extracted for dynamic beating profile quantification [10] / with permission of Optical Publishing Group.

synchronization between CMs. Contraction and relaxation time must be finely controlled without a time lag between signals of two CMs in the same sample. We will present the same state visually by analyzing the cross-correlation between two signals.

For the analysis of cell-to-cell synchronization, we obtained OPD images of CMs with DHM. Several single CMs from the OPD images were then extracted using a marker-controlled watershed algorithm (see Chapter 12 for more details). As shown in Figure 22.12, the extracted region contained mostly the nucleus of the

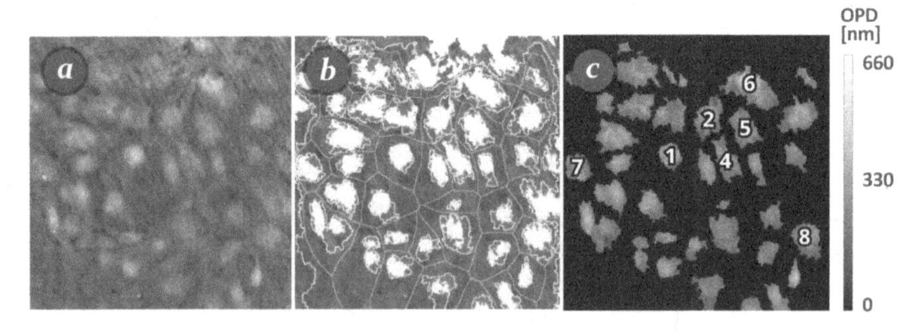

Figure 22.12 (a) Original OPD image of cardiac muscle cells obtained by DHM, (b) A cardiomyocyte image showing cell regions, internal markers, and external markers obtained using the marker-controlled watershed algorithm, (c) Final segmented cardiomyocyte image (some cells are labeled for further discussion). OPD scale applies to both (a) and (c) [10] / with permission of Optical Publishing Group.

cardiac muscle cell. Note that the segmented region also included a small portion of the cytoplasm within which the nucleus was enclosed. To quantify the beating profile of segmented single CMs, we used the variance of each OPD image frame subtracted from its successive frame. We compared contraction and relaxation points of CMs labeled in Figure 22.12c. Measured values were analyzed to demonstrate cell-to-cell synchronization. We found that the isolated CMs were rising and falling at the exact same time as the 3D representation (see Figure 22.13) and that positive peak-peak increases overlapped exactly, showing a highly synchronous population.

We performed a cross-correlation analysis to check the similarity and synchronization between two signals (two cells) using the equation

$$(f * g)[n] \stackrel{\text{def}}{=} \sum_{m=-\infty}^{\infty} f^*[m] g[m+n],$$

where f^* denotes the conjugate of f (first signal), g is the second signal, and n is the time lag between the two signals in the time domain (m) [11]. It is a useful tool to determine the time delay between two beat signals and analyze the signal synchronization. The maximum cross-correlation of functions indicates the time where the two signals are best aligned.

The argument that maximum cross-correlation can determine the point where the two signals are best matched is:

$$T_{delay} = \arg\max_{m} ((f * g)(m)).$$

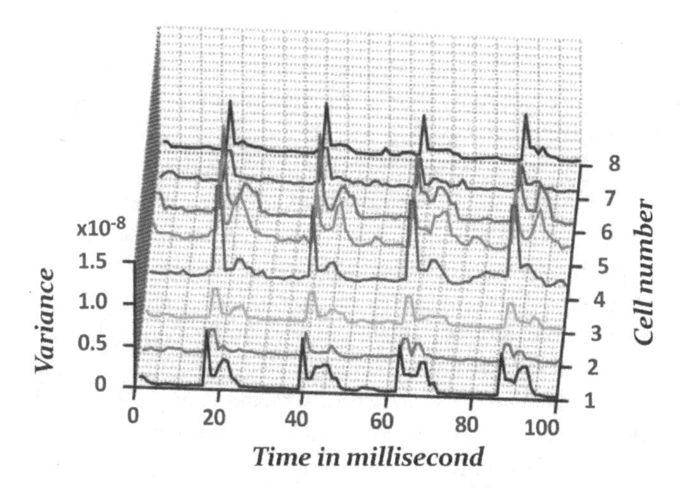

Figure 22.13 Rhythm strip comparison: a 3D representation of rhythm strips' synchronization detection [11].

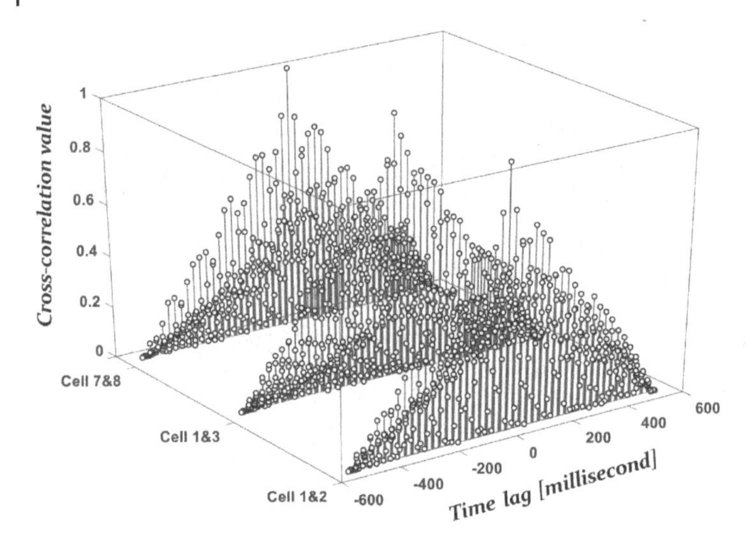

Figure 22.14 Cross-correlation between cells 1:2, 1:3, and 7:8 [11] / with permission of Optical Publishing Group.

Figure 22.14 shows cross-correlation between the two beating signals. We can see in this diagram that the two signals are at a maximum value when the time lag is zero, indicating that they are perfectly synced in time.

22.5 Conclusions

In this chapter, we introduced an automated FCN-based platform for CM nucleus extraction and beating pattern characterization at the single-cell level. We demonstrated that the CM nucleus section (denoted as ROI) from an OPD image can sufficiently reflect the beating pattern to characterize CMs at the single-cell level. We designed a new FCN-based model to discriminate the CM nucleus section from other sections of CMs using pixel classification techniques. Our FCN-based model outperformed the U-Net model for segmentation mask prediction with a high pixel classification accuracy (over 99%). The predicted mask was further used to quantify the dynamic contraction-relaxation of a single CM. We obtained multiple OPD images of CMs under different circumstance including shape, size, and orientations. We extracted several individual CMs from an OPD image using the predicted segmentation mask. Multiple parameters associated with the dynamic beating profile of each CM after removing the non-ROI were measured at the single-cell level. Single CM beating profile quantification was precisely performed using contraction-relaxation peak detection and multiple auxiliary points. Experimental results demonstrate that our method is robust for CM nucleus extraction and subsequent CM beating profile characterization at the single-cell level. Furthermore, our results showed that our method could quantitatively investigate cell-to-cell synchronization in the cardiovascular system. We believe that our models can be used to study cardiomyocyte disorders at the single-cell level.

References

1 Ossola, D., Amarouch, M., Behr, P. et al. (2015). Force-controlled patch clamp of beating cardiac cells. *Nano Lett.* 15 (3): 1743–1750.

2 Brüggemann, A., Haarmann, C., Rapedius, M. et al. (2017). Characterization of iPS derived cardiomyocytes in voltage clamp and current clamp by automated patch clamp. *Biophys. J.* 112 (3): 236a.

3 Huebsch, N., Loskill, P., Mandegar, M. et al. (2015). Automated video-based analysis of contractility and calcium flux in human-induced pluripotent stem cell-derived cardiomyocytes cultured over different spatial scales. *Tissue Eng. Part C Methods* 21 (5): 467–479.

4 Grespan, E., Martewicz, S., Serena, E. et al. (2016). Analysis of calcium transients and uniaxial contraction force in single human embryonic stem cell-derived cardiomyocytes on microstructured elastic substrate with spatially controlled surface chemistries. *Langmuir* 32 (46): 12190–12201.

5 Ahola, A., Kiviaho, A.L., Larsson, K. et al. (2014). Video image-based analysis of single human induced pluripotent stem cell derived cardiomyocyte beating dynamics using digital image correlation. *Biomed. Eng.* 13 (1): 1–18.

6 Bazan, C., Barba, D., Blomgren, P., and Paolini, P. (2011). Image processing techniques for assessing contractility in isolated neonatal cardiac myocytes. *J. Biomed. Imaging* 16: 729732.

7 Nitsan, I., Drori, S., Lewis, Y. et al. (2016). Mechanical communication in cardiac cell synchronized beating. *Nat. Phys.* 12 (5): 472–477.

8 Jaferzadeh, K., Rappaz, B., Kuttler, F. et al. (2019). Marker-free automatic quantification of drug-treated cardiomyocytes with digital holographic imaging. *ACS Photon.* 7 (1): 105–113.

9 Abassi, Y., Xi, B., Li, N. et al. (2012). Dynamic monitoring of beating periodicity of stem cell-derived cardiomyocytes as a predictive tool for preclinical safety assessment. *Br. J. Pharmacol.* 165 (5): 1424–1441.

10 Ahmadzadeh, E., Jaferzadeh, K., Shin, S., and Moon, I. (2020). Automated single cardiomyocyte characterization by nucleus extraction from dynamic holographic images using a fully convolutional neural network. *Biomed. Opt. Express* 11: 1501–1516.

11 Moon, I., Ahmadzadeh, E., Jaferzadeh, K., and Kim, N. (2019). Automated quantification study of human cardiomyocyte synchronization using holographic imaging. *Biomed. Opt. Express* 10: 610–621.

12 Litjens, G., Kooi, T., Bejnordi, B. et al. (2017). A survey on deep learning in medical image analysis. *Med. Image Anal.* 42: 60–88.

13 Garcia, A., Escolano, S., Oprea, S. et al. (2018). A survey on deep learning techniques for image and video semantic segmentation. *Appl. Soft Comput. J.* 70: 41–65.

14 Ronneberger, O., Fischer, P., and Brox, T. (2015). U-net: convolutional networks for biomedical image segmentation. *Lect. Notes Comput. Sci.* 9351: 234–241.

15 Chen, L., Bentley, P., Mori, K. et al. (2018). DRINet for medical image segmentation. *IEEE Trans. Med. Imaging* 37 (11): 2453–2462.

16 Huang, G., Liu, Z., Maaten, L., and Weinberger, K. (2017). Densely connected convolutional networks. In: *2017 IEEE Conference on Computer Vision and Pattern Recognition (CVPR)*, Honolulu, HI, USA (21–26 July 2017), 2261–2269. New York (NY): IEEE.

17 Khened, M., Kollerathu, V., and Krishnamurthi, G. (2019). Fully convolutional multi-scale residual DenseNets for cardiac segmentation and automated cardiac diagnosis using ensemble of classifiers. *Med. Image Anal.* 51: 21–45.

18 S. Jastrzębski, D. Arpit, N. Ballas, V. Verma, T. Che, and Y. Bengio. (2018). Residual Connections Encourage Iterative Inference. arXiv. https://doi.org/10.48550/arXiv.1710.04773

23

Automatic Quantification of Drug-treated Cardiomyocytes with DHM

23.1 Introduction

There is a high demand for innovative drug-discovery technologies. The past decade has seen a great expansion of label-free imaging platforms and methods in various stages of the drug discovery process including target engagement determination and drug safety assessments. Moreover, with the advent of human-induced pluripotent stem cell (HiPSC) technology, substantial attempts have been made to use HiPSCs to screen for new drugs and to test candidate drugs for toxicity. In particular, HiPSC-derived CMs can help us elucidate the molecular and cellular mechanisms of cardiac arrhythmias in patients. They can also serve as robust platforms for developing new drugs for clinical therapy. Therefore, it is highly relevant to develop label-free imaging systems capable of monitoring the effects of drug candidates on HiPSC-derived cells.

DHM can provide quantitative imaging of cell structures and dynamics in a non-invasive manner. This is of tremendous importance since it can enable us to accurately visualize live cells without disturbing them as drug-mediated cellular effects are assessed. The unique capability of DHM to visualize live cells without the need for scanning or contrast agents is particularly well suited for label-free, high-content screening [1–3]. Therefore, invaluable information regarding the morphology (contraction–relaxation) and dynamics of the quantitative redistribution of materials within CMs can be non-invasively obtained by DHM.

In this chapter, we will introduce an automated method to quantitatively assess the effects of E-4031, a class III anti-arrhythmic compound, and isoprenaline, which is a treatment for slow heart rates and heart-block abnormalities, on CMs by analyzing a set of dynamic cell parameters derived from phase imaging [4]. These cell parameters, which can help us measure mechanical and beating properties of CMs, can be used to quantitatively characterize effects of these

Artificial Intelligence in Digital Holographic Imaging: Technical Basis and Biomedical Applications, First Edition. Inkyu Moon.

compounds (E-4031 and isoprenaline) by monitoring control and drug-treated CMs. We believe that our label-free DHM is efficient for quantifying and assessing the specific effects of promising candidates on relevant CM functions within the framework of drug development.

23.2 Materials and Methods

23.2.1 Cell Preparations and Experimental Conditions

HiPSC-CMs (iCell CMs, catalog number CMC-100-010-001) were obtained from Cellular Dynamics Int. (Madison, WI, USA) and cultured according to the manufacturer's instructions (2×10^4 cells per well in 96-well plates pre-coated with 0.1% gelatin) and grown for 14 days. Measurements were achieved with a Chamlide WP incubator system for 96-well plates (LCI, South Korea) set at 37 °C and 5% CO_2 with high humidity. For the control condition, a sequence of CM images was recorded before treatment. Each drug was then added to individual wells at several concentrations and the CMs imaged again after a 15-minute incubation period. Experiments were performed in triplicate (three wells of about 50 cells with one 330×330 μm field acquired per well). For confirmation, cell viability was also determined using a Presto Blue fluorescence assay. Automated image analysis for toxicity assessment was performed with CellProfiler software 3.1.9 [5].

Figure 23.1a shows a recorded hologram of a control CM sample. Since the hologram was recorded in an off-axis configuration, a Fourier transform of the hologram could separately represent the bandwidth of the real image, virtual image, and zero-order noise as shown in Figure 23.1b. A spatial filter to cover only the bandwidth corresponding to the real image was used (see Figure 23.1c). Figures 23.1d and 23.1e show the amplitude and phase images of CMs reconstructed from filtered holograms. Note that phase and optical path difference (OPD) values can be exchanged with the equation $OPD = \frac{\lambda \times \phi}{2\pi}$, where ϕ denotes the phase value.

23.2.2 Cardiomyocyte Beating Signal Extraction

Several image pre-processing techniques were used to accurately quantify dynamic parameters of control and drug-treated CMs (see Figure 23.2a). The procedure is similar that used in Chapter 21. We used a signal fitting approach and a 3D median filter, which allowed us to enhance the accuracy of measurements. The filter ($2 \times 2 \times 2$) was applied to the whole stack to reduce spatial and temporal noise. According to Chapter 21, during the first step of our image processing method, temporal OPD images of CMs were arithmetically averaged. The resulting

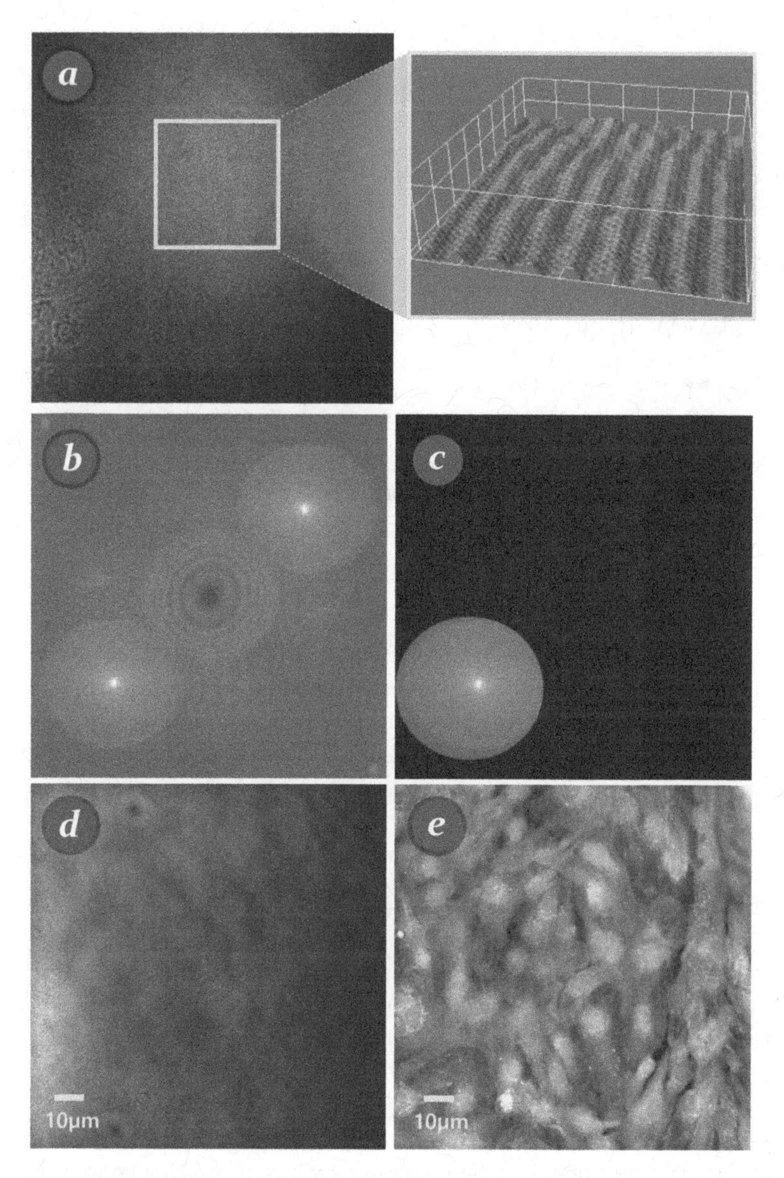

Figure 23.1 (a) A recorded hologram of a control cardiomyocyte sample (inset shows a portion of the hologram in 3D. Interference patterns between the reference wave and object wave were spatially recorded), (b) A spectrum of the off-axis hologram (three bandwidths were isolated). (c) Spatial filtering can preserve the bandwidth of a real image. (d) Amplitude image after numerical reconstruction. The contrast in the amplitude image is poor since cells are transparent. (e) Phase image after numerical reconstruction and phase unwrapping. The phase image provides high-contrast data for quantitative analysis [4] / with permission of American Chemical Society.

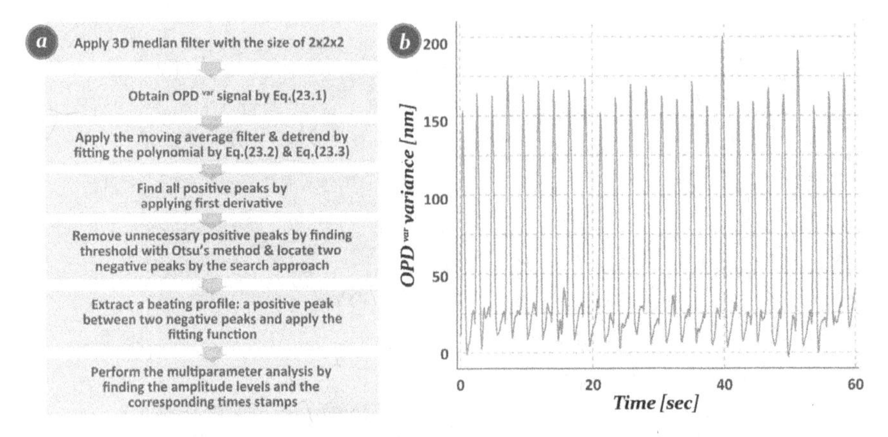

Figure 23.2 (a) General scheme of the proposed method to study cardio cells, and (b) Beating profile of a cardio sample at control conditions (beat rate = 26 beats per minute [bpm]).

averaged OPD image was then subtracted from each image in the sequence. Finally, the spatial variance between successive images was computed. This parameter, being sensitive to any redistribution of dry mass within CMs, can monitor various aspects of the cardiac beat over time. The corresponding equation is defined as

$$OPD_i^{var} = var\left[OPD_i(x,y) - \overline{\text{OPD}}\right], \tag{23.1}$$

where OPD_i is the ith OPD image from total samples and $\overline{\text{OPD}}$ is the average of total OPD images in the stack. Specifically, OPD^{var} (OPD variance) represents the time course of cell dry-mass redistribution occurring during the CM contraction–relaxation cycle. As a result, the OPD^{var} signal contains information about the beating pattern, such as the contraction and relaxation durations, resting duration, and beating period. Figure 23.2b shows a beating profile of control CMs after applying Eq. (23.1).

We will introduce a new approach to evaluate dynamic parameters corresponding to both a low CM beat rate (less than 20 bpm for which the resting period corresponds to a flat shoulder) and a high CM beat rate (greater than 30 bpm) characterized by a short resting time. For this purpose, the high-frequency noise of the OPD^{var} signal was removed by applying a moving average filter with a span of 6 (unweighted mean of 6 OPD^{var} points). This filtering procedure can efficiently reduce the noise that appears in the flat region of the signal corresponding to the

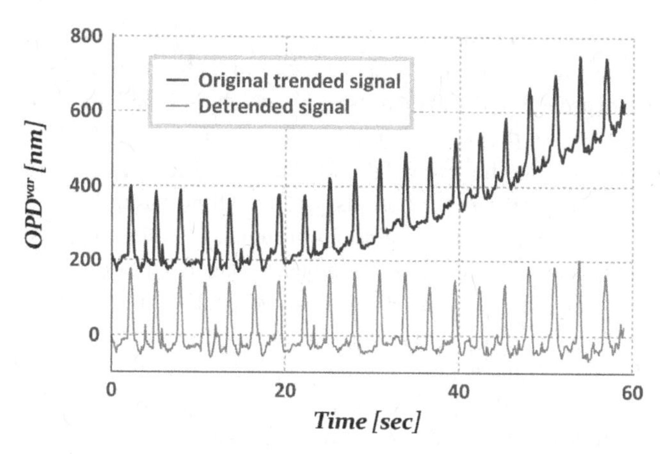

Figure 23.3 A trended cardiomyocyte signal and the detrended signal (beat rate = 21 bpm).

shoulder area (the resting duration) of the beating profile. Furthermore, it can provide less-noisy positive peaks for the peak-finding procedure.

We will also address the increasing or decreasing trends in OPD^{var} signal observed in some samples, which are probably due to experimental drift that can occur during DHM recording periods or unfocused particles moving across the ROI. The trended signal can compromise the peak-detection step described in Chapter 21 (a first-order derivation followed by Otsu's thresholding method to identify a positive peak threshold). The trended signal can be de-trended by subtracting the fitted 5-degree polynomial from the de-trended signal (see Figure 23.3). The fitted polynomial of degree 5 is defined as

$$f(x) = \sum_{i=1}^{6} b_i x^{6-i}, \tag{23.2}$$

where b_i is the coefficient of the polynomial obtained with a least-square criterion. The de-trended cardio beating profile is thus determined from

$$OPD^{de} = OPD^{var} - f(x), \tag{23.3}$$

where OPD^{de} denotes the de-trended signal considered for further analysis.

23.2.3 Cardiomyocyte-related Parameter Measurements

In Chapter 21, we presented the fitting scheme to cover specific parts of the beating pattern, namely the rising and falling portions as well as the flat resting area of each beating pattern. Here, we will focus on the rising–falling portion of the beating pattern since it represents the contraction–relaxation dynamics of CMs, which is specifically affected by E-4031 and isoprenaline drugs. The rising–falling portion of the beating pattern is asymmetrical due to the presence of CM-membrane ion channels and transporters as well as the mechanisms by which their activities are sequentially orchestrated during depolarization and repolarization [3, 6, 7]. The new polynomial function that fits the rising–falling portion makes it possible to characterize contraction and relaxation periods, resting time, and beating period/rate by allowing us to calculate the amplitudes of signals at different levels. The rising–falling portion is extracted from the two adjacent negative peaks with a positive peak between them. The location of the positive peak corresponds to either the point where the first derivative vanishes or the largest point between two neighbors. In the event of undesirable positive peaks, a general threshold value could be determined to remove peaks below the threshold. To do so, the Otsu's thresholding method is applied. All positive peaks below the threshold are excluded. There are cases in which two or three positive peaks are located very close to each other due to fluctuations of the beating signal near the actual positive peak. In this case, only the peak with the highest value is considered. The two negative peaks are detected by a search strategy around the positive peak. In this manner, the span of search is selected to cover an area larger than the rising–falling section. The search is initiated by computing the difference between adjacent elements of the OPD^{var} signal. The first element of the difference signal that exceeds a pre-determined threshold is the left-side negative peak. The right-side negative peak is also located in the same manner. Eventually, only one positive peak between two negative peaks is available for each beating profile. After extracting the rising–falling portion of each beating profile, a 9-degree polynomial function is fit to the rising–falling signal. The degree 9 polynomial is defined with the equation:

$$f(x) = \sum_{i=1}^{10} b_i x^{10-i}.$$

To find the b_i, the least-squares criterion is implemented. The fitted function yields a quantitative characterization of the beating profiles from which a set of highly relevant parameters to characterize the drug effects can be derived. Specifically, we can calculate the contraction and relaxation time, resting time, time of full width at half maximum (TFWHM), and the rising and falling slopes (see Table 23.1). The time corresponding to 10%, 50%, and 90% of the amplitude fitted signal (see Figure 23.4) can be calculated by solving the fitted polynomial equation. Specifically, 10% and 90% of the amplitude maximum are used to calculate the time duration of the contraction–relaxation process. TFWHM, the width of the beat profile amplitude at half weight, is an important parameter regarding

Table 23.1 Description of cardiomyocyte dynamic parameters and values for one control sample (au = arbitrary unit).

Parameter	Description	Values (Mean ± STD)
Contraction period (Rising duration)	The time difference between Amp_{10} and Amp_{90}. (T_3–T_1)	200 ± 20 [ms]
Relaxation period (Falling duration)	The time difference between Amp_{90} and Amp_{10}. (T_6–T_4)	290 ± 30 [ms]
Contraction–relaxation period	The time difference between T1 and T6. (T_6–T_1)	650 ± 50 [ms]
TFWHM	The time difference between two Amp_{50} (T_5–T_2) points corresponding to full width at half maximum (TFWHM).	410 ± 50 [ms]
Resting time	The time difference between the beating period and contraction–relaxation time.	1700 ± 100 [ms]
Beating period	The time between two adjacent positive peaks.	2380 ± 100 [ms]
Beating rate (beats per minute)	The total number of positive peaks in 60 seconds.	29 ± 2 BPM
Rising slope (Contraction slope)	The amplitude difference at (T_3, T_1).	75 ± 11OPD [nm]
Falling slope (Relaxations slope)	The amplitude difference at (T_6, T_4).	76 ± 11 OPD [nm]
Maximum contraction speed	Rising slope divided by the time interval (T_3–T_1).	38 ± 3 (au)
Maximum relaxation speed	Falling slope divided by the time interval (T_6–T_4).	25 ± 3 (au)
Average mass displacement	The total area under the OPD variance signal between T1 and T6.	350 ± 60 (au)

the action potential analysis of contractile CMs. It can provide information about the depolarizing and repolarizing phases of CMs [7, 8]. All the parameters mentioned above are measured at the single-beat level. Population average and standard deviations are calculated. Table 23.1 gives a description of calculated parameters that are important for the analysis of drug candidates' effects. To compute an average mass movement, the numerical integration of the OPD^{var} signal between T1 and T6 is evaluated.

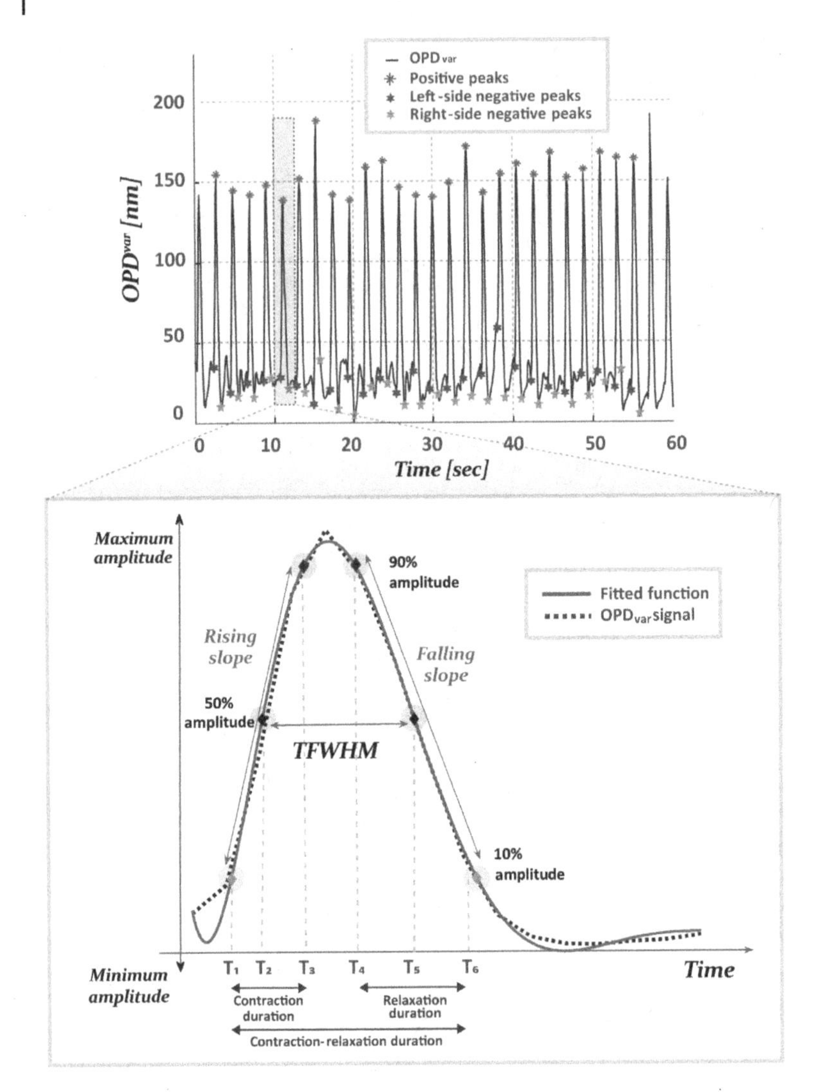

Figure 23.4 Representation of fitted functions on cardiac action potential and definitions of multiple parameters. The inset shows a sample in control conditions. The fitting function is similar for both the control and drug-treated samples. The fitting function should fit a single rising–falling portion for analysis.

23.3 Experimental Results and Discussion

23.3.1 Quantitative Analysis of Drug-treated Cardiomyocytes

Figure 23.5 shows the drug's effects on the beating activity of CM samples obtained using Eqs. (23.1) to (23.3). According to Figure 23.6, CMs exhibited an increase in beat duration in response to increasing concentrations of E-4031. However, their beat rate (defined as the number of positive peaks over a 60 s imaging interval) decreased (see Figure 23.5). Furthermore, E-4031 did not significantly change the contraction period, relaxation period, or the contraction–relaxation duration. Note that regardless of the E-4031 concentration, the two-sample Kolmogorov–Smirnov (KS) test indicated that the CM contraction period was significantly

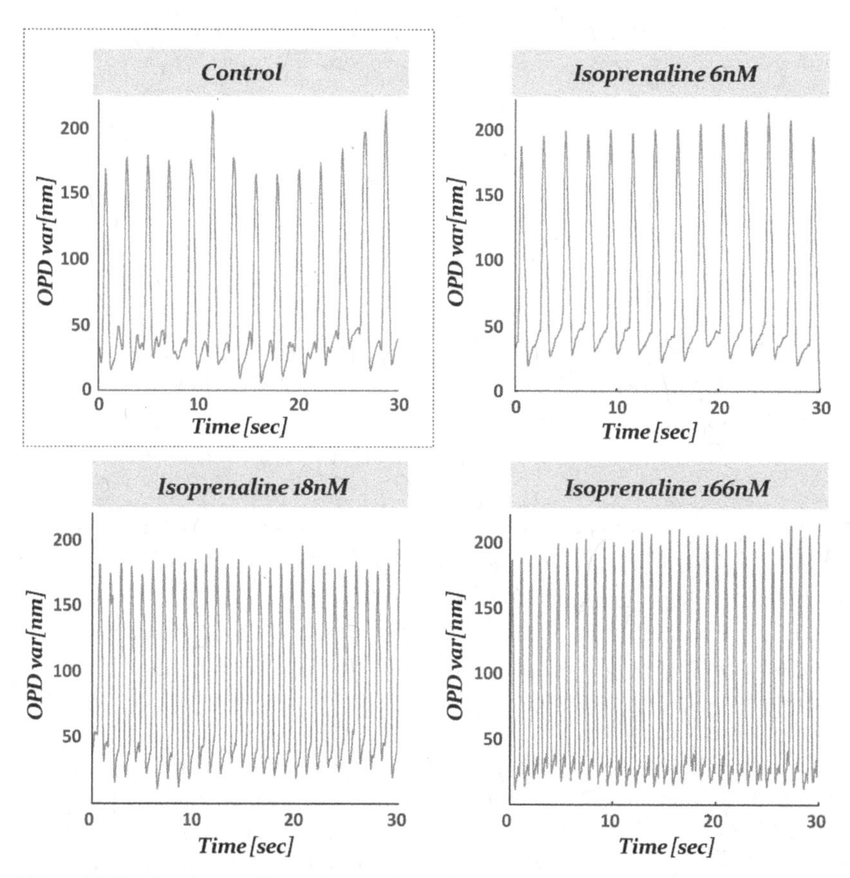

Figure 23.5 Beating profiles of control samples and samples treated with drugs (E-4031 and isoprenaline).

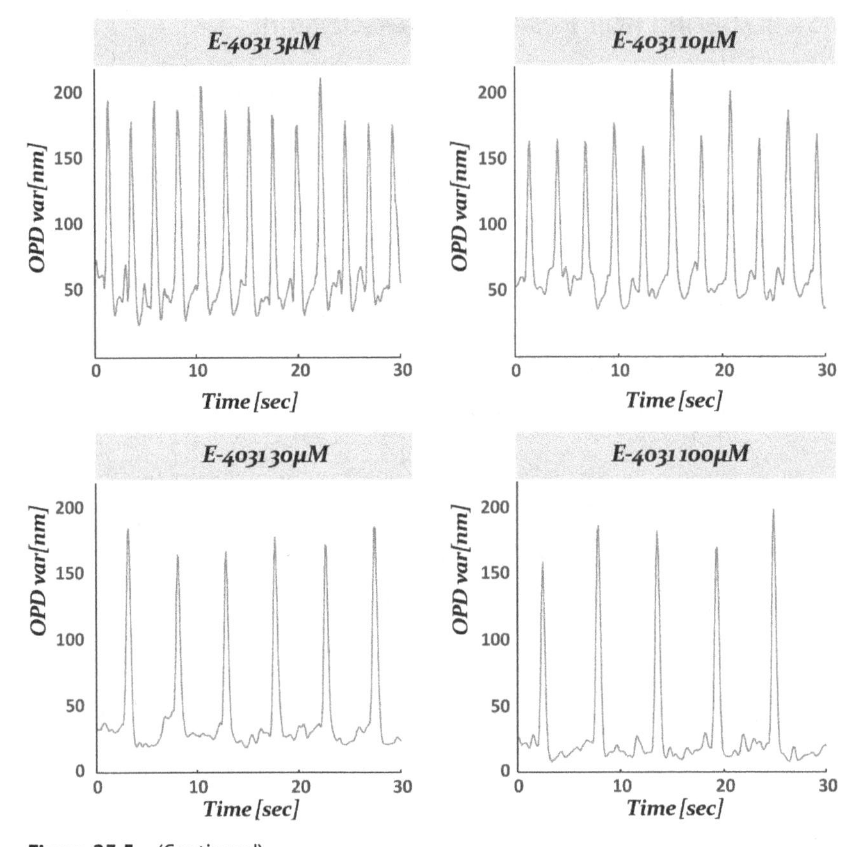

Figure 23.5 (Continued)

shorter ($p < 0.01$) than its relaxation period. On the other hand, resting duration prolonged with the increasing concentration of E-4031, which explained the increase in beat duration. However, a TFWHM decrease was observed for E-4031 at concentration of 3 μM or higher. Considering that the contraction–relaxation duration was not significantly modified by E-4031, these TFWHM decreases nevertheless reflected a modification of the dynamic contraction–relaxation response pattern.

Figure 23.7 shows that the average mass movement, the maximum relaxation speed, and the maximum contraction speed are increased by E-4031. However, these increases were already present with a low concentration of E-4031 at 3 μM. They did not show a clear dose-response relationship. Note that the maximum contraction speed was larger than the maximum relaxation speed according to the KS test ($p < 0.05$), consistent with previous reports [8–10]. These results for

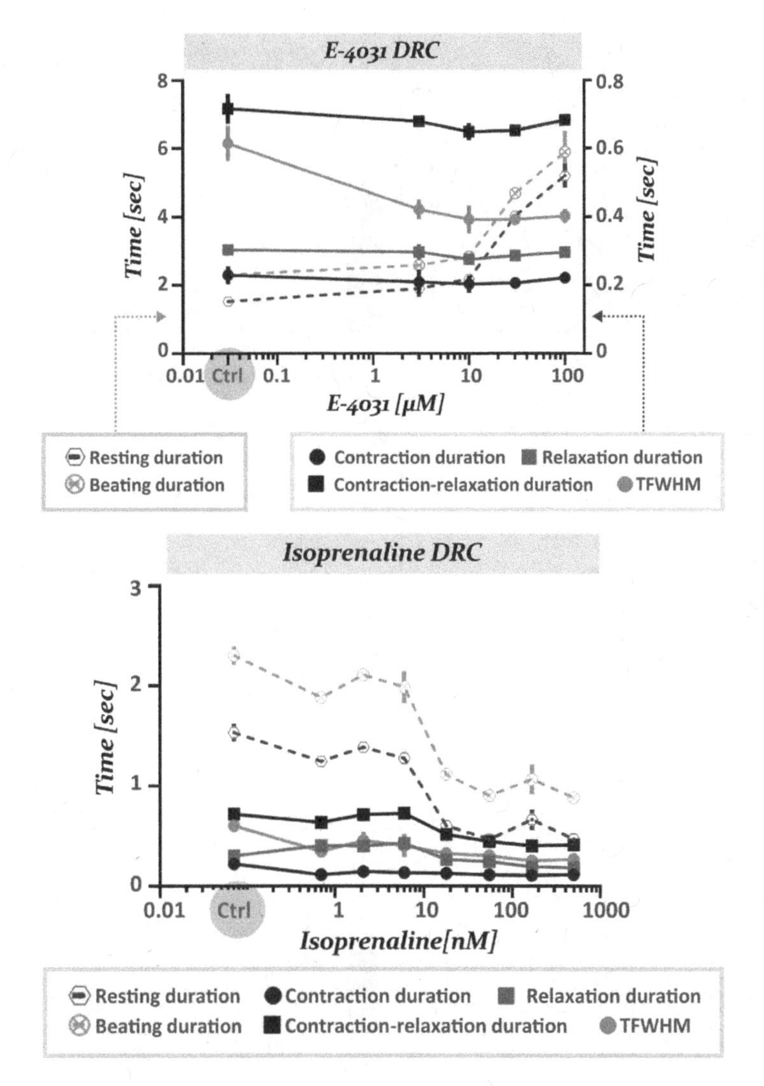

Figure 23.6 Multiple parameters in response to drug concentration. (Top) The effect of different concentrations of E-4031 on cardiac samples. (Bottom) The effects of different concentrations of isoprenaline on cardiac samples.

E-4031 align with its mode of action. E-4031 is a synthesized, class III antiarrhythmic toxin drug known to block hERG-type potassium channels by binding to open channels that can cause lethal arrhythmias. Specifically, it blocks the delayed rectifier potassium current in CMs, which increases the ventricular effective refractory period. This effect is compatible with our observations, particularly the decreased beat rate and increased resting duration.

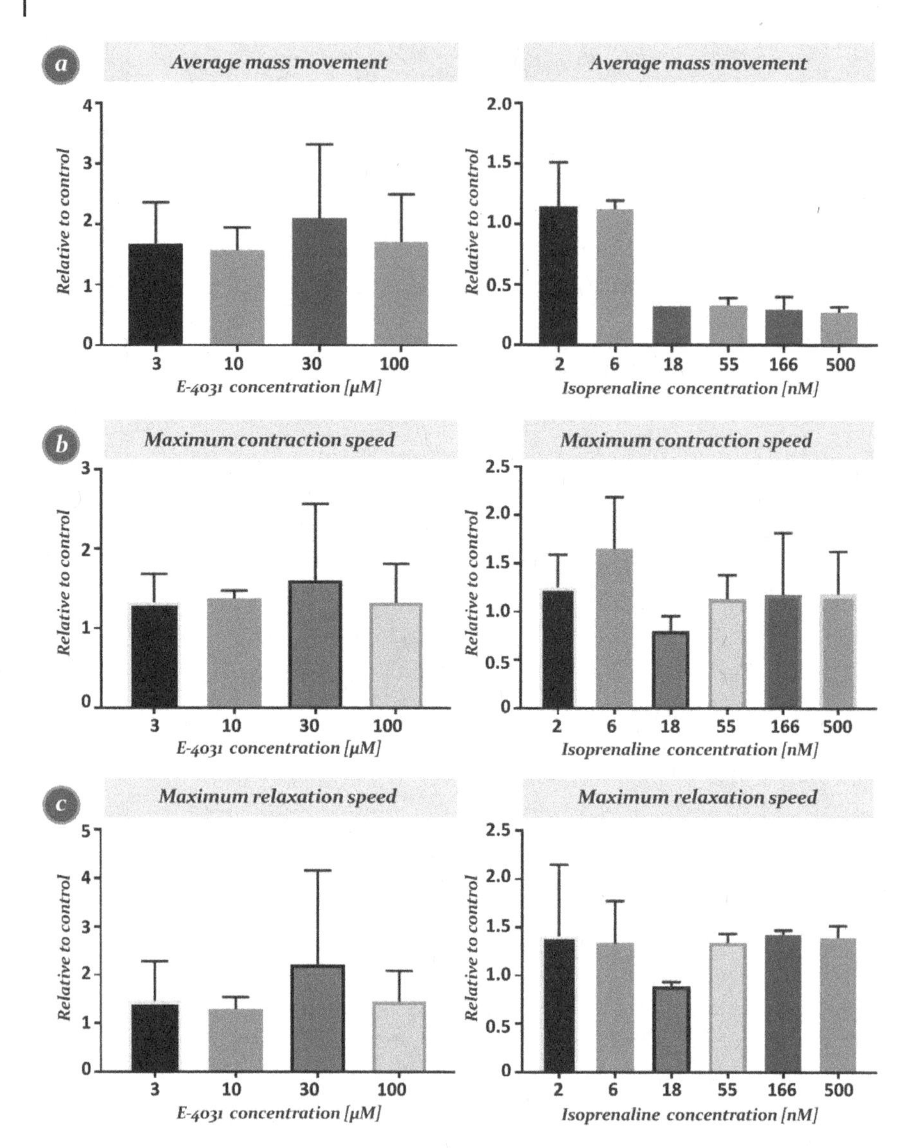

Figure 23.7 (a) Average mass movement relative to the control condition. (b) Maximum contraction speed relative to the control condition. (c) Maximum relaxation speed relative to the control condition [4] / with permission of American Chemical Society.

On the other hand, isoprenaline decreased both resting and beating durations in a dose-dependent manner (see Figures 23.5 and 23.6). There was also a trend toward a decrease in the contraction-relaxation duration. Similar to E-4031, we found that the contraction duration of isoprenaline-treated CMs was slightly shorter than the relaxation duration, as with previous findings [8]. As shown in Figure 23.6,the average mass movement was significantly increased for isoprenaline concentrations below 18 nM, but drastically decreased when samples were treated with higher concentrations of isoprenaline. The maximum contraction and relaxation speeds were significantly increased by isoprenaline. Although these speed increases were greater than those observed for E-4031 and compatible with an increase in contractile force, we did not observe a dose-dependent response. These observations were in good agreement with isoprenaline mechanisms of action, an agonist of cardiac β1 and β2 receptors, which increases ion movements through both sodium and calcium channels and leads to increased contractile force and beating rates characterized by a decreased refractory period.

Our method can measure multiple CM parameters in a label-free, noninvasive, and contactless manner. It can directly evaluate the mechanical contraction–relaxation of CMs by recording dry-mass changes during beating. Dry-mass redistribution can be related to the multi-parameter measurement of CMs and used to study the effects and cardiotoxicities of drugs.

23.3.2 Cardiotoxicity Assessment

We confirmed that DHM could monitor cell viability and toxicity in CMs. A parallel measurement of CMs treated with the toxic compounds doxorubicin and staurosporin showed both a decreased Presto Blue signal (a resazurin-based metabolic activity indicator, Figure 23.8a) and an increased average OPD (Figure 23.8b), which are indicative of cell death [11]. Both parameters were highly correlated (see Figure 23.8c) and fully translated morphological changes related to cell death (see Figure 23.8d).

We further obtained cell intensity (average OPD), cell count, and cell area data from all previous experiment OPD images by performing segmentation and cell-morphological parameter extractions. Figure 23.8e shows that all three parameters are correlated with Presto Blue data, showing that a feature analysis of CM OPD images obtained with DHM allows us to directly assess cell toxicity. This experiment further demonstrates that extensive information can be extracted from DHM without requiring invasive, costly, and/or time-consuming assays.

We, therefore, performed this extended image analysis to confirm that all the recorded parameter signals analyzed in previous figures were not perturbed by cell death or cell toxicity and that the monitored, beating CMs were in a healthy state throughout CM hologram acquisitions. As shown in Figure 23.9, we could clearly observe that all CM images from the sequence recording in presence of E-4031 or isoprenaline clustered in the same 3D space as the negative control (DMSO), while positive toxicity controls exhibited a very distinct pattern with higher cell intensities (average OPD) and decreased cell counts and areas.

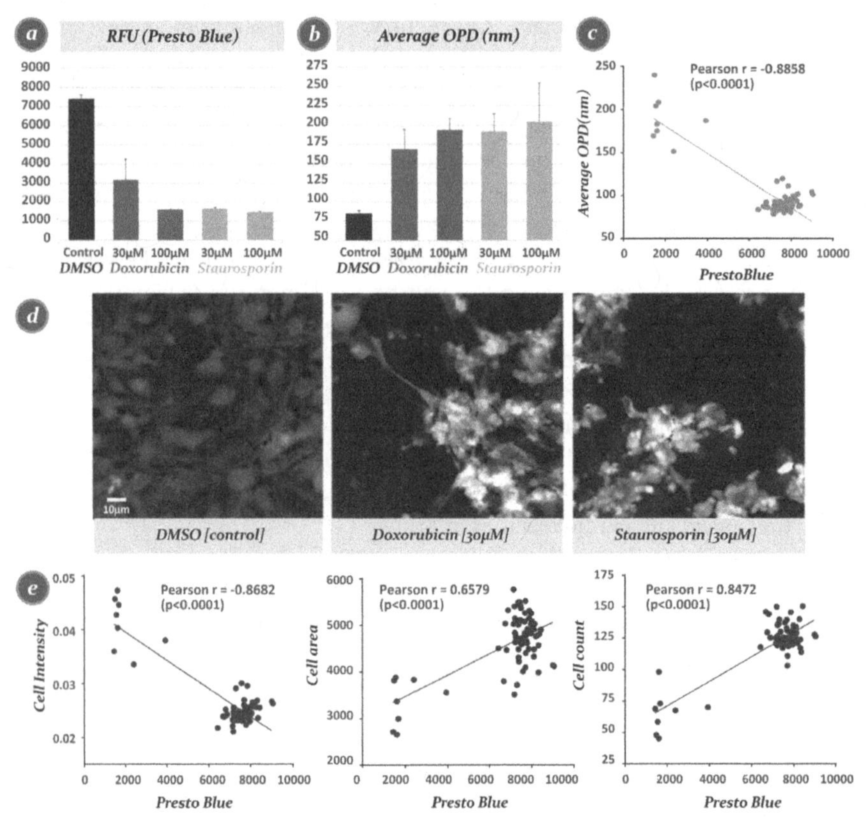

Figure 23.8 Cell viability assays with (a) Presto Blue or (b) DHM on cardiomyocytes treated with two dilutions of doxorubicin or staurosporin. (c) Correlation between Presto blue and DHM. (d) Selected DHM example images showing cardio-toxic conditions versus control condition. (e) Correlation between cell intensity, area, and count measured with CellProfiler and Presto Blue. Statistical analysis: two-tailed test for Pearson correlation with 95% confidence interval [4] / with permission of American Chemical Society.

23.4 Conclusions

In this chapter, we introduced DHM to directly and non-invasively measure the dynamics of drug-treated CMs by recording CM phase images. Our experimental results of E-4031-treated CMs showed that the beat rate decreased as the concentration of E-4031 increased as the result of an extended resting duration. We also observed that the contraction duration was shorter than the relaxation duration. Isoprenaline demonstrated its effect on CM samples by increasing both the contractile force and the beat rate as a result of the decreased resting period. These

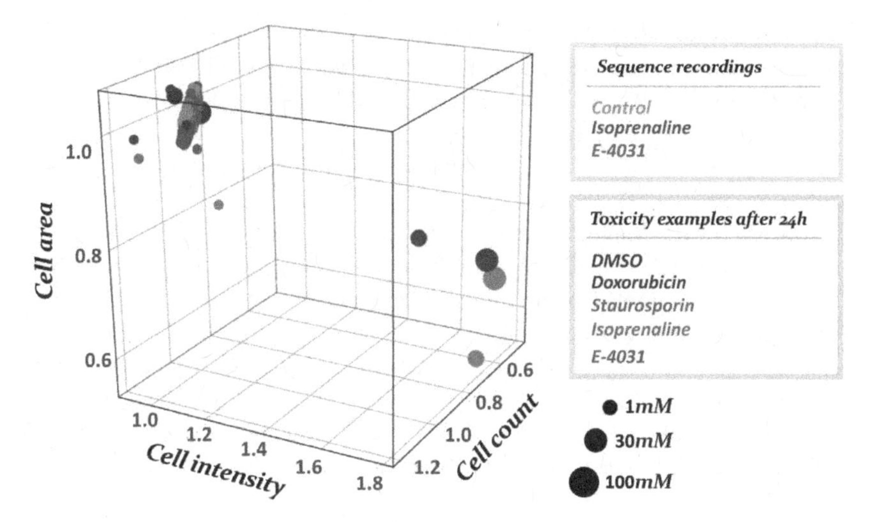

Figure 23.9 A 3D scatterplot of parameters extracted from image analysis using CellProfiler software. Cell count, cell intensity, and cell area are normalized by the corresponding control condition. Dots size is proportional to the compound concentration [4] / with permission of American Chemical Society.

findings demonstrate that our DHM system, with the development of proper analysis algorithms, can monitor specific drug-mediated effects on the dynamics of the CM contraction–relaxation process, which can reflect the underlying drug mechanisms of action at some extent. DHM, with a proper analysis of its phase image, thus represents a promising label-free approach for drug discovery.

References

1 Klabunde, R. and Richard, E. (2017). Cardiac electrophysiology: normal and ischemic ionic currents and the ECG. *Adv. Physiol. Educ.* 41 (1): 29–37.

2 Nerbonne, J. and Kass, R. (2005). Molecular physiology of cardiac repolarization. *Physiol. Rev.* 85 (4): 1205–1253.

3 Trenor, B., Cardona, K., Saiz, J. et al. (2017). Cardiac action potential repolarization revisited: early repolarization shows all-or-none behavior. *J Physiol.* 595 (21): 6599–6612.

4 Jaferzadeh, K., Rappaz, B., Fabien, K. et al. (2020). Marker-free automatic quantification of drug-treated cardiomyocytes with digital holographic imaging. *ACS Photonics* 7: 105–113.

5 Carpenter, A., Jones, T., Lamprecht, M. et al. (2006). CellProfiler: image analysis software for identifying and quantifying cell phenotypes. *Genome Biol.* 7 (10): R100.

6 Pinnell, J., Turner, S., and Howell, S. (2007). Cardiac muscle physiology. *Contin. Educ. Anaesth. Crit. Care Pain* 7 (3): 85–88.

7 Bers, D. (2002). Cardiac excitation-contraction coupling. *Nature* 415 (6868): 198–205.

8 Hayakawa, T., Kunihiro, T., Ando, T. et al. (2014). Image-based evaluation of contraction–relaxation kinetics of human-induced pluripotent stem cell-derived cardiomyocytes: correlation and com-plementarity with extracellular electrophysiology. *J. Mol. Cell. Cardiol.* 77: 178–191.

9 Nomura, F., Kaneko, T., Hattori, A., and Yasuda, K. (2011). On-chip constructive cell-network study (II): on-chip quasi-in vivo cardiac toxicity assay for ventricular tachycardia/fibrillation measurement using ring-shaped closed circuit microelectrode with lined-up cardiomyocyte cell network. *J. Nanobiotechnol.* 9 (1): 39.

10 Yamazaki, K., Hihara, T., Taniguchi, T. et al. (2012). A novel method of selecting human embryonic stem cell-derived cardiomyocyte clusters for assessment of potential to influence QT interval. *Toxicol. in Vitro* 26 (2): 335–342.

11 Kühn, J., Shaffer, E., Mena, J. et al. (2013). Label-free cytotoxicity screening assay by digital holographic microscopy. *Assay* 11 (2): 101–107.

24

Analysis of Cardiomyocytes with Holographic Image-based Tracking

24.1 Introduction

DHM is a promising tool with a remarkable role in live-cell analysis by providing quantitative phase images well-suited to study cell behavior without labels. In Chapter 23, we showed that cardiomyocyte (CM) responses to pharmacological compounds could be quantitatively characterized by monitoring dry-mass changes. However, developing a high-throughput screening system to monitor the beating behavior of human-induced pluripotent stem cell–derived (HiPSC) at the single-cell level remains essential. Obtaining functional signals that reveal detailed information about pharmacological effects on CM contractility at the single-cell level has several challenges. Optical flow-based motion-tracking methods allow the monitoring of CM contractile activity without needing a physical contact. In particular, the Farneback dense optical-flow algorithm uses single-pixel displacement detection to track moving objects. Furthermore, a combination of DHM and optical flow-based analysis can enable us to observe CMs for a long time at the single cell level. Using motion waveforms obtained from single CMs, the motion speed measurement and dynamic parameters of CMs, including the contraction, relaxation, beating, and resting periods, can be automatically quantified at the single cell level.

In this chapter, we will introduce a novel platform to integrate DHM and Farneback dense optical flow for single-CM contractile motion characterization [1]. First, we obtained phase images of CMs with DHM and performed motion tracking at the single-cell level using the Farneback dense optical-flow method that can detect high-resolution contractile centers. In this way, the CM contractile motion speed, which reflects changes in CM morphology, can be measured, then the motion waveform further characterized by computational algorithms to CMs. Multiple temporal parameters including contraction period, relaxation period,

Artificial Intelligence in Digital Holographic Imaging: Technical Basis and Biomedical Applications, First Edition. Inkyu Moon.

beating period, and resting period are then measured. Since our method occurs at the single-cell level, the synchronization of CMs can also be qualitatively assessed. We analyzed the pharmacological effects of isoprenaline (166 nM) and E-4031 (500 μM) in CM motion speed compared to control conditions at the single-cell level to evaluate the usefulness of our method for cardiotoxicity screening. We also quantified multiple parameters of CM dynamics using several whole-slide images for tens of cells (or beats). Furthermore, we validated our platform using speed measurements of fixed CMs versus live CMs, single-CM synchronization testing, and noise sensitivity analyses.

24.2 Materials and Methods

24.2.1 Cardiomyocyte Preparation and Imaging Conditions

HiPSC-CMs were obtained from Cellular Dynamics Int. (Madison, WI, USA) and cultured according to the manufacturer's instructions for 14 days before recording a hologram. For drug-treated CMs, a sequence was recorded for the control conditions before the drug was added. After a 15-minute incubation, CM images were recorded again. The images were acquired at three sampling frequencies of 10, 25, and 50 Hz. CMs were fixed using a 4% formalin solution (Sigma-Aldrich) and incubated for 15 minutes at room temperature. These cells were then washed three times with phosphate-buffered saline for 10 minutes at room temperature. A total of 1500 CM images were taken at a sampling frequency of 50 Hz before and after cell fixation.

24.2.2 Single-cardiomyocyte Motion Tracking with Farneback Optical Flow

Steps of single-CM motion tracking and motion waveform generation using the Farneback optical flow are shown in Figure 24.1a. In the first step, our algorithm generated a hierarchy of resolution levels from the original OPD image using Gaussian pyramids, with each level having a lower resolution than the previous level (see Figure 24.1b). Figure 24.1c shows an optical flow for the pixel displacement estimation in two successive image frames, where $I(x,y,t)$ is the pixel position in the reference frame and $I(x + dx, y + dy, t + dt)$ is the pixel displacement in the subsequent frame (current frame). The Farneback algorithm is a dense optical flow that performs motion tracking in multi-resolution levels [2]. Tracking procedures start with the lowest resolution and continues to the highest one. As a result, the displacement of two local patches in consecutive image frames is determined by approximating the neighborhood of each pixel with a quadratic polynomial [3]. The tracking is refined at each resolution level by starting from the lowest

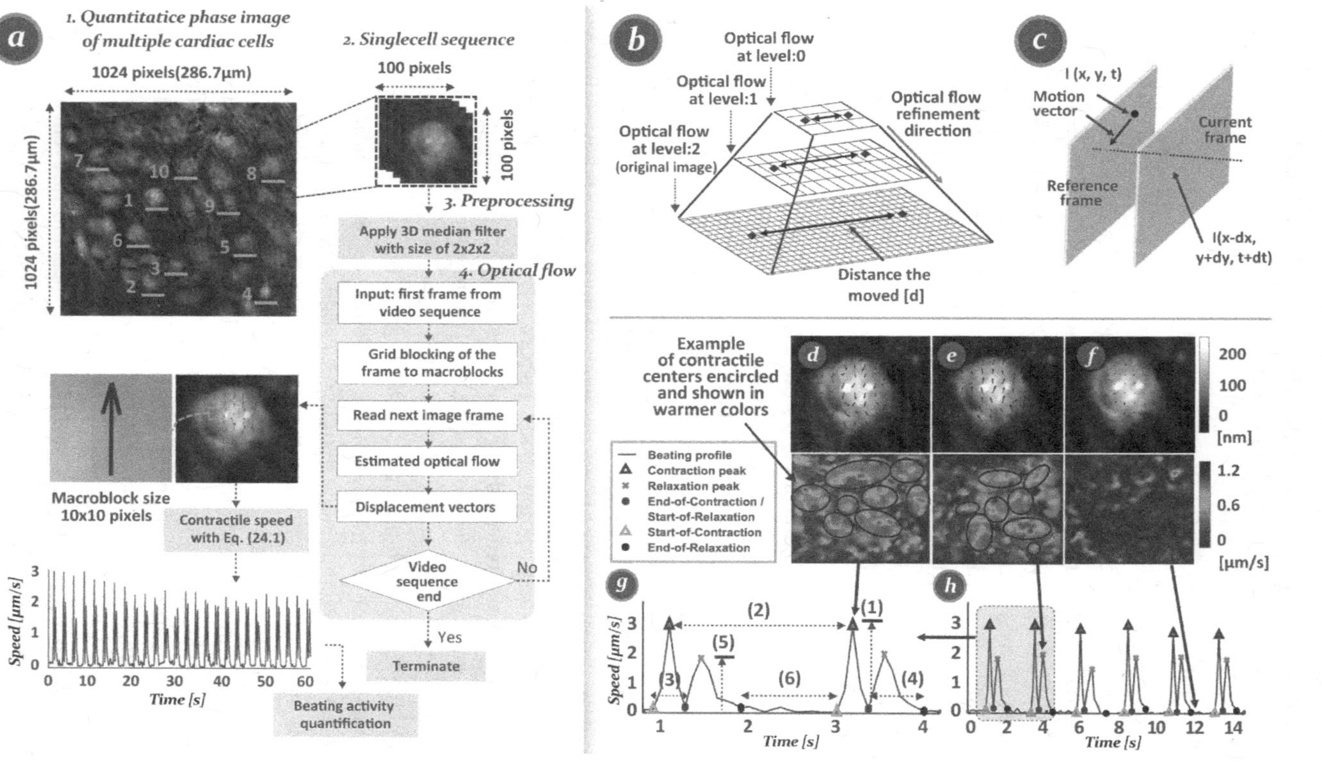

Figure 24.1 Overview of the workflow to motion track single CMs and beating profile quantification. (a) Quick steps of single-CM motion tracking using the Farneback optical flow and motion waveform generation. (b) Multi-resolution leveling of an image at three levels. The image resolution at each level was downsized. (c) Diagram of the optical flow for pixel displacement estimation. (d) Close-up phase image of a single CM with superimposed motion vectors on the CM image for contraction. (e) Relaxation state, (f) resting state (first row), and corresponding heat map generated from the absolute motion (second row) with encircled contractile centers. Contractile centers refer to regions in which contractions are maximized. (g) Quantification description as explained in Section 24.2.3. (h) Beating activity profile of a single cell [1] / with permission of Elsevier.

resolution level and moving to the highest resolution. Large displacement can be detected since the detected tracking points at each level are base points for the next. For details of the Farneback optical flow, please refer to [3].

24.2.3 Workflow for Beating Signal Quantification

Ten single CMs were manually extracted from different parts of the CM phase image (see Figure 24.1a). The extracted cell area mostly includes the nucleus region, assuming the nucleus part is the center of the motion. During CM beating activity, an array of motion vectors is generated by repeating the optical flow as shown in Figure 24.1a, in which the motion direction and action potential (AP) speed of each CM are computed. The motion waveform of the CM is generated by computing the AP speed using the equation

$$Speed_{\mu m/s} = \frac{Displacement}{Time} = \frac{\sqrt{(dx)^2 + (dy)^2}}{Time}, \quad (24.1)$$

where dx and dy denote displacements in x and y directions, respectively. They are estimated by the Farneback algorithm, which represents CM motion in the x and y directions. The time in Eq. (24.1) is the time between two consecutive frames. Once the motion waveform is generated, the contraction peak, relaxation peak, and several auxiliary points for quantifying CM physiological behavior can be calculated with the automated peak identification method. A single CM, with superimposed motion vectors on the image that refer to the motion directions for different bating statuses, are shown in Figures 24.1d, 24.2e, and 24.2f, which correspond to contraction, relaxation, and resting beating status, respectively. During contraction and relaxation, motion vectors indicated opposite directions. In contrast, motion vectors in the resting status indicated no motion. Heat maps generated from the absolute motion for each beating status are also shown (see the second row in Figures 24.1d–f). The heat map represents specific regions of the CM, which is a center of contraction. More contractile centers are observed on the heat map when the CM is in the contraction mode than in relaxation. In contrast, the heat map shows almost no contractile centers in the resting state, leading to an almost zero value for the motion speed.

The measurement of temporal motion speed can reveal details about the beating activity of the CM sample (see Figure 24.1g). Thus, we reconstructed important temporal parameters such as the contraction period, relaxation period, resting period, and beating period using the accurate peak detection and average period selection described in Chapter 22 (see Figure 24.1h). Figure 24.1g details the obtained, averaged temporal parameters related to CM mechanical events. To compute characteristics of CM activity based on motion speed signal, single beating profiles were extracted, which required two main contraction–relaxation peaks and three auxiliary points. The three stages of contraction are the contraction start, contraction end, and relaxation end. Figure 24.1g shows the representation of these points. Details of accurate peak detection along with auxiliary points are described in Chapter 22. Dynamic parameters measured for each isolated CM are described as follows: (1) the maximum contraction speed (see #1 in Figure 24.1g), the average amplitude of contraction peaks; (2) the beating period (see #2 in Figure 24.1g) is the time between two adjacent contraction peaks; (3) the contraction period (see #3 in Figure 24.1g), the average time between the start of contraction and end of contraction points; (4) the relaxation period (see #4 in Figure 24.1g), the average time from the start of relaxation to the end of relaxation; (5) the maximum relaxation speed (see #5 in Figure 24.1g), the average amplitude of relaxation peaks; and (6) the resting period (see #6 in Figure 24.1g), the average time between the end of relaxation and the next contraction start.

24.3 Experimental Results and Discussion

Single CM motion characterization, beating profile quantification results, and synchronization analysis are shown in Figure 24.2. A heat map analysis of the absolute motion was used to monitor contractile centers (see Figures 24.2a–c, second row). Four single CMs are provided as examples. Figure 24.2d shows that the maximum contraction speed is larger than the maximum relaxation speed and the contraction period is shorter than the relaxation period. During the CM contraction–relaxation beating activity, motion vectors indicate opposing directions. However, during the CM resting state, motion vectors are weak, indicating that the CM is nearly immobile (Figures 24.2a–c, first row). Quantification results revealed that all single CMs had a similar average beating rate and average beating period. The beating activity occurred at regular intervals (see Figure 24.2e). Note that there was a cell-to-cell variation in the magnitude of the speed value, whereas the CM beating rate and other properties related to physiological aspects of the CM (contraction period, relaxation period, and so on) were almost the same.

Furthermore, we performed a reliable synchronization analysis since our method could analyze CMs at the single cell level, Figures 24.2f and 24.2g show synchronization results for all extracted single CMs. As shown in Figure 24.2f, the temporal activity (contraction–relaxation) beats at the same frequency during different periods. As shown in Figure 24.2g, which presents the results of a cross-correlation evaluation between the beating activity signals of single CMs, the maximal cross-correlation value is on the zero time-lag, demonstrating that single CM signals are perfectly synced in time (see Figure 24.2g).

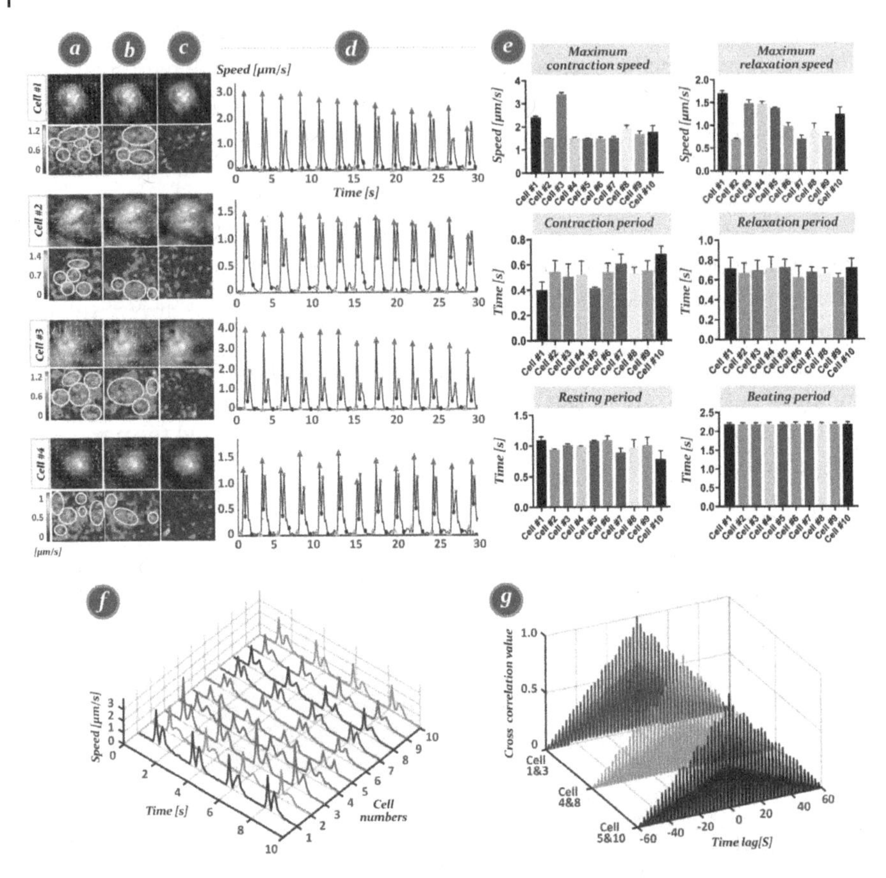

Figure 24.2 Single-CM motion characterization and synchronization analysis. Motion vectors superimposed on a single-CM image representing the motion direction for (a) contraction, (b) relaxation, and (c) resting beating cycle shown in the first row. The corresponding contractility heat map is demonstrated in the second row. Contractile centers are circled and shown in warmer colors on the heat map. (d) CM motion waveform derived from dry-mass redistribution speed calculation. (e) Results of CM motion waveform quantification parameters. (f) A 3D representation of single-CM beating activity synchronization. (g) Cross-correlation analysis between different pairs of individual CMs [1] / with permission of Elsevier.

24.3.1 Verification of the Proposed Motion Tracking Method

24.3.1.1 Whole-slide Image Motion Characterization

To demonstrate the robustness of our CM motion characterization method, we investigated five whole slide images with multiple CMs obtained at different sampling frequencies (10, 25, and 50 Hz). Sample #1, with superimposed motion vectors for the contraction beating status, is displayed in Figure 24.3a and its

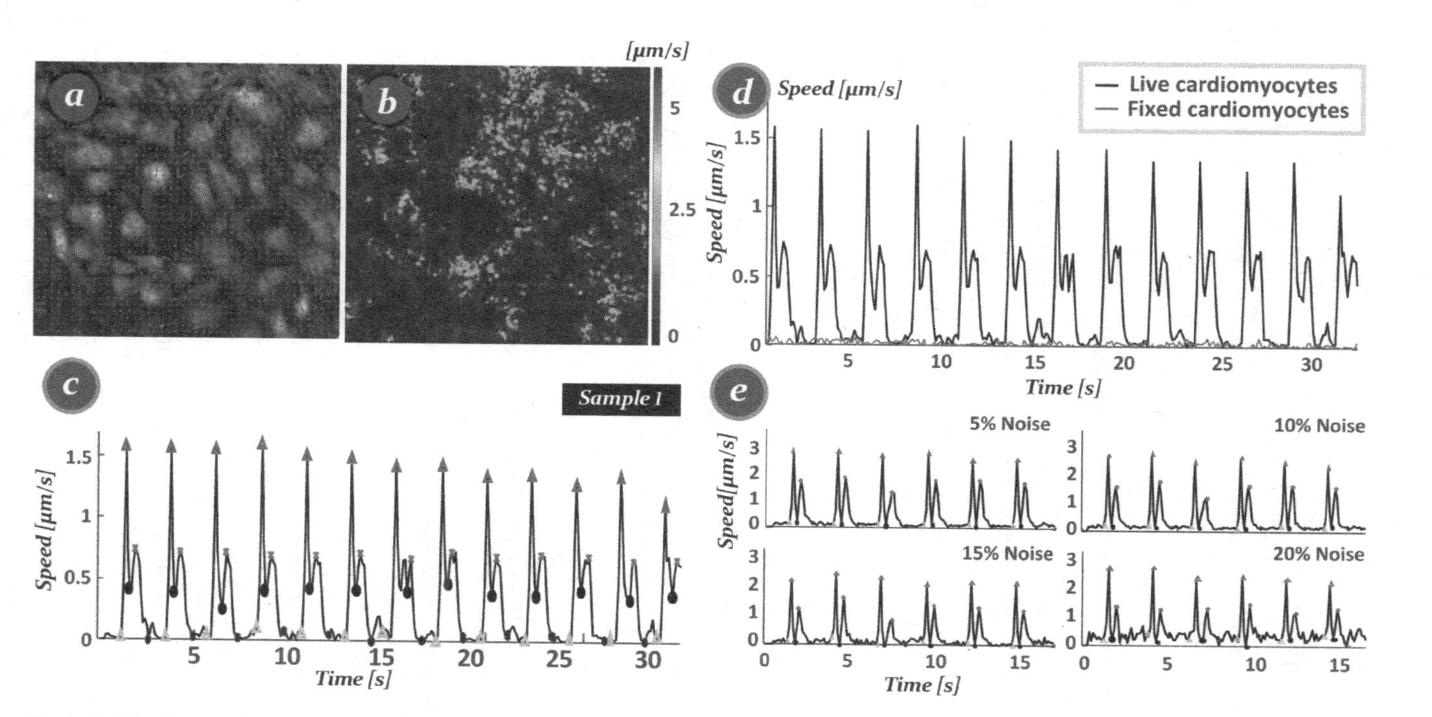

Figure 24.3 Proposed method validation. (a) Whole-slide image of multiple CMs with superimposed motion vectors for the contraction state, (b) the corresponding heat map, and (c) beating activity profile (maximum contraction speed: 1.4 μm/s; maximum relaxation speed: 0.8 μm/s; contraction period: 0.58 s; relaxation period: 0.63 s; beating period: 2.1 s; resting period: 0.75 s). (d) Contractile speed measurement of fixed versus live CMs. (e) Image of single-CM #1 with Gaussian noise ranging from 5% to 20%. The beating profile regularity was constant for all noise percentages.

corresponding heat map is presented in Figure 24.3b. Figure 24.3c shows a beating activity profile with the detected contraction–relaxation peaks and auxiliary points. It was noted that the average contraction–relaxation motion speed of the whole image with tens of CMs was affected by the motionless region, where there were few cells. These results showed decreased average speeds compared to the results of the single-cell analysis.

24.3.1.2 Speed Measurement of Fixed Cardiomyocytes Versus Live Cardiomyocytes

Our method was verified by measuring the speed of fixed CMs versus live CMs. Results are shown in Figure 24.3d. The amplitude of the speed of the fixed CMs fluctuated around zero. The amplitude was substantially less than both contraction and relaxation peaks but similar to the amplitude of the resting mode.

24.3.1.3 Noise Sensitivity Analysis

Noise might have several sources. To demonstrate the robustness of our method for CM motion characterization in noisy images, we artificially applied Gaussian noise ranging from 5% to 20% on the CM #1 image (see Figure 24.3a). We then quantified the motion waveform generated by our method (see Figure 24.3e). The regularity of the CM beating profile remained constant all at different noise levels. The quantified dynamic beating properties or parameters were almost constant.

24.3.2 Monitoring Pharmacological Effects of Compounds on Cardiomyocytes

24.3.2.1 Whole-slide Image Analysis of Drug-treated Cardiomyocytes

We performed additional experiments to examine the effect of pharmacological compounds on CM motion activity. We investigated the effects of an adrenergic receptor agonist (isoprenaline) and an hERG channel blocker (E-4031) on CM contractile speed using our automated method. We treated multiple CMs with 166 nM of isoprenaline or 500 μM of E-4031 and compared the beating activity parameters to those obtained under control conditions (see Figures 24.4a and 24.5a). Compared to controls, isoprenaline increased CM beating frequency by raising the contractile speed and shortening the resting period, in line with previous findings [4]. CMs responded to the E-4031 drug by decreasing their contractile speed and prolonging resting period, thus slowing down their beating frequency, which reflect the drug's mode of action.

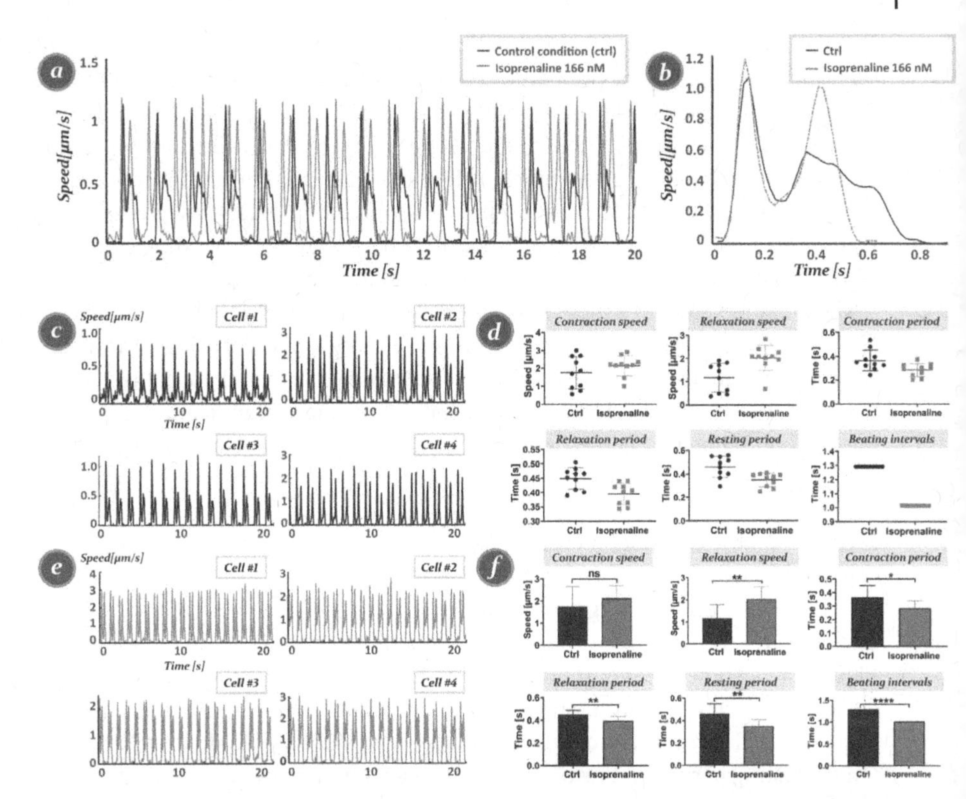

Figure 24.4 Single-CM contractile motion analysis in control (ctrl) and drug-treated conditions in response to treatment with 166 nM of isoprenaline. (a) Whole-slide image contractile motion analysis in control and drug-treated conditions, (b) Single-beat contractile motion comparison under control and drug-treated conditions. (c) Motion waveforms of single CMs #1 to #4 extracted under control conditions. (d) Quantification result comparison under control conditions (dark grey points) versus drug-treated conditions (grey points) at the single-cell level. (e) Single CMs #1 to #4 extracted after isoprenaline treatment. (f) Average of each quantification parameter for all extracted single cells in control (dark grey bars) versus drug-treated conditions grey bars. All statistical comparisons were carried using an unpaired Student's *t*-test. Statistical differences between the two groups with *p*-value <0.05 were considered statistically significant. We averaged each quantification result for all extracted single CMs (n=10) in control conditions and isoprenaline-treated conditions. Bars represent the mean and standard deviation for each quantification parameter.

24.3.2.2 Single-cell Motion Characterization of Isoprenaline-treated Cardiomyocytes

Figure 24.4a shows the contractile speed of the whole-slide image with tens of CMs in both control and isoprenaline-treatment conditions. Figure 24.4b demonstrates an example of a single-beat profile of the whole-slide image and a comparison between control and drug-treated conditions. The contraction–relaxation speed of CMs increased during isoprenaline treatment compared to control conditions. Examples of single-CM beating profiles of cell numbers 1 to 4 extracted from control (blue) and isoprenaline-treated (yellow green) conditions are presented in Figures 24.4c and 24.4e. Single-cell parameter quantification results are compared in Figure 24.4d. A summary of isoprenaline effects is presented in Figure 24.4f. We averaged each quantification result for all extracted single CMs (10 single cells) under control and isoprenaline-treated conditions.

The unpaired Student's *t*-test analysis showed that single CMs responded to isoprenaline with a significant increase in the relaxation speed during contraction speed, although it was not significantly increased compared to control CMs. The resting period was significantly reduced, causing a significant increase in the beating frequency. These findings are in line with previous reports [5, 6].

24.3.2.3 Single-cell Motion Characterization of E-4031-treated Cardiomyocytes

It was reported that treating CMs with E4031 can significantly decrease the heart rate due to the loss of intracellular potassium ions (K^+) in CMs [6, 7]. Figure 24.5a shows the contractile motion of a whole-slide image with multiple CMs under control and E-4031-treated conditions. After drug treatment, a decrease in the contraction–relaxation speed was observed, along with a prolonged resting period that resulted in a lower beat frequency compared to control conditions. The quantification of beating parameters showed a prolonged relaxation period. An example of a single-beat profile comparison is shown in Figure 24.5b.

The effects of E-4031 on CM contractile speed were tested at both the whole-slide and single-cell levels. Single-CM beating patterns for cell numbers 1 to 4 are given in Figures 24.5c and 24.5e. A comparison of the beating profile quantification parameter was performed at the single-cell level as shown in Figure 24.5d. We observed a contractile response of CMs to the E-4031 drug. The average of all single-CM contractile speeds decreased compared to control conditions. The average resting period for all the extracted single CMs treated with E-4031 was nearly the same and showed a significant prolongation compared to control conditions. As a result, there was a significant decrease in the beating period. The motion profiles of some single CMs showed irregular beating patterns after drug treatment. The averaged results of each quantification parameter for all extracted single CMs are presented in Figure 24.5f. We averaged each quantification result for all extracted single CMs (10 single cells) under control and E-4031-treated conditions. These results are in good agreement with previous findings [4].

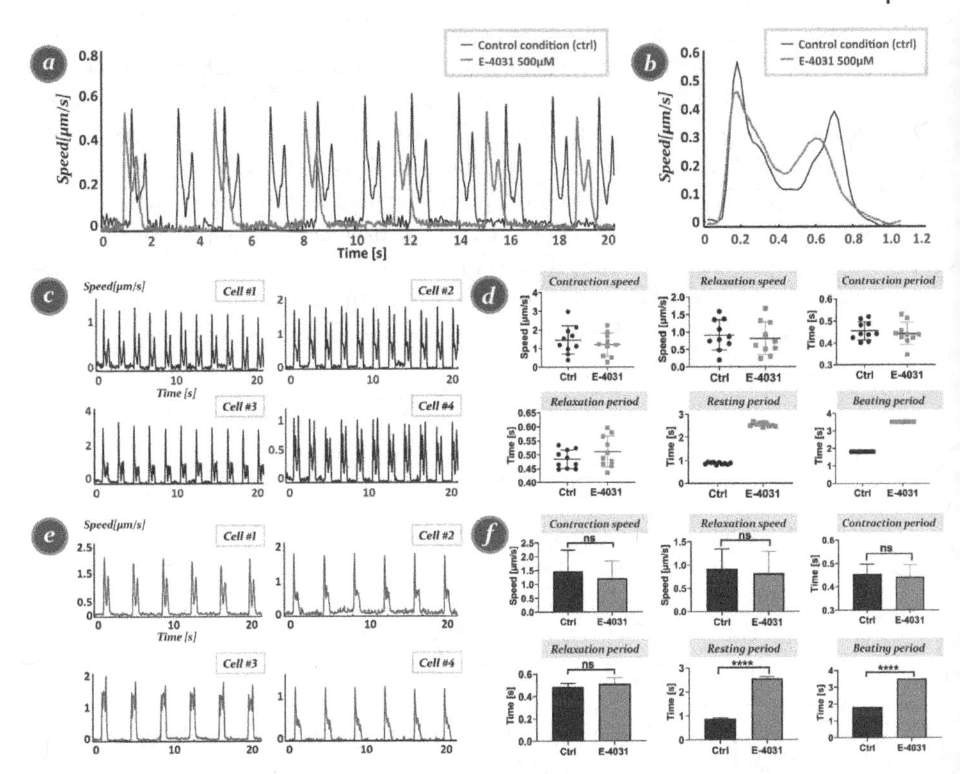

Figure 24.5 CM contractile motion analysis under control (ctrl) and drug-treated conditions (500 μ M E-4031). (a) Whole-slide image contractile motion analysis for control versus 500 μM E-4031 conditions along with the beating profile. (b) Single-CM beat contractile motion comparison between control and drug-treated conditions. (c) Motion waveforms of the single CMs, cells #1 to #4 extracted under control conditions and (d) a comparison of quantified results at the single-cell level under control conditions (dark grey) versus drug-treated conditions (grey). (e) Single CM cells #1 to #4 extracted after 500 μM E-4031 treatment. (f) The average of each quantification parameter for all extracted single cells under control conditions (dark grey) versus drug-treated (grey) conditions. All comparisons were conducted using an unpaired Student's t-test and a p-value < 0.05. We averaged the quantification results for each extracted single CM (n=10) under control and E-4031-treated conditions. Bars represent the mean and standard deviation for each quantification parameter of all extracted single CMs.

24.4 Conclusions

We presented a motion characterization platform for single HiPSC-CMs using the optical-flow method combined with DHM for cardiotoxicity applications. We obtained detailed information about CM functionality by generating CM beating activity profiles based on single-cell speed calculations. We quantified beating profiles for every extracted single CM using our automated peak identification method. We validated our CM characterization platform by measuring the contractile speeds of fixed versus live CMs with a noise sensitivity analysis. Furthermore, we validated the applicability of our platform for cardiotoxicity screening at the single-cell level by comparing the effects of E-4031 and isoprenaline chemicals on multiple CM beating activity-related parameters to those under control conditions. Our platform demonstrated that it could efficiently and accurately reveal detailed quantification results about the pharmacological effects of a drug on _single CMs. Our findings offer insights into the contractile motion of single HiPSC-CMs and a deeper understanding of their kinetics at the single-cell level for cardiotoxicity screening and predictive toxicology.

References

1 Ahamadzadeh, E., Jaferzadeh, K., Park, S. et al. (2022). Automated analysis of human cardiomyocytes dynamics with holographic image-based tracking for cardiotoxicity screening. *Biosens. Bioelectron.* 195: 113570.

2 Plyer, A., Le Besnerais, G., and Champagnat, F. (2016). Massively parallel Lucas Kanade optical flow for real-time video processing applications. *J. Real-Time Image Process* 11: 713–730.

3 Farnebäck, G. (2003). Two-frame motion estimation based on polynomial expansion. *Lect. Notes Comput. Sci.* 2749: 363–370.

4 Hayakawa, T., Kunihiro, T., Ando, T. et al. (2014). Image-based evaluation of contraction-relaxation kinetics of human-induced pluripotent stem cell-derived cardiomyocytes: correlation and complementarity with extracellular electrophysiology. *J. Mol. Cell. Cardiol.* 77: 178–191.

5 Isobe, T., Honda, M., Komatsu, R., and Tabo, M. (2018). Conduction and contraction properties of human iPS cell-derived cardiomyocytes: analysis by motion field imaging compared with the guinea-pig isolated heart model. *J. Toxicol. Sci.* 43: 493–506.

6 Luo, C., Wang, K., and Zhang, H. (2017). Modelling the effects of quinidine, disopyramide, and E-4031 on short QT syndrome variant 3 in the human ventricles. *Physiol. Meas.* 38: 1859–1873.

7 Dempsey, G., Chaudhary, K., Atwater, N. et al. (2016). Cardiotoxicity screening with simultaneous optogenetic pacing, voltage imaging and calcium imaging. *J. Pharmacol. Toxicol. Methods* 81: 240–250.

25

Conclusion and Future Work

In this book, we introduced applications of deep learning in digital holographic cell imaging and DHM-based phenotypic analysis methods. Basically, DHM can provide quantitative phase images if the exact distance between the sensor plane and the reconstruction plane is correctly provided. This process requires an iterative diffraction calculation, which is computationally time consuming. We presented a deep-learning convolutional neural network with a regression layer as the top layer to estimate the best reconstruction distance. Experimental results obtained using microsphere beads and RBCs showed that the proposed method could accurately predict the propagation distance from a filtered hologram. Additionally, our approach could be used at the single-cell level for cell-to-cell depth measurement and cell adherent studies.

Conventional, numerical phase unwrapping techniques can connect wrapped phases to recover the optical path length of a target object. However, these methods are computationally time consuming. We introduced a new deep-learning model that can automatically reconstruct unwrapped, focused phase images by combining digital holography and a generative adversarial network (GAN) for image-to-image translation. Compared with numerical phase unwrapping methods, the proposed GAN model overcomes the difficulty of accurate phase unwrapping due to abrupt phase changes. It can perform phase unwrapping at twice the rate of numerical methods. We showed that the proposed model could be generalized for different types of cell images. It has a higher performance than recent U-Net models.

In addition, we showed that deep learning could eliminate the superimposed twin-image noise in phase images from Gabor DHM setups. This is achieved with a conditional generative adversarial model (C-GAN) trained by input–output pairs of noisy phase images obtained from Gabor DHM and the corresponding quantitative noise-free contrast phase image obtained by off-axis digital holography. Surprisingly, we discovered that our model could recover other elliptical cell lines

Artificial Intelligence in Digital Holographic Imaging: Technical Basis and Biomedical Applications, First Edition. Inkyu Moon.

not included in the training dataset. Additionally, some misalignments can be compensated for with the trained model. In particular, if the reconstruction distance is slightly incorrect, our model can still retrieve in-focus images.

Furthermore, we demonstrated the potential of novel approaches to study live RBCs by integrating DHM with deep learning, which achieved good segmentation and classification accuracy with a Dice coefficient of 0.94 and a high-throughput rate of about 152 cells per second. Moreover, our holographic image-based deep-learning models could be applied to identifying morphological changes that occur in RBCs during storage. These deep learning–based classification results were in good agreement with previous findings describing RBC-marker changes affected by storage duration. Therefore, we believe that our DHM-based phenotypic analysis has potential as a new, efficient tool for the automated assessment of RBC quality and storage lesions to ensure safe transfusions as well as the diagnosis of RBC-related diseases.

Finally, we introduced DHM as a non-invasive measure of the dynamics of cardiomyocyte mechanical contraction and relaxation by recording quantitative phase images. Our automated image-processing algorithms extracted a set of parameters from quantitative phase images that allowed us to characterize beating patterns that reflect the mechanical contraction–relaxation cycle. Our experimental results of E-4031-treated cardiomyocytes showed that the beat rate was negatively correlated with E-4031 concentration, mainly due to an extended rest period. We also found that the contraction period was shorter than the relaxation period under these conditions. Isoprenaline increased both the contractile force and the beat rate, resulting in a decreased resting period. These results stress that intelligent DHM, with the development of proper, automated analysis algorithms, can be used to monitor drug-mediated effects on the dynamics of cardiomyocyte contraction and relaxation, which may help uncover their underlying mechanisms of action. DHM with proper AI or deep learning–based analysis of its quantitative phase image might be a promising label-free approach for drug discovery.

In this book, we introduced new deep-learning DHM systems well suited to label-free high-content screening. However, these models require complete labels or paired datasets (holograms and corresponding phase images) during training, which are often not available in practice. Moreover, since hologram patterns differ significantly according to hologram recording conditions and target cells, it may be difficult to establish generalized models with hologram patterns, thus restricting their efficient application. To overcome these issues, unsupervised learning or self-supervised learning DHM models should be studied to develop live-cell imaging and analysis platforms that are more efficient in practice.

While we presently use black-box deep learning techniques, explainable deep learning models for DHM, with a proper analysis of its quantitative phase signal for theranostical purposes, could be used in the near future. We also suggested, here-in, a new multi-modal imaging technique using deep learning to better understand cellular and subcellular physiology. This is vital to improve studies on disease pathology, disease prevention, and disease treatment.

Index

Artificial Intelligence in Digital Holographic Imaging: Technical Basis and Biomedical Applications, First Edition. Inkyu Moon.

e

f

n

o

p

q